Biological Magnetic Resonance
Volume 11

In Vivo Spectroscopy

A Continuation Order Plan is available for this series. A continuation order will bring delivery of each new volume immediately upon publication. Volumes are billed only upon actual shipment. For further information please contact the publisher.

Biological Magnetic Resonance
Volume 11

In Vivo Spectroscopy

Edited by

Lawrence J. Berliner
Ohio State University
Columbus, Ohio

and

Jacques Reuben
Hercules Incorporated
Research Center
Wilmington, Delaware

PLENUM PRESS • NEW YORK AND LONDON

The Library of Congress has cataloged the first volume of this series as follows:

Library of Congress Cataloging in Publication Data

Main entry under title:

Biological magnetic resonance:

Includes bibliographies and indexes.

1. Magnetic resonance. 2. Biology—Technique. I. Berliner, Lawrence, J. II. Reuben, Jacques.

QH324.9.M28B56 574.19′285 78-16035
AACR1

ISBN (Vol. 11) 0-306-44276-0

A Division of Plenum Publishing Corporation
233 Spring Street, New York, N.Y. 10013

Printed in the United States of America

Contributors

Lawrence J. Berliner • Department of Chemistry, The Ohio State University, Columbus, Ohio 43210

Lizann Bolinger • Department of Radiology, University of Pennsylvania, Philadelphia, Pennsylvania 19104

C. Tyler Burt • Magnetic Resonance Center, University of Illinois at Chicago, Chicago, Illinois 60680

G. Herbert Caines • Department of Chemistry and Biochemistry, Sinsheimer Laboratories, University of California, Santa Cruz, California 95064

Lee-Hong Chang • Department of Pharmaceutical Chemistry, University of California, San Francisco, California 94143

Gabriel A. Elgavish • Division of Cardiovascular Disease, Department of Medicine and Department of Biochemistry, University of Alabama at Birmingham, Birmingham, Alabama 35294

Hirotada Fujii • Department of Chemistry, The Ohio State University, Columbus, Ohio 43210. *Present address*: Tokyo Metropolitan Medical Research Institute, Tokyo, Japan

Thomas L. James • Department of Pharmaceutical Chemistry and Department of Radiology, University of California, San Francisco, California 94143

Robert E. London • Laboratory of Molecular Biophysics, National Institute of Environmental Health Sciences, Research Triangle Park, North Carolina 27709

Robert E. Lenkinski • Department of Radiology, University of Pennsylvania, Philadelphia, Pennsylvania 19104

Sandra K. Miller • Frederick Philips NMR Research Center, School of Medicine, Department of Radiology, Emory University, Atlanta, Georgia 30322

Jan M. Rydzewski • Department of Chemistry and Biochemistry, Sinsheimer Laboratories, University of California, Santa Cruz, California 95064

Thomas Schleich • Department of Chemistry and Biochemistry, Sinsheimer Laboratories, University of California, Santa Cruz, California 95064 and Department of Pharmaceutical Chemistry, School of Pharmacy, University of California, San Francisco, California 94143

Barry S. Selinsky • Department of Chemistry, Villanova University, Villanova, Pennsylvania 19085

Preface

In vivo nuclear magnetic and electron spin resonance spectroscopy is concerned, *inter alia*, with the noninvasive observation of metabolic changes in living systems, including animals and humans. Typically, the physiological (or pathological) state of an organ or tissue is monitored. This multifaceted approach was developed during the 1980s. It is still a research technique, but will undoubtedly become a clinical tool. We are proud to present this volume (the eleventh of our series) in which some of the pioneers in this area summarize their contributions and review related literature.

Bolinger and Lenkinski describe a variety of localization methods suitable for clinical applications of NMR spectroscopy. Schleich, Caines, and Rydzewski summarize their contributions to approaches involving off-resonance rotating frame relaxation and critically compare these with other NMR techniques that may yield similar information. Chang and James outline their approach and share their experience with the technical aspects of ^{1}H and ^{31}P NMR spectroscopy and spatially localized spectroscopy in studies of brain ischemia.

Sodium plays an important role in living systems, a key aspect being the large gradient between intra- and extracellular concentrations of sodium that is maintained by a variety of transport mechanisms. Miller and Elgavish give us a comprehensive review of an important research tool in this area—^{23}Na NMR spectroscopy as aided by shift reagents.

While the NMR study of living systems with the naturally occurring nuclides ^{1}H, ^{31}P, and ^{23}Na has provided a wealth of information, one can ask new questions in areas such as pharmacology and physiology by using suitable labelling with other NMR-active probes. An example is ^{19}F, which is very sensitive in NMR experiments and has the advantage of absence of endogenous biological background. Selinsky and Burt provide an excellent review of the state of the art of *in vivo* ^{19}F NMR.

The massive proton background in biological systems can be circumvented by using deuterated tracers with ^{2}H NMR. London reviews the use of ^{2}H NMR for *in vivo* studies of cellular metabolism.

While *in vivo* NMR spectroscopy appears to be a maturing field, electron spin resonance techniques for *in vivo* studies are still in the developmental stages. In this volume (to the personal pride of J.R.) we have a synopsis of the pioneering work of Berliner and Fujii on the feasibility and applications of *in vivo* ESR and EPR imaging to animal systems.

As always, we welcome the comments and suggestions of our readers. In the past they have been instrumental in improving this series.

Lawrence J. Berliner
Jacques Reuben

Contents

Chapter 1

Localization in Clinical NMR Spectroscopy

Lizann Bolinger and Robert E. Lenkinski

Chapter 2

Off-Resonance Rotating Frame Spin–Lattice Relaxation: Theory, and *in Vivo* MRS and MRI Applications

Thomas Schleich, G. Herbert Caines, and Jan M. Rydzewski

Chapter 3

NMR Methods in Studies of Brain Ischemia

Lee-Hong Chang and Thomas L. James

Chapter 4

Shift-Reagent-Aided ^{23}Na NMR Spectroscopy in Cellular, Tissue, and Whole-Organ Systems

Sandra K. Miller and Gabriel A. Elgavish

Chapter 5

In Vivo **^{19}F NMR**

Barry S. Selinsky and C. Tyler Burt

Chapter 6

In Vivo ^{2}H NMR Studies of Cellular Metabolism

Robert E. London

Chapter 7

Some Applications of ESR to *in Vivo* Animal Studies and EPR Imaging

Lawrence J. Berliner and Hirotada Fujii

1

Localization in Clinical NMR Spectroscopy

Lizann Bolinger and Robert E. Lenkinski

1. INTRODUCTION

Over the past decade, there has been a rapid proliferation of MR imagers into the diagnostic medical community. These scanners are used daily to provide anatomically based images in which the contrast is derived from differences in the intrinsic NMR properties (e.g., spin density, relaxation times, rates of diffusion) of the chemical constituents (i.e., fat and water) of the tissues being imaged. In the diagnostic setting, these images are used to both visualize and characterize abnormalities. It has proven to be difficult to completely characterize abnormalities based on the signal intensities observed on MRI alone. There have been a number of approaches that are based on MRI-derived methods such as contrast-enhanced MR imaging, MR diffusion and perfusion imaging as well as MR angiography, which are directed at obtaining so-called "functional" rather than anatomically based contrast. The ability of NMR spectroscopy to monitor metabolic processes noninvasively in living systems provides an attractive complementary technique for the characterization of tissue. In addition to this diagnostic application, NMR spectroscopy also has the potential to monitor the metabolic consequences of therapeutic intervention in order to provide early indicators of the success or failure of the intervention. It is reasonable to expect that metabolic changes that occur in the course of therapy precede any anatomical changes that might be observed on MRI.

Lizann Bolinger and Robert E. Lenkinski • Department of Radiology, University of Pennsylvania, Philadelphia, Pennsylvania 19104.

Biological Magnetic Resonance, Volume 11: In Vivo Spectroscopy, edited by Lawrence J. Berliner and Jacques Reuben. Plenum Press, New York, 1992.

Precise spatial localization techniques are critical in implementing NMR spectroscopy (often referred to as MRS) in a clinical setting. These methods provide a means for the determination of metabolic parameters from well-defined regions of interest selected from anatomically based MR images. Various aspects of spatial localization methods have been reviewed in the literature (Aue, 1986; Narayana and Delayre, 1988). Rather than survey all of the methods that have been described, we will focus on those methods that are currently employed in the bulk of the clinical MRS studies described in the literature. In our discussion we will specifically discuss methods employed in studies of ^{1}H MRS and ^{31}P MRS.

Before discussing the specific problems, it is important to define the characteristics of an ideal localization method. The ideal method should be capable of sampling all regions of space simultaneously. The resulting spatially localized spectra should have high signal-to-noise ratios and should be free of any distortions that prohibit accurate and precise quantitation. The precision of spatial localization should be high. That is, the spectra obtained should be free of contamination with signals arising outside of the nominal region of interest. In the clinical setting, it is advantageous to be able to reference the spatially localized spectra to MR images. Since there are no ideal localization methods available, it is fair to say that all of the methods currently employed make some compromises in achieving the characteristics outlined above.

Selecting an appropriate spatial localization method for the region of tissue, nucleus, and disease state being examined, involves the careful consideration of a number of technical and practical issues. For example, some of the localization methods are specific for use with surface coils rather than whole volume coils. Other localization methods work exclusively with RF coils which provide uniform excitation profiles. Also, some of the intrinsic properties of the spin systems being sampled may have a large influence on the choice of localization method. Accordingly, we will discuss some of the properties of the nuclei within the molecules of interest that influence the choice of localization method. These include the intrinsic sensitivities of these nuclei, their relaxation behavior, and the effects of spin–spin coupling on the observed spatially localized spectra. There are also technical limitations present in all of the localization methods currently employed. In the following sections we will address some of these theoretical and practical issues that are important in understanding the relative merits of the localization methods currently employed in clinical MRS studies.

2. INTRINSIC PROPERTIES OF THE NUCLEI

2.1. The Relative Sensitivities of the Various Nuclei in Normal Tissue

The relative sensitivities of the various nuclei routinely monitored by MRS are given in Table 1. These numbers can be used in combination with the physiological concentrations of various compounds present to arrive at estimates of the volumes of tissue that will give rise to localized spectra with reasonable signal-to-noise ratio (S/N) in realistic acquisition times. As an example, we can calculate the volumes of tissue that can be currently sampled by MRS with reasonable S/N. Using the concentrations reported for creatine and phosphocreatine (PCr) measured by conventional wet chemical analysis, we can estimate the minimum voxel volumes achievable at 1.5 T by comparing the minimum volumes detectable in MRI at 1.5 T. In these estimates, we are assuming that there are no S/N losses associated with the localization methods employed and that RF coils of comparable quality to the coil employed in MRI are used in MRS. We use as our baseline MR images obtained at the highest spatial resolution possible with the extremity or head coil. These, in our experience, can be obtained at a field of view of 8 cm, slice thickness of 3 mm, and matrix size of 256 × 256, in about 13 min of acquisition time (repetition time of 2500 msec; echo time of 20 msec). The voxel size in this protocol is $(8/256) \times (8/256) \times (0.3)\ \text{cm}^3$. If we wish to achieve a comparable S/N in a comparable acquisition time for MRS, then we multiply this voxel volume by the appropriate sensitivity factors for the compounds of interest. As an example, we can employ the proton resonance of creatine, which occurs at 3.0 ppm and is present at about 10 mM in human brain (this resonance is actually the sum of the creatine and PCr present and thus represents the total creatine pool). This

TABLE 1
NMR Properties of Select Nuclei of Biomedical Interest

Nucleus	Spin	Natural abundance	Frequency	Physiological concentration	Relative sensitivity at 1.5 T[a]	Receptivity at 1.5 T[b]
H-1	1/2	100	63.89	110 M	1.00	1.00
H-2	1	0.015	9.80	110 M	2.4×10^{-6}	2.4×10^{-6}
H-1	Proton-containing compounds		63.89	1–10 mM	1.00	10^{-5}–10^{-6}
C-13	1/2	1.18	16.06	10 mM	2.5×10^{-4}	2.5×10^{-8}
F-19	1/2	100	60.08	10 mM[c]	0.85	8.5×10^{-5}
Na-23	3/2	100	16.89	80 mM	0.13	9.5×10^{-5}
P-31	1/2	100	25.85	10 mM[d]	8.3×10^{-2}	8.3×10^{-6}
K-39	3/2	93.1	2.99	45 mM	1.0×10^{-3}	4.1×10^{-7}

[a] Relative sensitivity at 1.5 T computed by multiplying the intrinsic sensitivity at constant field by the natural abundance of the nucleus.
[b] The relative sensitivity multiplied by the relative (to water protons) physiological concentration present. For reference, an entry for proton-containing compounds such as lactate is included.
[c] Introduced exogenously as a fluorinated drug like 5-fluorouracil.
[d] The concentration reported is an approximate average over tissues and different phosphorus-containing compounds. For example, the concentrations of ^{31}P-containing species are: skeletal muscle, PCr 30 mM, ATP 5 mM; cardiac muscle, PCr 10 mM; brain, PCr 5 mM, ATP 3 mM, and PDE 14 mM (taken from Bottomley, 1989).

calculation gives a minimum voxel volume of about 1 cm^3 for the detection of creatine using solvent-suppressed proton MRS. For computing the sensitivity of ^{31}P we can employ the resonance of PCr, which is present at about 30 mM in skeletal muscle and about 5 mM in brain. The minimum volume of tissue that will yield reasonable S/N for the ^{31}P resonance of PCr can range between 1 and 10 cm^3. Assuming cubic voxels, these volumes translate into linear voxel dimensions of about 1 cm for ^{1}H and about 1–3 cm for ^{31}P. These estimates can be extended to other nuclei by scaling with the appropriate factors provided in Table 1.

2.2. The Chemical Shifts and Homonuclear Coupling Constants of ^{1}H and ^{31}P Nuclei in Molecules of Biological Interest

The chemical shifts and coupling constants of some proton-containing compounds that have been observed in solvent-suppressed proton spectra in humans are given in Table 2. Only those compounds that resonate at frequencies lower than water are reported. This is the region of the spectrum where most investigators have focused their attention, although there has been some interest in the aromatic resonances observed *in vivo*. Some of the resonances, like the methyl groups of *N*-acetylaspartate (NAA), creatine, and choline, are singlets. It is interesting to note that many of the compounds exhibit *J*-coupled spectra which are non-first-order. That is, the value of J is comparable to the difference in chemical shifts between the coupled spins. At reasonably low fields like 1.5 T, these compounds can exhibit deceptively simple spectra. Examples of these sorts of systems are citrate, taurine, and myoinositol. Examples of complex spin systems are glutamate and glutamine.

The ^{31}P chemical shifts and homonuclear coupling constants (for ATP) for the resonances commonly observed *in vivo* are given in Table 3. Both the phosphomonoester (PME) and phosphodiester (PDE) peaks are composite peaks arising from more than one species. The precise resonance positions of some of the compounds are dependent on both the pH and ionic composition of the medium. In the case of P_i, the relationship between pH and chemical shift has been exploited to give an estimate of the pH of the tissue. The frequencies of the resonances of ATP vary with the concentration of Mg^{2+} in the tissue, with the β-ATP resonance exhibiting the largest perturbation on binding to Mg^{2+}. There has been increasing interest in using the frequency of the β-ATP resonance to estimate the concentration of Mg^{2+} present in the tissue.

As will become evident in some of the following sections, the values of both the chemical shifts and coupling constants impose technical and practical constraints on the localization methods employed. For example,

TABLE 2
The Proton Chemical Shifts and Coupling Constants for Compounds That Exhibit Coupled Proton Spectra *in Vivo*

Compound[a]	Proton	Spin system	δ (ppm from DSS)	J (Hz)
Alanine	α-CH	Quartet	3.76	$J_{\alpha\beta} = 7.2$
	β-CH_3	Doublet	1.46	
Aspartate	α-CH	$AB\underline{X}$	3.89	$J_{AB} = 17.7$
	β-CH_2	$\underline{AB}X$	A 2.79	$J_{AX} = 4.4$
			B 2.68	$J_{BX} = 7.8$
Citrate	α-CH1, γ-CH1	$\underline{A}B$	2.65	$J_{AB} = 15.9$
	α-CH2, γ-CH2	$A\underline{B}$	2.50	
Glutamate	α-CH	$AA'BB'\underline{X}$	3.75	
	β-CH_2	$AA'\underline{BB'}X$	2.09	
	γ-CH_2	$\underline{AA'}BB'X$	2.33	
Glutamine	α-CH	$AA'BB'\underline{X}$	3.77	
	β-CH_2	$AA'\underline{BB'}X$	2.14	
	γ-CH_2	$\underline{AA'}BB'X$	2.44	
Lactate	C2H	Quartet	4.11	$J_{2,3} = 7.1$
	C3 (methyl)	Doublet	1.32	
Myoinositol	H1, H3	Complex	3.54	
	H4, H6		3.61	
	H5		3.29	
	H2		4.07	
N-Acetyl aspartate	α-CH	$AB\underline{X}$	4.38	$J_{AB} = 15.3$
	β-CH_2	$\underline{AB}X$	A 2.66	$J_{AX} = 8.4$
			B 2.47	$J_{BX} = 5.0$
	N-acetyl CH_3			
N-Acetyl glutamate	α-CH		4.09	$J_{\alpha\beta 1} = 8.5$
	β-CH_2		1.86, 2.05	$J_{\alpha\beta 2} = 4.9$
	γ-CH_2		2.22	$J_{\beta\gamma} = 8.0$
	N-acetyl CH_3		2.01	
Taurine	*S*-CH_2	$\underline{AA'}BB'$	3.41	$J_{AB} = 6.4$
	N-CH_2	$AA'\underline{BB'}$	3.26	

[a] Taken from Seeholzer (1985).

TABLE 3
The ^{31}P Chemical Shifts and ^{31}P–^{31}P Coupling Constants for ATP

	δ (ppm from H_3PO_4)[a]	J (Hz)
γ-ATP	−5.07	$J_{\alpha\beta} = 16.7$
α-ATP	−10.31	$J_{\beta\gamma} = 15.4$
β-ATP	−18.57	

[a] Taken from Gorenstein (1984). The data shown were determined in a solution of ATP at pH 7.8 and a $MgCl_2$ concentration of 5 mM.

the range of chemical shifts determines the bandwidth of the method. For the complete proton spectrum, this bandwidth is about 10–12 ppm or about 700 Hz at 1.5 T. Interestingly, the bandwidth for ^{31}P is also about 700 Hz at 1.5 T (27 ppm). The excitation bandwidths of RF pulses are related to their B_1 strengths. For rectangular pulses this relationship is straightforward. In many localization methods, the amplitude of the excitation pulses are set in combination with the amplitudes of B_0 gradients to selectively excite a volume of tissue. In practice, the finite amplitude of the RF pulses may result in limited excitation bandwidths which can give rise to off-resonance effects in localization methods that rely on RF methods. The amplitude of the gradient employed can lead to slice mislocation artifacts in methods that employ switched B_0 gradients for spatial localization (see Section 3.4).

Coupled spin systems may exhibit a modulation of peak amplitude which depends on the parameters and method employed for spatial localization. This effect is often loosely referred to as "J-modulation" and is addressed in detail in a following section. The effects of J-modulation can complicate the determination of T_2 and may create difficulties in obtaining quantitatively reliable spectra. These effects can also be exploited to obtain edited spectra in which certain spin systems are selectively inverted. An example of this effect is the selective inversion of the methyl resonance of lactate in proton spectra obtained using the stimulated echo method with an interecho delay of 135 msec (see, e.g., Rothman *et al.*, 1985). The effects of J-modulation can be simulated exactly for first-order spin systems using the product operator formalism (see Section 4.3.3). Simulations for non-first-order spin systems may be treated more conveniently with formalisms like the density matrix.

2.3. The Relaxation Times of ^1H and ^{31}P Nuclei in Molecules of Biological Interest

The relaxation times for some of the proton resonances observed in different regions of normal brain (Frahm *et al.*, 1989b) are given in Table 4. The values for T_1 range from 0.9 sec for myoinositol in the occipital area to 1.7 sec for taurine also in the occipital area. A knowledge of these values is important in selecting the appropriate repetition times and/or flip angles. Some localization methods can be employed in combination with excitation pulses set at reduced flip angles according to the "Ernst angle" relationship (Ernst, 1966). Other methods require that the repetition time be set for nearly full recovery between successive excitations. The choice of the most appropriate repetition time can become relatively complex in methods where localization is achieved with the use of multiple pulses in combination with surface coils.

TABLE 4
The Proton Spin–Lattice and Spin–Spin Relaxation Times (msec) for Selected Compounds in Different Regions of the Brain Determined from Solvent-Suppressed Proton Spectra *in Vivo*

Compound[a]	Group	δ (ppm)	Insular area T_1	Insular area T_2	Occipital area T_1	Occipital area T_2	Thalamus T_1	Thalamus T_2	Cerebellum T_1	Cerebellum T_2
Choline	$N\text{-}(CH_3)_3$	3.2	1100	380	1150	330	1200	320	1500	410
Creatine/	$N\text{-}CH_3$	3.0	1750	250	1550	240	1750	200	1500	190
PCr	$N\text{-}CH_2$	3.9	1750	160	1050	190				
Myoinositol	CH	3.5	1350	130	900	110	1100	150	1850	130
Lactate	CH_3	1.3			1550	1200				
NAA	CH_3	2.0	1650	330	1450	450	1400	340	1700	300
Taurine	$N\text{-}CH_2$ $S\text{-}CH_2$	3.3			1700	270				

[a] Taken from Frahm *et al.* (1989b). The data shown were determined using a stimulated echo sequence.

The T_2 values given in Table 4 range between 110 msec for myoinositol in the occipital region and 410 msec for choline in the cerebellum. At the relatively short interecho times that can be used in some of the localization methods currently available (i.e., the stimulated echo method with an interecho delay of 10 msec) there will be a small amount of differential T_2 decay in these resonances (0.91 magnetization remaining for myoinositol and 0.97 magnetization remaining for choline). At longer interecho delays, these differences become more pronounced. These effects must be taken into account in any calculation of absolute concentration of compounds from proton spectra.

The values for the spin–lattice relaxation times for ^{31}P-containing compounds observed in different tissues are given in Table 5. Although some of the values are crude estimates based on relative changes in peak amplitudes obtained at two different repetition times, these values still indicate the range of repetition times that must be employed to yield fully relaxed spectra. The T_2 values for the metabolites observed in human brain are given in Table 6. The apparent short T_2 values obtained for the resonances of ATP have prompted investigators to obtain localized spectra with as short a delay as possible. In methods that employ spin-echo techniques, this means that the interecho delay is kept as short as possible. In methods that employ direct excitation to produce FID-like signals, the delay between the excitation and detection of the signal (the preacquisition delay) has generally been kept as short as possible. There are some reasons to believe that the T_2 values of the ATP resonances are actually substantially longer than those reported in Table 6. Posse *et al.* (1990) have shown that the

TABLE 5
Representative Spin–Lattice Relaxation Times (sec) of ^{31}P-Containing Compounds in Human Tissues Measured *in Vivo*

Tissue	PME	P_i	PDE	PCR	γ-ATP	α-ATP	β-ATP
Brain[a]	2.74	1.47	1.64	3.29	1.36	0.97	1.03
Liver[b]		0.4	1.4		0.4	0.7	0.4
Liver[c]	4	0.5	7				1
Muscle[a]		5.25		6.87	4.95	3.17	4.05
Muscle[b]		4.7		6.5	4.0	4.2	4.1
Brain[d]	4.0	2.5	2.0	3.0	0.7	0.7	1.0

[a] Taken from Luyten *et al.* (1989). These values were obtained using the ISIS method.
[b] Taken from Buchtal *et al.* (1989). These values were obtained using a 1D-CSI sequence.
[c] Taken from Cox *et al.* (1991). These values were estimated from saturation effects observed using a surface coil in combination with a 2D chemical shift imaging sequence.
[d] Taken from Merboldt *et al.* (1990). These values were obtained using a stimulated echo sequence.

resonance of the β-ATP is observable with relatively high S/N in a spin-echo sequence performed with an interecho delay of 45 msec where only about 10% of the signal should be present if the T_2 were 20 msec. Turner and Garlick (1984) have shown that there can be severe modulations of the amplitudes of the ATP resonances as a result of J-modulation. This observation has recently been made by van Zijl *et al.* (1989). We believe that the presence of J-modulation is the reason that the T_2 values of the ATP resonances are underestimated in the stimulated echo sequence. In a following section (Section 4.3.3), we treat the effects of J-modulation on an AX spin system using the product operator formalism.

In vivo, the resonance of β-ATP is close to a triplet with a coupling of about 16 Hz. With a spin-echo delay of about 32 msec ($1/2J$) the outer lines of the triplet will be inverted relative to the central line. There can be a net cancellation of signal if the spectral resolution is too poor to resolve these lines. At 63 msec ($1/J$), the resonances of the triplet should all be in phase. Similar modulations occur in the other two resonances of ATP. This discussion clearly indicates the importance of understanding the effects of J-modulation and the manifestations of these effects in spatial localization methods.

TABLE 6
Spin–Spin Relaxation Times (msec) for ^{31}P-Containing Compounds Measured in Human Brain

PME	P_i	PDE	PCr	γ-ATP	α-ATP	β-ATP
70[a]	80	20	150	30	30	20

[a] Taken from Merboldt *et al.* (1990). The data shown were obtained using a stimulated echo technique.

3. PRACTICAL ASPECTS OF SPATIAL LOCALIZATION

3.1. General Approaches to Spatial Localization Methods and Their Classification

The need for obtaining tissue-specific NMR spectra was recognized early in the field of *in vivo* NMR. This need led to the use of local RF coils called surface coils (Ackerman *et al.*, 1980), which sampled superficial tissue in relatively close proximity to the coil. In animal experiments, deeper-lying tissues and organs can be surgically exposed and sampled by surface coil methods. In the clinical setting, the goal is to be able to obtain NMR spectra from a user-selectable region or regions using spatial localization methods rather than surgery. There have been a large number of these methods described in the literature. Spatial localization methods fall into two very broad categories, single volume or voxel methods and multiple volume methods. These two categories can be thought of as obtaining a spectrum from an area of interest or obtaining a spectrum at every point in an image. The single volume methods can be further subdivided into three categories: selective excitation methods which excite the area of interest only and attempt to leave all of the other volumes untouched; selective deexcitation methods which excite all of the spins outside the volume and saturate or dephase them so that only the spins in the volume of interest are left; and encoding schemes which use multiple acquisitions and addition and subtractions of spectra in order to obtain the volume of interest.

The multiple voxel methods may also be further subdivided into two categories, those that use slice selection techniques only to obtain localization and those that use additional methods to sample multiple points in space. All of these techniques use some form of encoding that requires postprocessing in the spatial dimensions in order to obtain the set of localized spectra. Examples of both single voxel and multiple voxel methods will be discussed in detail in the following sections.

The NMR signal, $s(t)$, has three basic parameters into which information about the spins may be stored. These include the frequency (ω), amplitude (A), and phase (ϕ):

$$s(t) = \sum_{i=1}^{N} A_i \cos(\omega_i t + \phi_i) \tag{1}$$

Spatial localization can be achieved by making any one of these three parameters dependent on the locations of the spins within the magnet. In practice, this process is achieved by sampling the NMR signal in the presence of either an RF gradient (B_1 methods) or through the use of techniques similar to those employed in MR imaging where B_0 gradients are used for spatial selection and encoding.

Before reviewing the details of the methods employed for spatial localization, we believe that it is important to understand some of the common practical problems that are intrinsic to both the nucleus being studied and the general class of localization method being employed. Accordingly, in the following section we will discuss: (1) dynamic range issues and approaches to solvent suppression for proton spectroscopy, (2) off-resonance effects in RF localization methods, (3) slice mislocation artifacts in gradient localization methods, (4) eddy current effects in gradient localization methods, and (5) the effects of delayed acquisition on the appearance of the localized spectra.

3.2. Dynamic Range Issues and Approaches to the Suppression of Water in Proton Spectroscopy

The considerations here are similar in principle to discussions of solvent suppression in conventional high-resolution NMR spectroscopy. Most commercial clinical MR imagers have analog-to-digital converters with a dynamic range of 16 bits, 15 bits for digital representation of the voltage detected in the receiver and 1 sign bit. This means that 1 part in 2^{15} (32,768) can be digitized if the full 15 bits is available for representation of the signal. Since the concentration of protons in water is about 111 M in pure solutions and is only somewhat lower in tissue, this means that compounds that have a resonance whose intensity is equivalent to 3 mM protons (as singlets) are at the limit of digitization. Since this concentration range corresponds to the order of magnitude of the concentrations of many of the compounds present in human tissue, there is a need to selectively reduce the intensity of the water resonance. This selective reduction can be accomplished in a variety of ways. These include the use of pulses that do not excite the water resonance (Plateau *et al.*, 1983; Hore, 1983); selective excitation of the water resonance using a chemically shift selective (CHESS) pulse, or pulses, followed by dephasing gradients (Haase and Frahm, 1985); using a selective inversion pulse to invert the water signal and begin the localization sequence when water is at its null point; exploiting the fact that the T_2 of water in tissue is much shorter than the T_2 values of the compounds present by employing spin-echo sequences with relatively long interecho delays; or combinations of these methods. It is important to recognize that the choice of solvent suppression technique must be compatible with the localization scheme being employed. For example, it is difficult to combine the stimulated echo or STEAM sequence (Frahm *et al.*, 1985) with binomial pulses for water suppression.

In general, the efficiency of solvent suppression methods depends on B_1 and B_0 homogeneity over the volume of tissue being sampled. It is clearly more difficult to achieve both high RF and B_0 homogeneity over large volumes of tissue. These considerations can become quite important in multivoxel spatial encoding methods. One approach, which has been quite successful, is to employ a localization method to select a smaller field of view within the head, for example, for spatial encoding. This approach requires that the RF and B_0 homogeneity is optimized over the smaller region of interest. Although it is clear that the suppression of water should be as high as possible *in vivo*, the main motivation for achieving suppression ratios in excess of 100–200 is the flatness of the baseline in the spectral region of interest since a suppression of water by a factor of 128 gains 7 bits in the digitizer. Examples of simulated proton spectra obtained at suppression factors of 100, 500, 1000, and 10,000 are shown in Fig. 1 to illustrate this point.

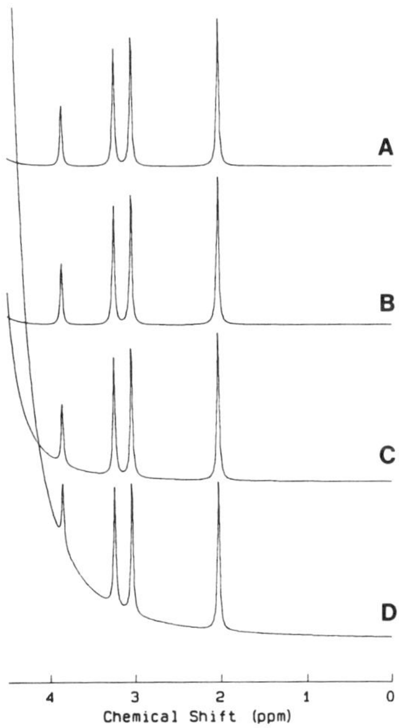

Figure 1. The upfield region of simulated spectra containing resonances at frequencies corresponding to *N*-acetylaspartate (2.0 ppm), creatine (3.0 ppm), choline (3.2 ppm), and myoinositol (3.85 ppm). The spectra shown were simulated with a linewidth of residual water of 10 Hz at suppression factors of 100, 500, 1000, and 10,000 in spectra A–D, respectively. The line shape of the water line was assumed to be Lorentzian and there were no distortions associated with eddy currents or water suppression included in the simulations.

3.3. Off-Resonance Effects in B_1 Localization Methods

All of the B_1 methods rely on the fact that the application of an inhomogeneous excitation RF pulse or pulses causes different parts of the sample to experience different flip angles. If the pulse or pulses have relatively narrow excitation bandwidths, the situation can become quite complicated since the flip angle may also depend on the difference in frequency between the observed resonances and the transmitter. Off-resonance effects in B_1 localization methods will cause compounds with differing chemical shifts to be sampled from different regions of the tissue. This occurs because the magnetizations at all chemical shifts are not rotated through the same flip angle. Resonances farther away from the transmitter frequency will be rotated through a flip angle that is less than that of a resonance directly at the transmitter frequency. Thus, the resonances that are farther away from the transmitter frequency will appear to be spatially displaced from their true location in the tissue. In order to counter this problem, it is necessary to generate relatively strong B_1 fields to ensure that all resonances within the spectral window of interest experience approximately the same flip angle. In multinuclear applications, this condition may be difficult to achieve since for a rectangular pulse the flip angle is given by

$$\Theta = 2\pi \cdot \gamma \cdot B_1 \cdot t_p \tag{2}$$

where Θ is the flip angle, γ is the magnetogyric ratio of the nucleus, B_1 is the strength of the RF field, and t_p is the length of the excitation pulse. In order to achieve the same excitation bandwidth at ^{31}P as for protons, the RF field must be (γ_H/γ_P) higher. This means that the transmitter power necessary is $(\gamma_H/\gamma_P)^2$ higher at ^{31}P than protons for the same excitation bandwidth.

3.4. Slice Mislocation Artifacts in Gradient Localization Methods

The core of many localization schemes lies in the principle of slice selection. This is accomplished by applying pulses that are selective in frequency in the presence of a B_0 gradient. The precise resonance frequency of a nucleus in the presence of a gradient applied along the x direction is given by

$$\omega = \gamma \cdot G_x \cdot x \tag{3}$$

where ω is the frequency of the nucleus, γ is its magnetogyric ratio, G_x is the strength of the gradient, and x is the position along the x axis. When this gradient is applied to a sample containing nuclei that resonate at slightly different frequencies, there will be a displacement of the sensitive volume for each of the different species. The displacement can be easily estimated using Eq. (3). For two species with a chemical shift difference of 3 ppm (e.g., fat and water) at 1.5 T, the displacement will be about 0.5 mm at a gradient strength of 10 mT/m (the maximum values routinely achievable on commercial whole body gradient sets). There is an additional constraint imposed on the gradient by the bandwidth of excitation of the RF pulse. The frequency bandwidth or excitation profile of the slice selection pulses is very important to the amount of signal obtained from the desired volume elements. For slice selection, the ideal profile in the frequency domain would be rectangular. For small excitation flip angles, an RF pulse that is amplitude modulated by a $(\sin x)/x$ or sinc function gives a rather good approximation to a rectangular frequency excitation profile. For large flip angles, say greater than 30°, the nonlinearity of the response of spins to the excitation must be compensated. This nonlinearity becomes severe at flip angles of 180° and results in an excitation bandwidth that is narrower than the 90° pulse. This property of 180° sinc pulses may lead to large artifacts when these pulses are used as refocusing pulses.

As pointed out above, the spatial location of a slice at each chemical shift will be different. The ability to overcome this problem is ultimately limited not only by the strength of the gradients but also by the RF power available. As the bandwidth of the RF excitation is increased, the strength of the gradient must also be increased to maintain the same slice thickness. The percent of overlap of the volumes derived from excitation of species with different chemical shifts is proportional to the bandwidth of the excitation pulse (BW) minus the difference in chemical shifts (ΔF) of the two resonances divided by the bandwidth of the pulse:

$$\% = \frac{\mathrm{BW} - \Delta F}{\mathrm{BW}} 100 \tag{4}$$

The size of the volume in one dimension (S) is then determined by the strength of the gradient (G) used during the application of the pulses:

$$S = \frac{\mathrm{BW}}{G} \tag{5}$$

The slice mislocation artifact is illustrated in Fig. 2 for a single slice selective localization method. All localization techniques that utilize slice selections for localization are susceptible to this problem. There is no slice mislocation

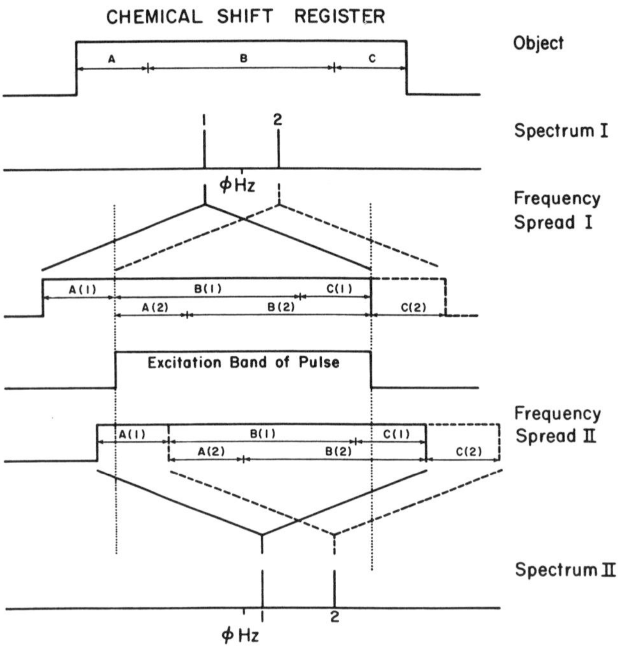

Figure 2. The so-called slice mislocation problem arises from the difference in frequencies of resonances associated with various chemical structures. For a single resonance at 0 Hz in the rotating frame, the sample in the presence of a gradient may give signal corresponding to the plot labeled "Object." If the chemical profile is equivalent to spectrum 1, then the object has two profiles, the solid line for peak 1 and the dashed line for peak 2. The amount of each peak observed is dependent on the frequency bandwidth of the excitation pulse. In general, for an object with profiles at each resonance that extend beyond the frequency limits of the pulse, the overlap in volumes is equivalent to that described in Eqs. (4) and (5). Region B of the object is present in the profiles of both resonances excited by the pulse although the regions of nonoverlap came from two different regions of the sample.

A difficulty arises if the resonances are shifted such that one of their profiles does not extend past the excitation bandwidth of the pulse. The resulting spectrum would show the intensity of resonance 2 to be significantly reduced as compared to resonance 1 since the amount of overlap has decreased. This situation also occurs in spectrum 1 if the two resonances are not uniformly distributed in the sample. In general, the best way to avoid these problems is to use a very broad excitation band and very large gradients.

artifact in the phase-encoded dimensions of the Fourier chemical shift technique.

3.5. The Effects of Eddy Currents in Gradient Localization Methods

The effects of eddy currents on MR images have been analyzed recently by Ahn and Cho (1991). The term *eddy current* is usually applied incorrectly to describe two different effects. The first effect is caused by the finite inductance of the gradient coils and is not related to any eddy current phenomena. This finite inductance causes a lag in the rise time of the gradient. Also, a residual magnetic field gradient persists after the current in the gradient coil is turned off. This residual gradient field decays with a time constant of tens of milliseconds. An example of this phenomenon is shown in Fig. 3. The effects of these residual gradients can be partially compensated by applying the appropriate corrections to the input waveform to the gradient. The applications of these corrections have become standard practice in all commercial whole body MR imagers. The second kind of effect is the true eddy current induced by interactions between the gradient coil and conducting elements within the magnet structure. The effect of these eddy currents is a time-dependent variation in the main magnetic field (B_0) which may not decay with a single exponential and may persist for much longer than the residual gradient field.

One of the earliest examples of the effects of eddy currents on proton spectra was described by Ordidge *et al.* (1987) and is shown in Fig. 4. There have been a number of remedies suggested for this problem. The example shown in Fig. 4 employed a method in which the phase of the FID obtained for a sample of water on resonance was applied as a correction to the FID

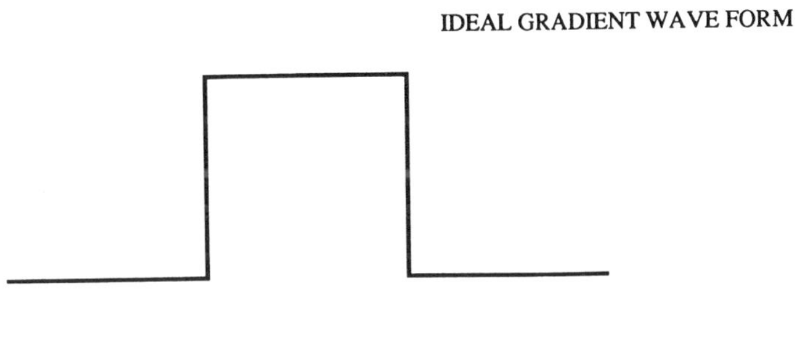

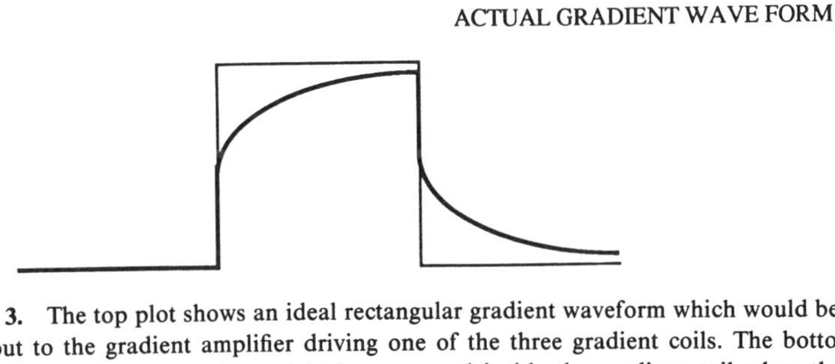

Figure 3. The top plot shows an ideal rectangular gradient waveform which would be fed as an input to the gradient amplifier driving one of the three gradient coils. The bottom plot shows an actual response that might be measured inside the gradient coil when the input waveform is distorted by the finite impedance of the gradient coil. Note that the gradient is initially underdriven and persists after the current driving the coil is turned off.

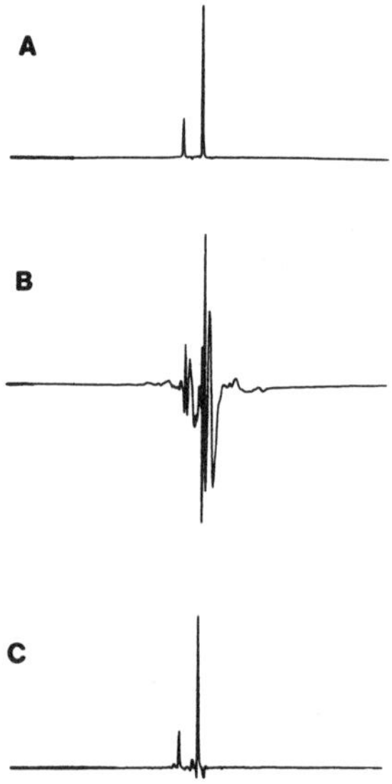

Figure 4. (A) A proton spectrum obtained from a 10-cm sphere filled with methanol in a 1.5-T Bruker whole body magnet. (B) The same spectrum as in A sampled 4 msec after the application of gradient pulses along the three orthogonal axes with a strength of 0.2 G/cm (0.02 mT/m). (C) The spectrum of B after correction using the measured time-dependent phase function. (Reproduced with permission from Ordidge *et al.*, 1987.)

obtained for methanol. The dramatic improvement obtained clearly demonstrates that it is possible to correct for spectral distortions produced by residual eddy currents using relatively simple algorithms (Ordidge and Creshull, 1986; Klose, 1990; Riddle *et al.*, 1992).

3.6. Distortions Caused by Delayed Acquisition

Although this problem has been discussed in detail in the literature (Allman *et al.*, 1988), it is important to review the origins of the spectral distortions caused by delayed acquisition. In all FID-like acquisitions (pulse-acquire), there is a delay between the time at which coherence is produced by the RF pulse and the time at which the magnetization is sampled. In high-resolution NMR this delay is shorter than a single dwell period since the data acquisition can be started as soon as the RF coil has recovered from the application of the excitation pulse. The presence of this short delay in the acquisition of the time domain signal, often referred to as the dead time, introduces the need for a frequency-dependent phase correction of the transformed, frequency domain spectrum. This first-order phase correction increases as the length of the delay increases at a rate of 180° for each dwell period during which data acquisition is delayed. If the delay is close to an integral number of dwell periods, the first-order phase shift can be removed by right shifting the time domain data by the appropriate number of points so that the start of data acquisition more closely corresponds to the time at which coherence was created. This method is equivalent to multiplying the FID with a window in which the first n points are zero and the remaining points are set to 1. This operation can be conveniently described by the convolution theorem as

$$\mathbf{FT}\{f(x)g(x)\} = F(x) * G(x) \tag{6}$$

where $f(x)$ and $g(x)$ are two functions, $F(x)$ and $G(x)$ are their Fourier transforms, and $*$ is the convolution operator. If we define $f(x)$ as the true FID and $g(x)$ the function where the first n points are equal to zero and all remaining points are set to 1, it is clear that the result will be a convolution of the spectrum with the Fourier transform of $g(x)$ in the frequency domain. The Fourier transform of $g(x)$ is a $\sin x/x$ function with the width of the function determined by n, the number of points set to zero. An example of this effect is shown in Fig. 5. As the number of missing points is increased, the presence of sinc-wiggles in the transformed spectrum becomes more pronounced. If the spectrum is phased after FT by conventional phasing routines (Cooper, 1983) rather than right shifting data points in the time domain, the identical sinc side lobes appear.

These distortions are not present if an echo rather than a delayed acquisition is acquired. Most of the localization methods employed for ^{31}P utilize FID-like acquisitions in order to avoid the losses in signal associated with the T_2 decay of the PME, PDE, and ATP resonances. The shortest preacquisition delay possible in methods employing slice-selective gradients on whole body scanners is of the order of several milliseconds. This delay amounts to several dwell periods. As a result, the spectra acquired with these delays require substantial first-order phase correction (180° per dwell period) and appear to sit in a rather deep well on a rolling baseline (see Fig. 5). Localized ^{31}P spectra exhibiting these distorted baselines have been published by a number of investigators. These distortions are also present in CSI methods which employ slice-selective methods in combination with phase-encoding techniques.

There are a number of remedies for these distortions. However, any component with a very short T_2 will be lost since it is impossible to recover the information contained in the points that were not acquired. The data acquired with a delay can be fit in the time domain using a variety of time-domain fitting algorithms. Many of these methods can provide good estimates of the component exponential decays even when the first few points of the FID are missing. Alternatively, the baseline of the frequency domain spectrum can be corrected using a sum of sinc functions each

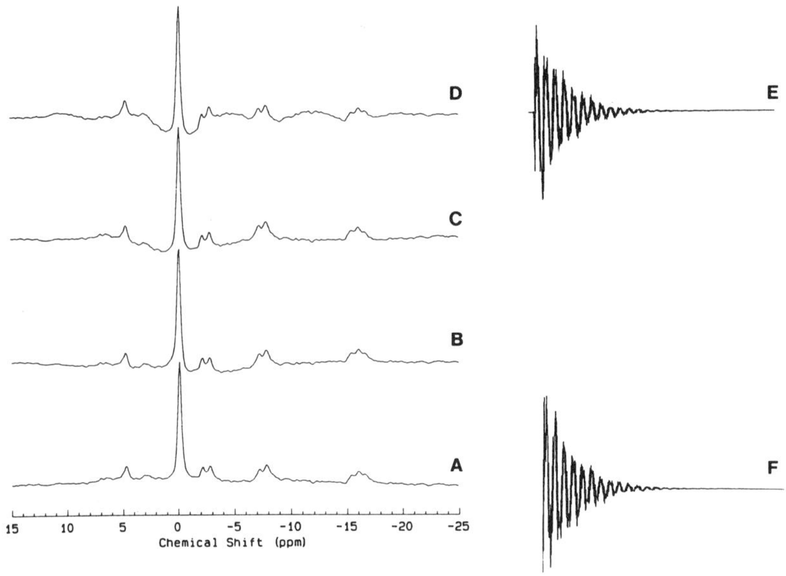

Figure 5. Phosphorus-31 spectra taken at 1.5 T with a 12.5-cm coil placed over the gastrocnemius muscle of a normal volunteer. The spectrum shown in A was obtained using a rectangular RF pulse followed by a short delay before data acquisition. The spectra shown in B–D are the phased Fourier-transformed results obtained from the FID corresponding to spectrum A (trace F) after the time domain data were left shifted by 4, 8, and 12 points, respectively. The FID shown in E was obtained by the inverse FT of spectrum D. Note the presence of the points set to zero in this FID.

centered at the resonance position of the observed lines. This approach is equivalent to an approximate "sinc deconvolution" in the frequency domain (Allman *et al.*, 1988). Another approach to this problem is to acquire the localized data as a full or half echo. More investigators have become aware of this point since there have been more reports dealing with the development of localization methods for ^{31}P which employ some sort of echo rather than direct FID-like acquisition.

4. SPATIAL LOCALIZATION METHODS

4.1. Single Voxel B_1 Methods

For many localization experiments, it is advantageous to use a surface coil instead of a volume coil in order to increase the amount of signal obtained. This is true whenever the volume of interest is close to the surface of the body such as the liver, skeletal muscles, and areas of the brain like the occipital lobe. Unlike the volume coils used in imaging, surface coils do not have a uniform B_1 field. In general, the flip angle of an NMR pulse will vary in space according to the B_1 profile of the coil. Because the surface coil has a varying B_1 field, the spatial extent of its excitation is limited and some degree of localization may be obtained when a simple single pulse experiment is performed with a surface coil. The B_1 profile of a surface coil is shown in Fig. 6. It is important to note that the usable B_1 field of the surface coil is not circularly symmetric. When placed with its axis perpendicular to the B_0 field, only the x and y components of the B_1 field are effective in the RF pulse. A slice in the xz plane of the B_1 profile of the coil does not look like a slice through the yz plane as shown in Fig. 7.

For many purposes, the localization provided by the B_1 field of the surface coil alone is not sufficient. Consider the liver for example. In order to obtain a spectrum of only liver tissue, the signal from the surface muscle must be eliminated. This can be done using some very simple methods which utilize the B_1 distribution of the surface coil. The first such method, termed depth pulsing (Bendall and Gordon, 1983), utilizes a Θ; $2\Theta[\pm x, \pm y]$ sequence in order to select a range of B_1 contours according to a $\sin^3 \Theta$ distribution around the region in space corresponding to a $\pi/2$ pulse. The location of the volume of interest is chosen by making sure that the $\pi/2$ flip angle occurs in the center of this region. Figure 8 shows a surface plot of the excitation distribution in a plane for a single pulse and the depth pulse which gives a $\sin^3 \Theta$ distribution. The power of the pulse has been chosen so that a flip angle of $\pi/2$ occurs at 15 cm from the plane of the coil. The coil has a radius of 20 cm. The signal close to the coil has been suppressed by this sequence. By adding more 2Θ pulses, the excitation band can be made narrower, Θ; $(2\Theta[\pm x, \pm y])_n = \sin^{2n+1}(\Theta)$.

Pekar *et al.* (1985) introduced a method called harmonically analyzed sensitivity profile (HASP) by which the same B_1 sensitivity profile may be obtained but without the necessity to phase cycle. The result can be obtained by acquiring data sets with flip angles of Θ, 3Θ, 5Θ, ... and combining these according to the coefficients needed to obtain $\sin^{2n-1} \Theta$. For example, $\sin^3 \Theta$ can be expanded to $3 \sin \Theta - \sin 3 \Theta$. Thus, two FIDs can be acquired, one with a flip angle of Θ, once again the power of the pulse is set so that the $\pi/2$ flip angle is at the depth desired, and one with a flip angle of 3Θ, i.e., three times the pulse length of the $\pi/2$ choice. These can be multiplied by the coefficients 3 and -1 and added. Alternatively, three times more summations of the Θ FID could be acquired than the 3Θ FID and the two results subtracted. This will provide the same B_1 profile as shown in Fig. 8 for the depth pulse. For $\sin^5 \Theta$, 10 times Θ, minus five times 3Θ and one 5Θ data acquisitions are required. Thus, the sine series describing $\sin^{2n-1} \Theta$ is used to obtain the correct coefficients.

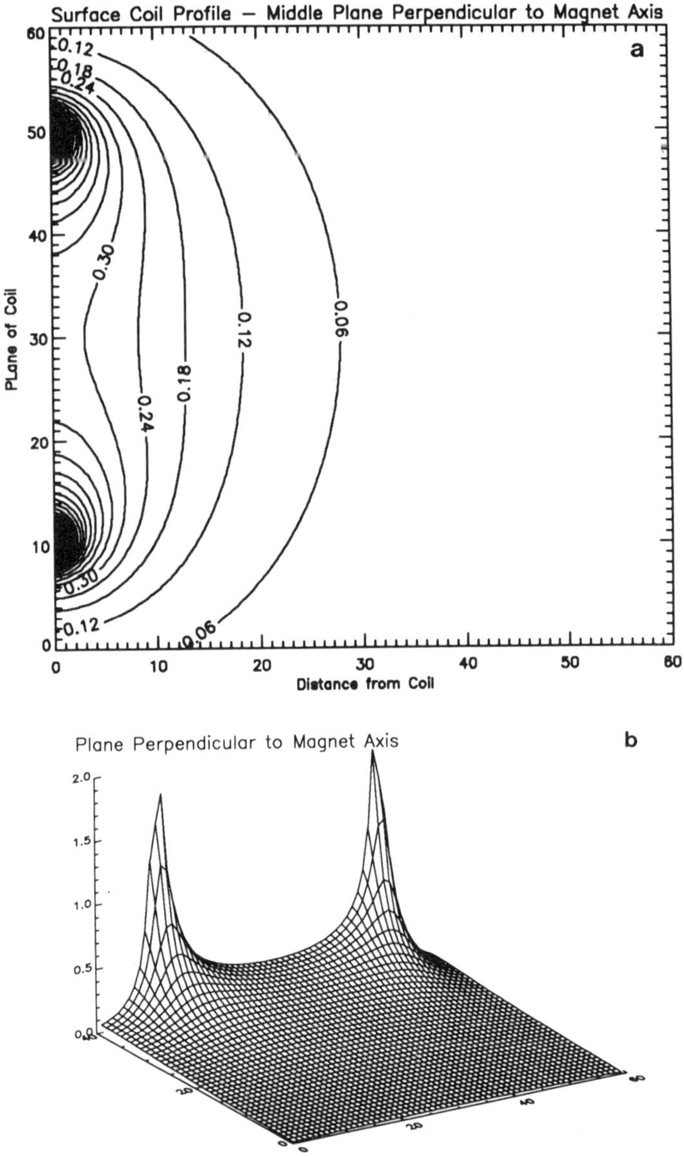

Figure 6. Computer calculations of the NMR usable B_1 field of a single circular coil using a Biot–Savart law. The 20-cm circular coil is approximated by 64 straight line segments. The B_1 field is calculated over a 40 by 60 by 60 cm volume. The main magnetic field (B_0) is assumed to be in the z direction. Therefore, only the x and y components of the B_1 field contribute to the RF pulse. The plot is of the square root of the sum of the squares of the x and y components. The surface coil lies in the yz plane and the plots (a, contour plot; b, 3D surface plot) show the plane across the diameter of the coil in the xy plane. The two plots show the isocontours of the B_1 field generated by the coil.

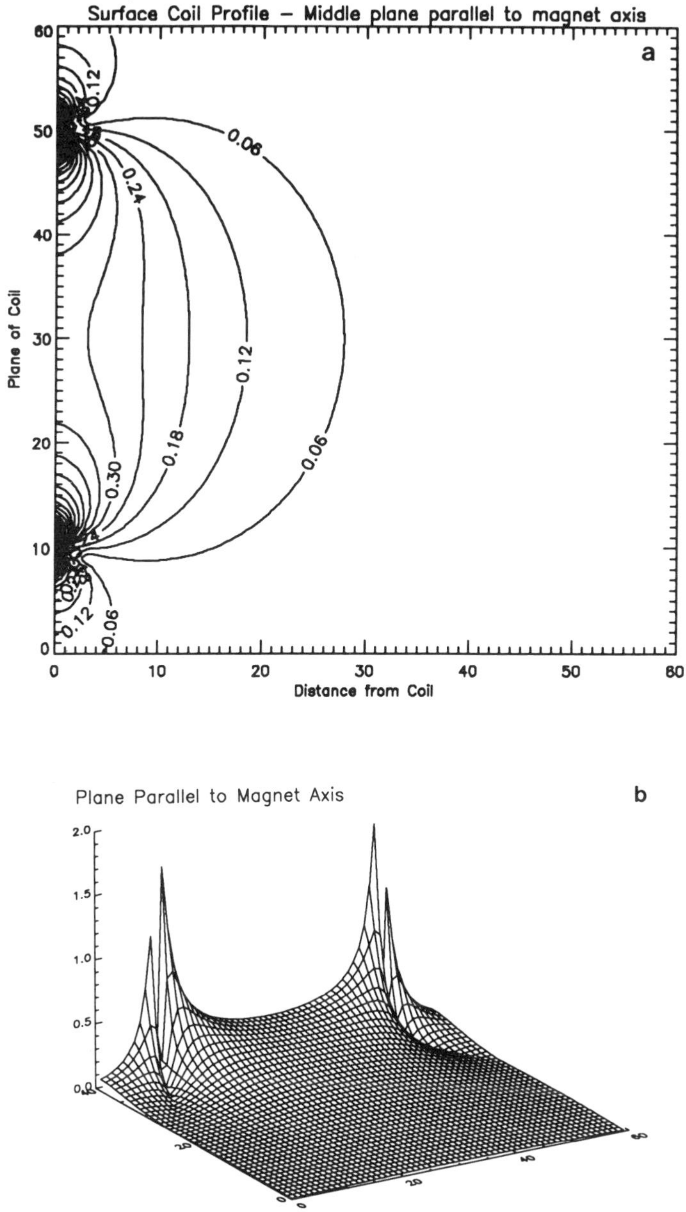

Figure 7. The plots of the B_1 field of the coil as described in Fig. 6 showing the *xz* plane (a, contour plot; b, 3D surface plot). It should be noted that the main magnetic field (B_0) along the *z* axis breaks the circular symmetry of the coil causing the two planes (*xy* and *xz*) to have different B_1 profiles.

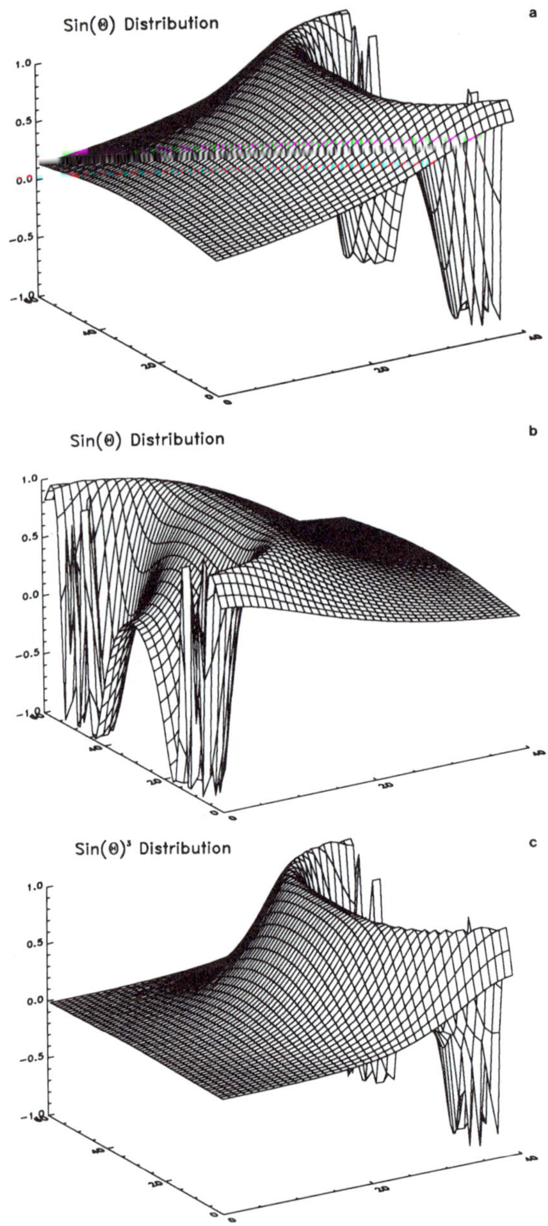

Figure 8. The excitation profile of the surface coil defined in Figs. 6 and 7 was obtained by choosing a point 15 cm along the axis of the surface coil and defining this point as the location for the $\pi/2$ pulse. The B_1 profile calculated in plot 7 was then converted to a flip angle profile using the equation $\Theta = \gamma B_1$. The normal excitation profile of a surface coil in the xy plane then looks like a (view remote from the coil) and b (view from behind the coil). The surface shown in c and d show the $\sin^3 \Theta$ profile. The $\sin^5 \Theta$ profile is shown in e and f. These profiles can be produced using the B_1 localization schemes outlined in the text.

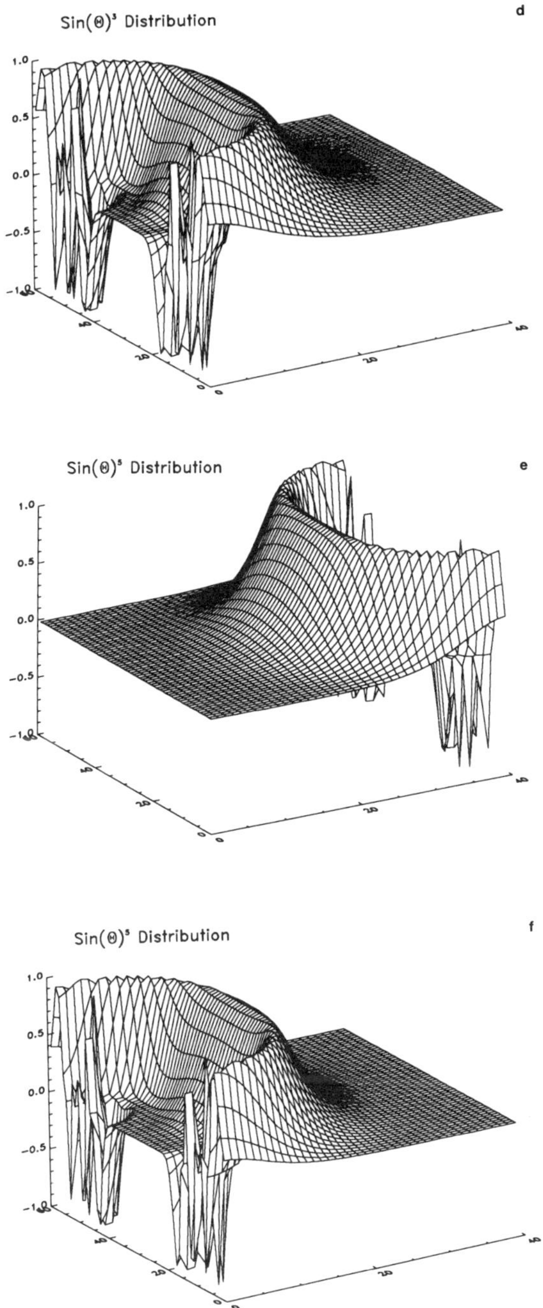

Figure 8. *(Continued)*

A variation on this theme called fast HASP can be implemented which allows all of the necessary data to be acquired in one experiment. The sequence consists of a series of Θ pulses. An FID is collected after each pulse and the acquisition followed by a gradient to dephase any transverse magnetization. Thus, the next pulse flips only those spins which were left along the z axis by the preceding pulse. The same B_1 localization obtained in the slower form of HASP may also be achieved in the fast form.

The slower form of HASP has been generalized into a Fourier sine series approach, called Fourier series windowing (FSW), as described by Garwood *et al.* (1985). In this approach a window function or shape of the desired flip angle distribution is chosen and the discrete sine transform of this shape is obtained:

$$W(\Theta) = \pm C(nT)\sin(\Theta nT) \qquad (7)$$

The Fourier coefficients $C(nT)$ are the weighting functions needed for the combination of the data obtained at different flip angles. For $W(\Theta) = \sin^{2n-1}\Theta$, the same result is obtained as the slow HASP experiment. However, shapes different from powers of sine functions are obtainable with FSW.

4.2. Multiple Voxel B_1 Methods

HASP and FSW are single volume methods which are related to a B_1 imaging scheme called rotating frame imaging. The concept of using a coil where B_1 varies linearly in space in order to obtain spatial information was first proposed by Hoult (1979) and was called rotating frame zeugmatography. Styles *et al.* (1985) made the technique practical for use with surface coils by first designing a double coil system which allows one to use the linear B_1 region of the surface coil for excitation. The two coils consist of a large coil used for excitation and a smaller coil which is placed in such a manner that it may receive signal only from the region excited with a linear B_1 by the large coil. An example of this coil assembly is shown in Fig. 9. Using these coils, the flip angle is varied in a stepwise fashion and the resulting data Fourier transformed in two dimensions. This results in a one-dimensional spatial image with spectra obtained at various depths along the axis of the surface coil. This process is illustrated by example in Fig. 10. The other two dimensions are limited by the size of the small coil used to receive the signal. In this way, the spatial information is encoded into the amplitude of the observed signal. This results in the loss of possible signal since some of the magnetization remains along the z axis and is not detected. It is possible to convert this method into a phase encoding technique by applying a second pulse which has a relative phase of 90° to

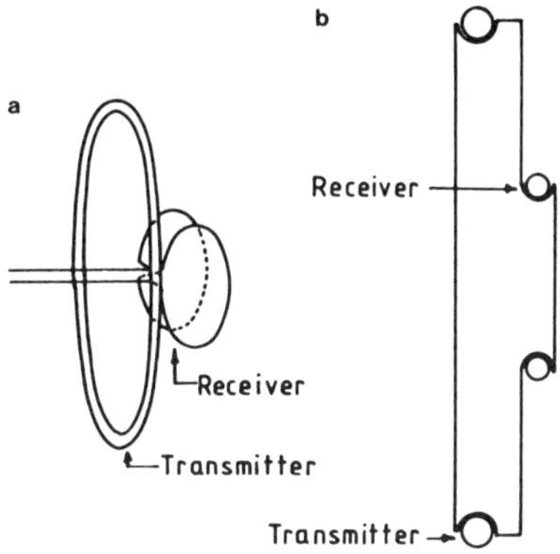

Figure 9. Diagrams of two different arrangements of double coils used for the rotating-frame spectroscopic imaging protocol. (a) The probe used for imaging the rat kidney (5-cm-diameter circular transmit coil, elliptical pair of receiver coils, 2 cm major diameter, 1 cm minor diameter). The two receiver coils are separated by 1 cm. The receiver coil assembly is 2 cm from the center of the transmit coil. (b) The coil assembly used for human studies. The transmit coil has a diameter of 15 cm, the receiver coil has a diameter of 6.5 cm, and the two coils are 2.5 cm apart. (Reproduced with permission from Styles *et al.*, 1987.)

the first pulse and is close to a $\pi/2$ flip angle (Blackledge *et al.*, 1987). Most of the signal is now rotated from the *xz* plane into the *xy* plane increasing the effective S/N of the experiment. Using this method, which is the phase-modulated form of rotating frame imaging, the spatial information is distributed in the phase of the *xy* magnetization. Again a two-dimensional Fourier transform of the data gives an image in which spectra are localized to various positions along the axis perpendicular to the plane of the coils.

One of the major problems with this method is the presence of off-resonance effects caused by the necessity of generating relatively uniform

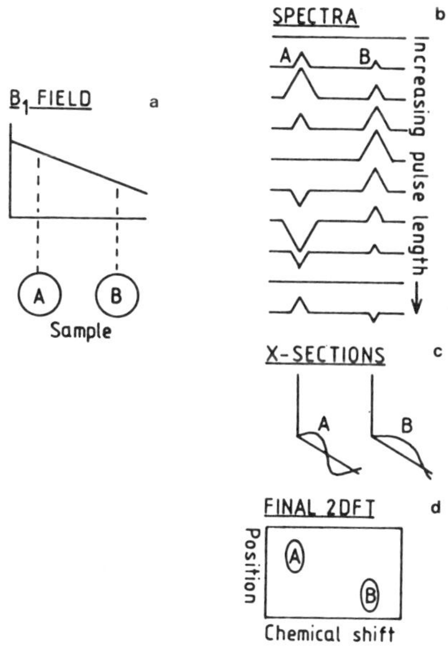

Figure 10. A schematic representation of the rotating-frame spectroscopic imaging protocol. (a) A two-compartment phantom placed in a gradient B_1 field. (b) Set of spectra obtained with increasing pulse length. (c) Cross sections through the data matrix of b at positions corresponding to the centers of the two spectral lines. (d) The final data matrix obtained after a second Fourier transform has been performed on the data in b. (Reproduced with permission from Styles *et al.*, 1987.)

B_1 fields over the entire spectral region of interest. This consideration is simply one of transmitter output power. For ^{31}P at 1.5 T, the total excitation bandwidth necessary is about 750 Hz for 30 pppm. Therefore, if the transmitter is placed at the center of the spectrum, the β-phosphate of ATP will resonate about 375 Hz from the zero reference. For a $\pi/2$ pulse of 200 μsec, this resonance will be flipped with an effective field which is 1.04 times stronger than the field on resonance. Thus, the β-phosphate of ATP will be moved in the spatial direction to a B_1 of the surface coil which is 1.04 times stronger. Thus, it is quite common to observe a decrease in the intensity of the β-NTP resonance in spectra observed at deeper slices using the rotating frame method. This effect is similar to the slice mislocation artefact experienced with B_0 slice selection methods.

Narayana and Delayre (1988) have pointed out that the most simple application of this method may cause smearing in the spatial dimension caused by incomplete recovery of longitudinal relaxation of the magnetization. This is true only if the encoding is left as an amplitude modulation in the *xz* plane. Since different amounts of magnetization are inverted on each encoding, incomplete T_1 relaxation between encodings would cause such smearing. For the phase-modulated form of the rotating frame experiment, almost all of the magnetization ends up in the transverse plane prior to the acquisition. Thus, no matter which encoding is being performed, all spins relax longitudinally from the same starting point and this smearing is eliminated. If the amplitude-modulated experiment must be performed, then a remedy proposed by Garwood *et al.* (1984) may be used. This includes a set of preparation pulses along each orthogonal axis followed by a delay before the evolution pulse is applied. This may be necessary if the two-coil method of Styles *et al.* (1985) is not used.

The advantages of this method are that it is possible to obtain spectra from relatively thin slices in reasonable acquisition times. The technical requirements for implementing this method primarily involve construction of the coil assembly, which is relatively straightforward. The rotating frame method can be combined with frequency-selective saturation pulses to produce localized estimates of the rates of transfer of saturation.

4.3. Localization Methods Based on the Use of Switched B_0 Gradients

There have been a large number of localization methods developed which employ switched B_0 gradients to localize particular regions in space. Some of these methods are most commonly employed in obtained multinuclear NMR spectra in clinical settings. These methods can be subdivided into four separate categories: single and multivoxel methods most commonly used for localized ^{31}P NMR spectroscopy, single and multivoxel methods most commonly used for proton NMR spectroscopy. In the following section we will discuss the most commonly used methods in each of these categories as well as other methods which illustrate important aspects of the approaches to spatial localization. In these discussions we will make a somewhat arbitrary distinction between methods which can be extended to sample more than one region of tissue (by methods like slice-interleaving) and spatial encoding methods. The former kinds of methods will be discussed along with the single voxel methods from which they are derived. The latter will be discussed separately as part of spatial encoding methods. We will outline two different approaches to spatial encoding methods as well as some of the problems associated with these approaches. We will also illustrate by example how the use of particular parameters in a given localization method can influence the appearance of the spectra obtained by using the product operator formalism to calculate the appearance of an A_nX spin system sampled by the stimulated echo sequence.

4.3.1. Single Voxel Methods Commonly Used for Localized ^{31}P NMR Spectroscopy

The simplest method using B_0 gradients is the DRESS (depth-resolved surface coil spectroscopy) method described by Bottomley *et al.* (1984) shown in Fig. 11. The DRESS method is an example of a method in which only the desired spins are excited and sampled. A gradient is applied in the plane perpendicular to the plane of the surface coil in the presence of a frequency-selective RF pulse. This operation excites a slice which is parallel to the plane of the coil. The advantages of this method are its simplicity and the fact that the flip angle can be adjusted within the selected slice. The thickness of the slice is determined by the bandwidth of the excitation pulse and the strength of the gradient. There may be a significant

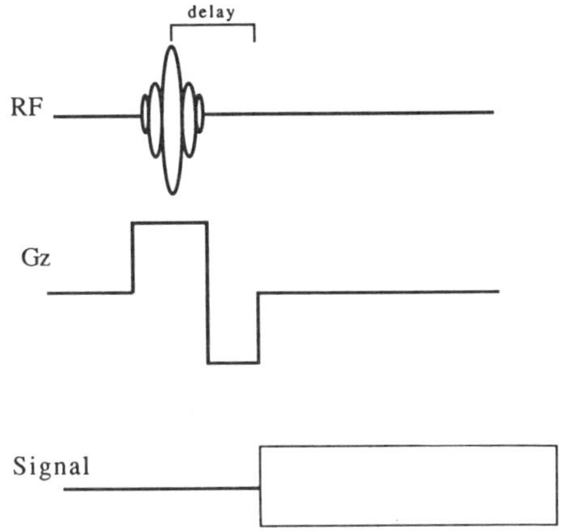

Figure 11. The pulse sequence for depth-resolved surface coil spectroscopy (DRESS). There is a short delay between the center of the RF pulse, shown as a sinc pulse, and the start of data acquisition.

slice mislocation artifact present when the gradient amplitude is set to excite a rather thick slice. The localization profile depends heavily on the geometry of the surface coil in two dimensions and there is most often a preacquisition delay of several milliseconds. It is of course possible to sample several slices in an interleaved manner (SLIT-DRESS, slice-interleaved DRESS; Bottomley *et al.*, 1985) during the normally relatively long TR time.

FROGS (fast rotating gradient spectroscopy; Sauter *et al.*, 1987) is an example of a method in which unwanted spins are removed during a preparation period prior to excitation of the desired spins. The pulse sequence for FROGS is shown in Fig. 12. This method is commonly applied in combination with a surface coil to sample tissue like liver which lies just below the most superficial layer of tissue within the patient. The first operation in the pulse sequence is identical to the slice selection process in the DRESS method except that the slice location is set to be the slice closest to the surface coil. This step is followed by a dephasing gradient, or gradients, along one, or both, of the orthogonal axes. This step destroys any coherence of the transverse magnetization of the spins excited by the first operation. The combination of the two operations is similar to the "homo-spoil" procedure employed in high-resolution NMR. The process is repeated *n* times until no coherent magnetization remains in the most superficial slice. After a short settling delay, primarily to allow recovery from residual eddy currents, the desired magnetization is sampled with a "hard" nonselective excitation pulse. This pulse may be set to any flip angle desired. The advantages of this method are similar to those of the DRESS method with the additional feature that there is no preacquisition delay. The localization is poorer than with DRESS since all of the spins except those in the most superficial tissue are sampled.

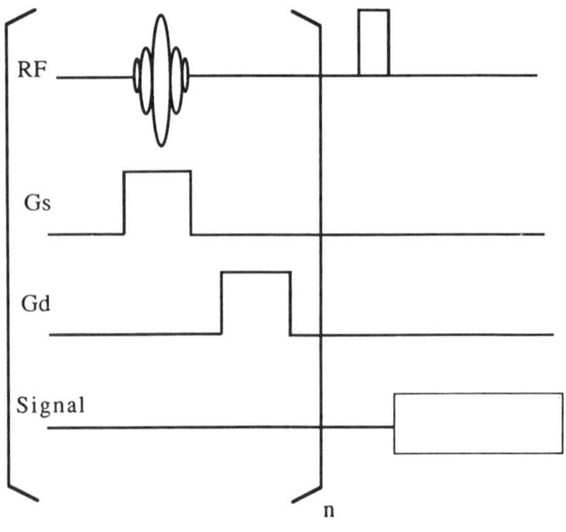

Figure 12. The pulse sequence for fast rotating gradient spectroscopy (FROGS). The symbols Gs and Gd denote the slice select and dephasing gradients, respectively.

An interesting method in which unwanted spins are removed during a preparation period prior to excitation of the desired spins is the method of volume-selective excitation (VSE) proposed by Aue *et al.* (1984). The pulse sequence for VSE is shown in Fig. 13. This method involves the use of a selective 45°, nonselective −90°, 45° spin sandwich. Since the selective 45° pulse is applied in the presence of a gradient, it affects only the spins within the slice. The nonselective −90° pulse then rotates all spins by −90°. The width of this pulse must be short because it is applied in the presence of a gradient and must excite all spins within the entire object. The last selective 45° pulse resets the full longitudinal magnetization of the spins within the slice. The pulse "sandwich" has the following effects: the magnetization of the spins within the desired slice are left aligned along the magnetic axis while the magnetization of all other spins is rotated into the transverse plane. The transverse coherence of the unwanted spins is destroyed by the other gradient pulses. When this operation is repeated along three different axes, the only magnetization that remains lies within the three-dimensional cubic voxel in space experiencing the three successive null operations. After a short settling delay, again primarily to allow recovery from residual eddy currents, the desired magnetization is sampled with a "hard" nonselective excitation pulse. As in the FROGS method, this pulse may be set to any flip angle desired. The VSE method can be used as described with a volume coil that has good B_1 homogeneity. Modifications of the VSE method for use with surface coils have been described (Mueller *et al.*, 1985a, b). The VSE method gives well-localized spectra with no preacquisition delay. The major drawback of this method is the need for sufficient RF power to deliver a broadband 90° pulse over the entire object. Since this pulse is applied in

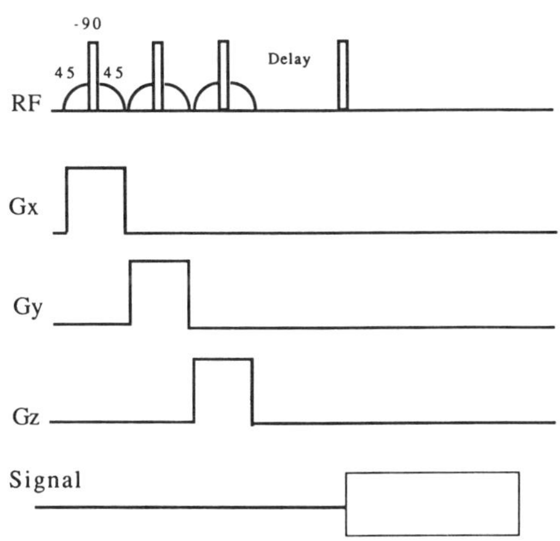

Figure 13. The pulse sequence for volume-selective excitation (VSE).

the presence of a gradient, the frequency bandwidth of a large object like the human body is quite large resulting in the need for very short 90° pulse widths. This requirement is difficult to meet with the RF coils used on humans. Since the method depends on the destruction of coherence of unwanted spins, it is not possible to modify VSE for multivoxel applications.

Haase (1986) described a method called LOCUS (localization of unaffected spins) which is essentially a three-dimensional analogue of FROGS. Even though this method was used only in phantoms, it offers intriguing possibilities as a localization method as described or as a preparation procedure which can be combined with other methods. The pulse sequence for LOCUS is shown in Fig. 14. A series of selective 90° excitation pulses are applied in the presence of a gradient. The carrier frequency of each of these pulses (four as shown in Fig. 14 for example) is set so that a single slice of spins was left unaffected (unexcited). The coherence of the transverse magnetization of the excited spins is destroyed by the successive gradients applied. After this process is repeated along each of the orthogonal axes, only spins in a block volume element remain unaffected. The location and size of this volume element depend on carrier frequencies of the RF pulses and the strengths of the gradients employed. As in the VSE method the preparation period is followed by a short delay to permit the settling of residual eddy currents followed by a nonselective pulse which can be set to any flip angle desired. The LOCUS method requires high B_1 homogeneity and was described for use with volume coils. There can be significant slice mislocation artifacts introduced which depend on the

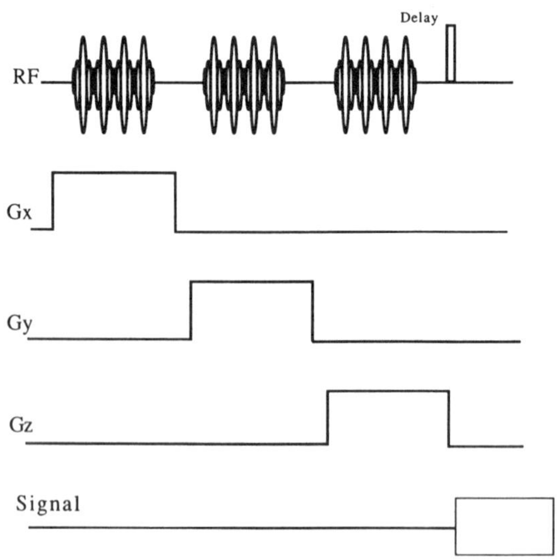

Figure 14. The pulse sequence for localization of unaffected spins (LOCUS).

strength of the slice selection gradients applied during the preparation period. Haase also pointed out that there can be recovery of magnetization from outside the region of interest if the time required to play out the pulse train is significant relative to the T_1's of the species being sampled. As was the case for VSE, this method cannot be modified easily to give multivoxel capabilities unless LOCUS is employed as the volume selection method prior to spatial encoding. It is interesting to note that the RF power deposition reached in LOCUS is about that employed in conventional multislice, multiecho MR imaging.

Ordidge *et al.* (1986) described a method called ISIS (image-selected *in-vivo* spectroscopy) in which the signals acquired during operations involving selective inversions of regions of the sample are combined to give localized spectra from a single volume in space. The pulse sequence for ISIS is shown in Fig. 15. The eight difference experiments along with the prescription for addition or subtraction of the results are given in Table 7. As will become evident later, this method has similarities to the Hadamard encoding method which will be discussed in Section 4.3.4. The ISIS method has no preacquisition delay because the magnetization is sampled by a broadband 90° pulse after each combination of inversion pulses is applied. The ISIS method is an elegant combination of binary spatial encoding with storage of magnetization along the longitudinal rather than transverse plane. It produces good localization under the appropriate acquisition conditions. The addition and subtractions depend on the fact that the magnetization recovers fully between each application of inversion

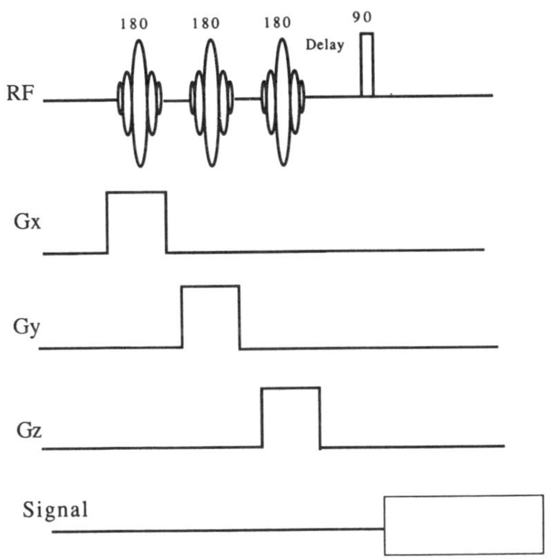

Figure 15. The pulse sequence for image-selected *in vivo* spectroscopy (ISIS).

TABLE 7
The Combinations of RF Pulses Employed in the ISIS Sequence Together with the Addition/Subtraction Required

	Inversion pulse			
Operation	G_x	G_y	G_z	Add/subtract
1	Off	Off	Off	Add
2	On	Off	Off	Subtract
3	On	On	Off	Add
4	On	On	On	Subtract
5	Off	On	On	Add
6	Off	On	Off	Subtract
7	On	Off	On	Add
8	Off	Off	On	Subtract

pulses. It is relatively easy to see that incomplete recovery will result in mislocation artifacts and poor definition of the final volume element. This problem can be overcome with the use of relatively long repetition delays. Modifications to ISIS in which the shape and number of the final volume element(s) sampled have been described (Ordidge *et al.*, 1988). Although it is possible to obtain MR images of the final ISIS voxel(s) by carrying out the eight steps in an imaging sequence, this procedure is rarely carried out. Because the ISIS voxel(s) are defined by the addition and subtractions of eight separate experiments, the region of interest is usually shimmed using a different localization method. The ISIS method is easily implemented and can be referenced to MR images in a straightforward manner. It is the most widely used single voxel method in clinical ^{31}P NMR spectroscopy.

4.3.2. Single Voxel Methods Commonly Used for Localized Solvent-Suppressed ^1H NMR Spectroscopy

There has been a great deal of recent interest in obtaining solvent-suppressed proton spectra in humans. The two most popular methods that are currently being employed are PRESS (point-resolved spectroscopy; Bottomley, 1984; Ordidge *et al.*, 1987) and STEAM (stimulated echo acquisition mode; Haase *et al.*, 1985; Granot, 1986; Kimmich and Hoepfel, 1987, Frahm *et al.*, 1987, 1989a; Moonen *et al.*, 1989). Both of these methods are direct excitation methods. Before describing the details of these two methods, we will describe another method called SPARS (spatially resolved spectroscopy; Luyten and den Hollander, 1986; Jensen *et al.*, 1987) which is a method similar to VSE in that the coherence of the unwanted spins is destroyed and the longitudinal magnetization of the desired spins is preserved. The pulse sequence for SPARS is shown in Fig. 16. The volume of

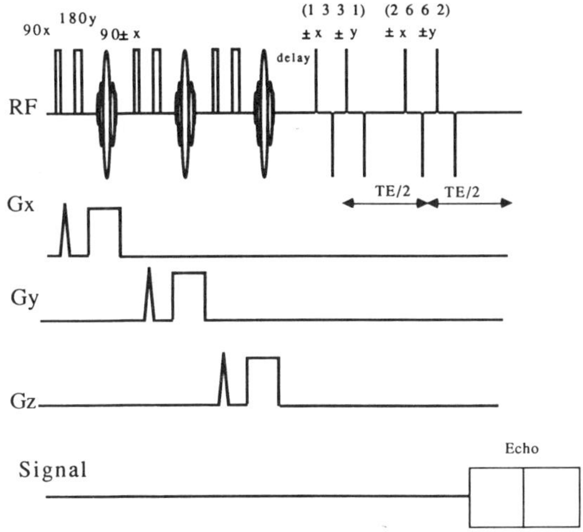

Figure 16. The pulse sequence for spatially resolved spectroscopy (SPARS).

interest is localized by performing essentially the same operation on each of the three orthogonal axes. Each of these operations involves the generation of a spatially nonselective spin echo (90° pulse followed by a 180° pulse) followed by a selective 90° "reset" pulse in the presence of a gradient. There is a gradient switched on between the nonselective 90° and 180° pulse to compensate for dephasing of spins by the slice-selective gradient. There are no gradients applied during the nonselective pulses because the durations of the pulses necessary to generate the required flip angles are much longer than 100 μsec. The excitation bandwidths of these pulses are too narrow to excite the entire volume of the object (head or body) being sampled. The net result of one of these operations is that the magnetization of the spins in the slice selected by the final "reset" pulse is returned to the z axis, while all of the magnetization left in the transverse plane after the spin echo is dephased by the slice-selective gradient. This operation can be repeated along the two other orthogonal planes to select a cubic element in space. There is only longitudinal magnetization remaining in this volume and this magnetization can be sampled by a nonselective pulse. Usually a short delay is included before the application of this final pulse to allow for the decay of residual eddy currents. In Fig. 16 the particular SPARS sequence employs a 1-3-3-1, 2-6-6-2 pair of pulses to generate a solvent-suppressed spin echo. However, any set of sampling pulse or pulses can be combined with the spatial selection portion of the sequence. Another feature of the sequence shown in Fig. 16 is the phase alteration of the selective 90° "reset" pulse in each three pulse operations. With this phase alteration the desired magnetization is reset along the $+z$ or $-z$ direction. An eight-step phase cycling scheme can be employed in which the acquired signals are either added or subtracted as the phases of the three "reset" pulses are cycled.

The PRESS method shown in Fig. 17 is a direct excitation method which employs a double spin echo. A volume in space is localized by applying three selective pulses each along one of the orthogonal axes in the presence of the appropriate gradients. The flip angle of the last two pulses is set to be 180° over the volume of interest. Any deviations in the flip angles of these pulses will cause undesired transverse magnetization which is dephased by the application of so-called "TE-crushers" applied around each of the selective 180° pulses. One of the problems with the first implementations of PRESS was the poor quality of the 180° refocusing pulses available. The development and implementation of more sophisticated, higher-quality refocusing pulses have dramatically improved the performance of the PRESS method. The magnitude and duration of the crusher gradients depend on the B_1 homogeneity of the RF coil and the B_0 homogeneity over the sample. Solvent suppression can be achieved as shown in Fig. 17 by the application of frequency-selective CHESS pulses each followed by a dephasing gradient. Alternatively, the CHESS pulses can be replaced by a selective inversion pulse followed by a delay chosen so that the water is passing through a null (the WEFT method) when the localization part of the sequence is started. One of the characteristics of this sequence is that the minimum echo delay achievable is quite long (more than 30 msec)

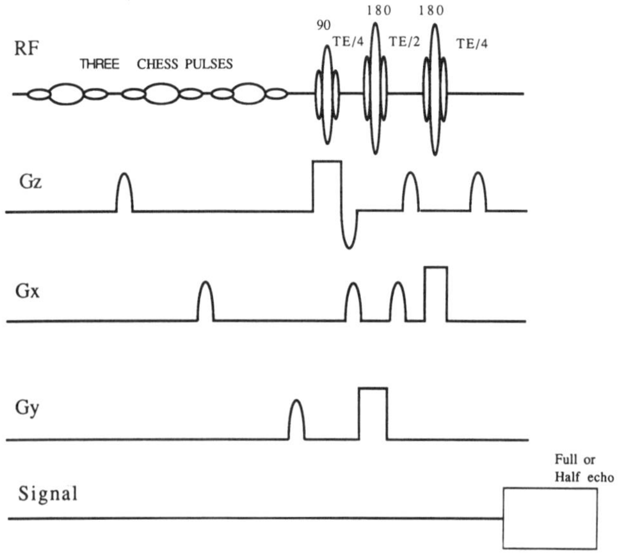

Figure 17. The pulse sequence for point-resolved spectroscopy (PRESS).

on most clinical scanners. This feature may be a serious drawback in detecting resonances with short T_2's or resonances that arise from complex coupled spin systems. The ability to obtain spectra with short echo delays may be very important *in vivo* since it has been shown in animals that the information content of spectra obtained at TEs of about 15 msec is much higher than in spectra obtained with long TEs (van Zijl *et al.*, 1989).

Because the PRESS sequence employs 180° refocusing pulses, there is no mechanism present for creating multiple quantum coherences in coupled spin systems (there are still effects due to J modulation present). This is in contrast to the STEAM sequence where localization is achieved with three selective 90° pulses. In some cases this feature of PRESS is an advantage since there are no manifestations of multiquantum coherences in the observed spectra. However, the presence of these multiquantum effects can be exploited to selectively edit spectra arising from specific coupled spin systems such as lactate. Also at short echo delays the PRESS sequence may be sensitive to the presence of residual eddy currents particularly when the volume of interest is chosen off isocenter in the magnet.

The STEAM sequence shown in Fig. 18 is used to obtain proton spectra from a single localized volume. This technique is based on the stimulated echo sequence which was first discussed in the literature by Hahn (1950). The sequence contains three 90° pulses, each of which is slice selective in one of three perpendicular directions. The stimulated echo occurs only in the volume which was excited by each of the pulses. Gradients are appropri-

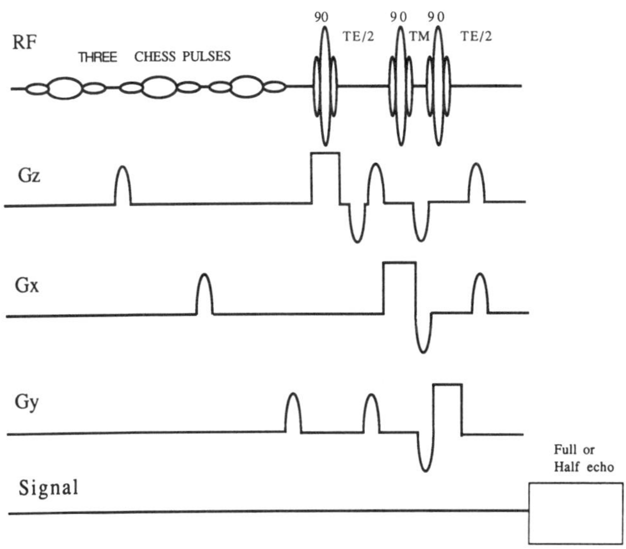

Figure 18. The pulse sequence for stimulated echo acquisition mode spectroscopy (STEAM).

ately placed to dephase any spins which are excited by the pulses but are not contained in the desired volume. Solvent suppression is achieved by the three CHESS pulses each followed by dephasing gradients which are applied before the three slice-selective 90° pulses. There have been a number of different reports dealing with the optimization of the CHESS pulses, the order of the dephasing gradient pulses and their durations (see, e.g., Moonen *et al.*, 1990; Kreis *et al.*, 1991). Using some of these approaches, one can obtain excellent solvent suppression (suppression factors greater than 1000) even at short echo times. Because good slice profiles can be achieved using simple sinc pulses for the slice-selective 90° pulses, it is possible to implement STEAM with relatively short echo times. For example, Griffey and Flamig (1990) have reported a version of STEAM called VAPOR which has a minimum echo time of 10 msec. As with PRESS, at short echo delays the STEAM sequence may be sensitive to the presence of residual eddy currents particularly when the volume of interest is chosen off isocenter in the magnet.

There have been several reports which have employed simultaneous multifrequency pulses to excite two different volume elements in a STEAM sequence (Muller *et al.*, 1989; Hafner *et al.*, 1990). This approach may have problems in shimming and solvent suppression in both volumes simultaneously, particularly when the two volumes selected are far apart. In spite of this potential problem, Hafner *et al.* (1990) were able to obtain high-quality spectra from a human volunteer using a modification of the STEAM sequence.

4.3.3. A Product Operator Treatment of an AX Spin System Acquired Using the STEAM and PRESS Sequence

The two techniques primarily being used to obtain localized proton spectroscopy are STEAM and PRESS. Since most metabolites of interest have coupled proton resonances, the effects of spin–spin coupling on the intensity of these resonances need to be understood. Such an understanding will allow accurate quantification of the spectra. The difficulties arise from modulation of the observed signal due to spin–spin coupling and the creation of zero and multiple quantum coherences which are converted to observable signal at the end of the pulse sequence. These effects can seriously distort the spectrum. In this section the results of the Cartesian product operator analysis of both of these sequences are given for an AX spin system. The results provide the information necessary to choose delay times and flip angles such that the intensities of the desired signals are maximized and all others are minimized. The analysis for the STEAM sequence has been published in detail by Kimmich *et al.* (1989). Their results are presented here along with a discussion of the implications and the complementary results for the PRESS sequence.

The "general" pulse sequence analyzed is the following:

$$(\pi/2)_x\text{-}\tau_1\text{-}(\beta)_y\text{-}\tau_2\text{-}(\delta)_x\text{-}\tau_3\text{-acquisition}$$

The first pulse is taken to be $\pi/2$ which greatly simplifies the rest of the analysis. For the STEAM sequence, the second and third pulses are also $\pi/2$. Kimmich *et al.* (1989) left these two pulses as variables in order to explore the use of various flip angles to filter signals. Here, we will consider only the results obtained when these pulses produce a $\pi/2$ flip and refer the reader to the above paper for the other results. For the PRESS analysis, the second and third pulses are expected to produce flips of π. Also in the STEAM sequence, a gradient is applied during τ_2 which eliminates all transverse magnetization and all multiple quantum coherence that is present at that time. The desired magnetization should be aligned along the z axis during this interval. This magnetization is unaffected by the gradient. The gradient does not affect the zero quantum coherence either since it evolves with the difference of two frequencies. For the PRESS sequence, no gradient is applied during τ_2 and the delay times are related, $\tau_2 = 2\tau_1$ and $\tau_3 = \tau_1$.

For both sequences, the coupling between the two spins is assumed to be weak, i.e., $2\pi|J_{12}| \ll |\Omega_1 - \Omega_2|$ where J_{12} is the spin-spin coupling constant and Ω_1 and Ω_2 are the chemical shift offsets of the two spins. Field gradients for the selective pulses are all refocused so that Ω_1 and Ω_2 are assumed to be constant during the entire pulse sequence. Relaxation is ignored and in order to avoid extremely complicated expressions, the RF pulses are assumed to be nonselective in chemical shift inside the selected slice and all spins outside the slice are ignored.

The results for the STEAM sequence follow:

$$\begin{aligned} s_{0q}(\tau_3) &= a_1[I_{1x}\cos\Omega_1\tau_3 + I_{1y}\sin\Omega_1\tau_3 - I_{2x}\cos\Omega_2\tau_3 - I_{2y}\sin\Omega_2\tau_3] \\ a_1 &= a_0\left[\cos(\Omega_1 - \Omega_2)\tau_2\cos\frac{\Omega_1+\Omega_2}{2}\tau_1(\sin)\frac{\Omega_1-\Omega_2}{2}\tau_1\right] \\ a_0 &= \sin\pi J_{12}\tau_1\sin\pi J_{12}\tau_3 \end{aligned} \tag{8}$$

$$\begin{aligned} s_{1m}(\tau_3) &= -c_1[I_{1x}\cos\Omega_1\tau_3 + I_{1y}\sin\Omega_1\tau_3] - c_2[I_{2x}\cos\Omega_2\tau_3 + I_{2y}\sin\Omega_2\tau_3] \\ c_1 &= c_0\sin\Omega_1\tau_1 \\ c_2 &= c_0\sin\Omega_2\tau_1 \\ c_0 &= \cos\pi J_{12}\tau_1\cos\pi J_{12}\tau_3 \end{aligned} \tag{9}$$

These equations are in the same form as the equations in Kimmich *et al.* (1989) but they have been simplified due to the $\pi/2$ pulses. The terms have been divided into the observable magnetization which was aligned along the z axis during τ_2, s_{1m}, and that which evolved as zero quantum coherences during τ_2, s_{0q}. Both of these signals arise from the localized volume since they result from all three pulses. The desired signal is only the magnetization which was along the z axis during τ_2, s_{1m}.

At the time of the stimulated echo, $\tau_3 = \tau_1$, s_{1m} reduces to

$$\begin{aligned} s_{1m}(\tau_3 = \tau_1) = -1/4[&I_{1x} \sin 2\Omega_1\tau_1 + I_{1y}(1 - \cos 2\Omega_1\tau_1) \\ &+ I_{2x} \sin \Omega_2\tau_1 + I_{2y}(1 - \cos 2\Omega_2\tau_1)] \\ &\times (1 + \cos 2\pi J_{12}\tau_1) \end{aligned} \tag{10}$$

There are two important points to be made about this result. First, unless something special is done, the observed signal is modulated in frequency unlike a true spin echo. In other words, the excitation of the spins is not uniform across the spectral width. At any one frequency, the entire signal may be available while at other frequencies no signal will be obtained. This arises in the following way. Prior to the second $\pi/2$ pulse, all of the magnetization is in the xy plane and is dispersed due to chemical shift evolution. If a pulse is now applied along the y axis, only that component of each magnetization vector which is along the x axis is stored along z during τ_2. All of the y components will then be dephased during τ_2. Thus, the x component which is now stored along the z axis is modulated in amplitude as $\sin \Omega\tau_1$. Unfortunately, when the z magnetization is brought into the xy plane by the third pulse, this nonuniform excitation still persists in the I_{ny} terms. This problem may be addressed in two ways. First, a gradient may be placed before the second pulse to dephase the spins as a function of space as well as chemical shift. This gradient pulse should be of sufficient strength to cause exactly a $2\pi\, n$ phase shift across the volume of interest. Now the spins at any one chemical shift will be dispersed in the xy plane. Upon application of the second pulse, exactly half of the spins at any one frequency will be placed along the z axis. The spatial dispersion of the spins must be refocused after the third pulse which is done by applying the same strength gradient at this time. A second approach to this problem is to develop a two-step phase cycle which causes the frequency modulation to be a sine function on one scan and cosine function on the next. The two data sets can then be combined to eliminate the frequency dependence. A possible phase cycle which would accomplish this is

$$\pi/2(x) - \pi/2(x) - \pi/2(x)$$

$$\pi/2(x) - \pi/2(y) - \pi/2(-y)$$

Once the frequency dependence has been removed from Eq. (10), only the modulation of the signal by J exists, i.e. $1/2(I_{1y} + I_{2y})(1 + \cos 2\pi J_{12}\tau_1)$. This is true even if the spin–spin coupling is unresolved, only here the spectral consequences are potentially severe. Only at certain delay times, $\tau_1 = 1/J$, will $\cos 2\pi J\tau_1$ be 1 as shown in Fig. 19. At this delay value, the observed signal will be similar to that obtained for uncoupled spins. Also the signal which arises from the zero quantum coherence during τ_2 is zero at this delay time as the following equation shows since it is modulated as $(1 - \cos 2\pi J_{12}\tau_1)$.

$$\begin{aligned} s_{0q}(\tau_3) = {} & 1/4(1 - \cos 2\pi J_{12}\tau_1) \cos(\Omega_1 - \Omega_2)\tau_2[1/2\{I_{1x} \sin 2\Omega_1\tau_1 \\ & - I_{1y}(1 - \cos 2\Omega_1\tau_1) + I_{2x} \sin 2\Omega_2\tau_1 - I_{2y}(1 - \cos 2\Omega_2\tau_3)\} \\ & - (I_{1x} \cos \Omega_1\tau_1 \sin \Omega_2\tau_1 + I_{1y} \sin \Omega_1\tau_1 \sin \Omega_2\tau_1 \\ & + I_{2x} \cos \Omega_2\tau_1 \sin \Omega_1\tau_1 + I_{2y} \sin \Omega_2\tau_1 \sin \Omega_1\tau_1)] \end{aligned} \quad (11)$$

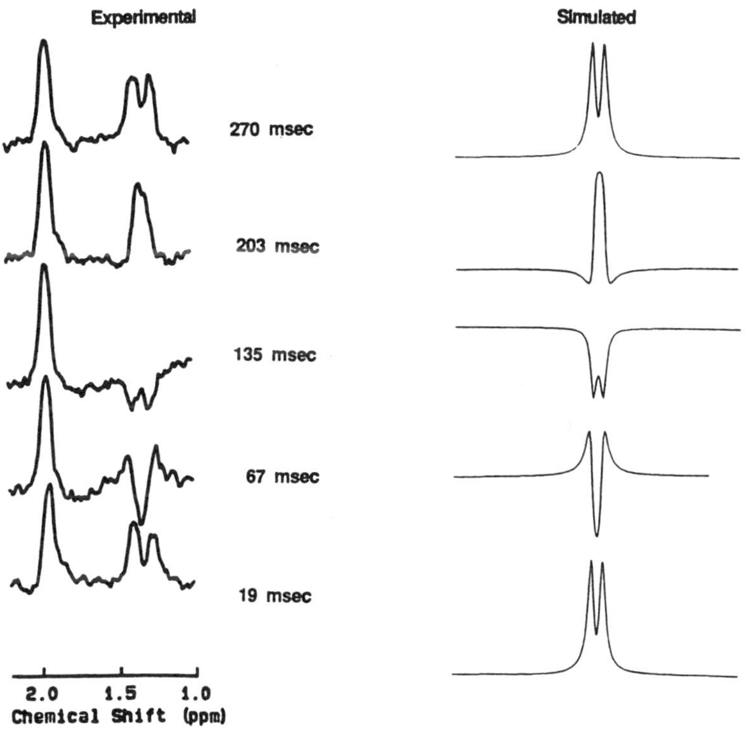

Figure 19. The experimental and simulated spectra for N-acetylalanine (10 mM) in water obtained using the STEAM sequence at the echo times indicated. The simulations were carried out using the product operator formalism described in the text.

Unlike a normal spin-echo sequence, the stimulated echo sequence never becomes negative due to J-modulation unless there is interference from the zero quantum term. For instance at $\tau_1 = 1/2J$, the s_{1m} will be zero and the only signal observed comes from zero quantum coherence developed during τ_2. Under some conditions of offset frequency and τ_2 this can be a negative signal with up to one-half of the value of the s_{1m} term.

Many parameters cause a modulation of the stimulated echo signal in a nonintuitive way and therefore can affect the signals being observed at τ_1 delays other than $1/J$. One of the most important things to note is that the amplitude of this signal will vary with the length of the τ_2 delay and the difference in the offset frequencies of the two spins due to the term, $\cos(\Omega_1 - \Omega_2)\tau_2$. These two parameters do not affect s_{1m}, but they can be adjusted to eliminate the zero quantum contribution in the STEAM experiment. Thus, it is possible to eliminate the zero quantum contribution at times other than $1/J$.

The equations which result from an analysis of the PRESS sequence are much simpler than those for STEAM.

$$I_{1x} \cos 4\pi J\tau_1 - 2I_{1y}I_{2z} \sin 4\pi J\tau_1 + I_{2x} \cos 4\pi J\tau_1 - 2I_{2y}I_{1z} \sin 4\pi J\tau_1 \tag{12}$$

The simplification comes from the use of π rotations for the second and third pulses. These pulses do not create any multiple quantum coherences. However, there is still modulation of this signal due to spin–spin coupling. If we choose τ_1 correctly, $\tau_1 = n/4J$ where n is an integer, then only the I_{1x} and I_{2x} terms survive. Thus, the PRESS sequence allows localization without interference from multiple quantum terms. Unfortunately, the analysis of both the PRESS and the STEAM sequence demonstrate a difficulty when these sequences are applied to coupled spin systems. In general, there will be more than one coupled spin system in the spectrum and it quickly becomes difficult to meet the criteria for eliminating the unwanted signals of all spin systems at once.

4.3.4. Multivoxel Methods Based on Spatial Encoding Techniques

The most common method of spatial encoding using B_0 gradients is Fourier chemical shift imaging (CSI) (Brown *et al.*, 1982; Haselgrove *et al.*, 1983; Maudsley *et al.*, 1983). Various aspects of this method have been reviewed by Buxton *et al.* (1988). The mathematical basis for CSI was elegantly presented by Brown *et al.* (1982) in the original paper describing this method. The CSI method can be performed in one, two, or three dimensions. The method can be divided into three parts: excitation of magnetization, evolution, and sampling. After an excitation pulse, the application of a phase encoding gradient causes the spins to precess at different Larmor frequencies which depend on their spatial positions, the strength and duration of the gradient. This information is stored in the relative phase angle that each spin has accumulated during the application of the gradient. The gradient is switched off and the spins are then sampled directly in an FID-like acquisition or can be refocused into an echo by the application of a refocusing pulse. All of the spatial information is encoded into the phase of the signal by applying gradients along the appropriate spatial axes. The spatial resolution along any axis is given by the inverse of the maximum value of the applied gradient or the field-of-view divided by the number of different phase encoding steps applied. The spatially registered spectra can be generated from the acquired data set by performing the appropriate number of Fourier transforms in the spatial dimensions followed by a Fourier transform of the spatially registered time domain signals. Since the spatial localization is encoded into the phase of the signal, and not the amplitude, there are no localization errors caused by relaxation effects. Also, there is no slice mislocation artifact present except when using slice selection and in this case the artifact occurs only in the direction of the slice selection. This technique may be used with a surface coil since the flip angle does not need to have any particular value. An example of a pulse sequence in which a phase encoding gradient is applied along one axis is shown in Fig. 20. This particular example has been referred to as a one dimensional or 1D-CSI experiment. For consistency, this experiment should be referred to as a 2D-CSI experiment since the higher spatial order CSI methods such as 3D- and 4D-CSI derive their designations from the total number of Fourier transforms that must be applied to the raw data set in order to obtain spatially registered spectra. Examples of 3D- and 4D-CSI sequences are shown in Figs. 21 and 22, respectively. These sequences are shown as direct acquisition CSI methods. The analogous spin-echo versions of the 1D-, 3D-, and 4D-CSI methods are similar to those shown

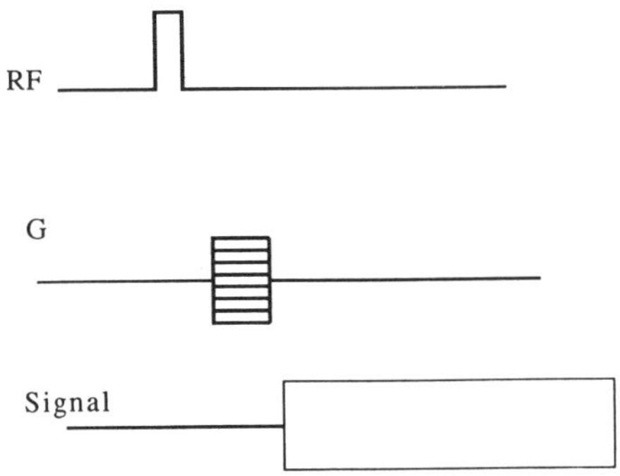

Figure 20. The pulse sequence for one-dimensional chemical shift imaging (1D-CSI). In this particular sequence, eight phase encoding steps are indicated.

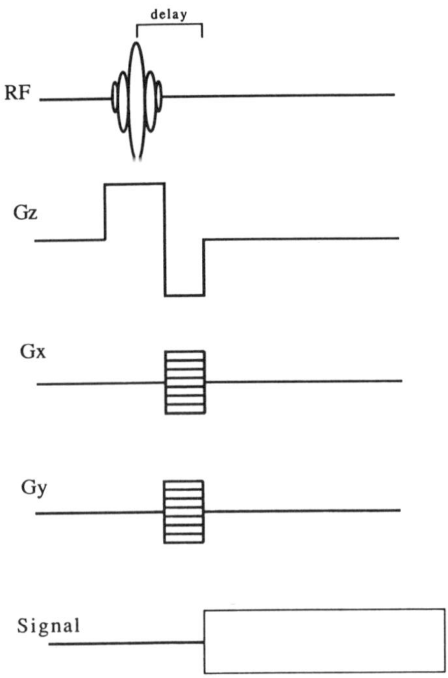

Figure 21. The pulse sequence for three-dimensional (two spatial and one frequency) chemical shift imaging (3D-CSI). In this particular sequence, eight phase encoding steps are indicated along each of the *x* and *y* axes. This results in the acquisition of 64 FIDs. As in the DRESS method there is a short delay between the center of the RF pulse and the start of data acquisition.

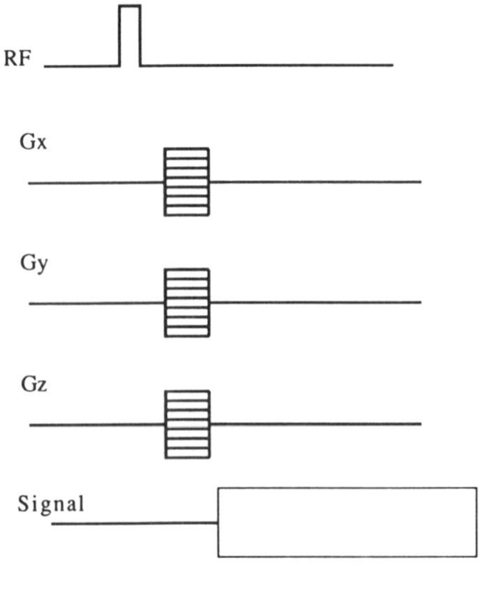

Figure 22. The pulse sequence for four-dimensional chemical shift imaging (4D-CSI). In this sequence, eight phase encoding steps are indicated along each of the three gradient axes. This will result in the acquisition of 512 FIDs.

in Figs. 20, 21, and 22 with the addition of a refocusing pulse. The 1D- and 4D-CSI sequences employ a nonselective excitation pulse followed by phase encoding gradients along one and three axes, respectively. In these two sequences the preacquisition delay can be kept relatively short. The total acquisition time for the 4D-CSI sequence is given by

$$\mathrm{TT} = \mathrm{NEX} * \mathrm{TR} * \mathrm{N}_x * \mathrm{N}_y * \mathrm{N}_z \tag{13}$$

where NEX is the number of averages per phase encoding steps, TR is the delay between successive acquisitions, and N_x, N_y, and N_z are the number of phase encoding steps applied along each of the axes. The total acquisition time would be 68 min for 16 steps acquired along each direction with a repetition time of 1 sec and 1 average per step. This time can be reduced significantly by decreasing the spatial resolution of the acquisition. As will be shown in the following section, under these conditions there may be significant truncation artifact present caused by the sparse sampling of the spatial dimensions.

One of the main advantages of the CSI method is the fact that it is a very efficient encoding method in terms of the S/N per encoding step. All of the spins within the encoding volume are sampled fully at every phase encoding step. Also the CSI methods can all be performed at reduced flip angles for excitation which can increase the S/N of the acquisition (Ernst, 1966).

The examples shown for the CSI methods are widely used in multivoxel studies of ^{31}P. The 1D-CSI method is commonly used in combination with surface coils to obtain localized spectra from several slices. In these applications the gradient direction is selected to sample slices parallel to the plane of the coil. The other two methods are more commonly used in combination with volume coils like a head coil or extremity coil to obtain localized spectra from 3D volume elements. The direct acquisition methods are most commonly used for ^{31}P since it is generally believed that the echo delays in the spin-echo methods are too long to adequately sample resonances with relatively short T_2's.

Applications of CSI methods to solvent-suppressed proton spectroscopy have involved combining phase encoding methods with single voxel localization techniques like STEAM and PRESS. In these implementations, the STEAM and PRESS sequences can be thought of as spatial preselection methods in which a portion of an object is selected for phase encoding. These STEAM and PRESS CSI methods involve the incorporation of the appropriate number of phase encoding gradients into the sequences shown in Figs. 17 and 18. In most of the applications of CSI to solvent-suppressed proton spectroscopy the slice-selective pulses and gradients are set to excite a large box typically of dimensions of about 5–10 cm with a thickness of around 1–2 cm. Since it is only the spins within this box that are sampled, the application of phase encoding gradients along the two long axes of the box subdivides the box into $N \times N$ voxels. One of the major motivations for preselecting a box to encode is the fact that it is difficult to shim a large volume of tissue well enough to suppress the water resonance in all parts of the sample.

It is important to note that the successful reconstruction of a Fourier chemical shift image is based on the assumption that the sum of the signal inside a voxel can be treated as concentrated at a point in the center of the voxel. This in effect places all of the signal intensity on a lattice point in k space and perfect reconstruction of such an object would be possible. However, this assumption has effects on the localization obtained by this technique. First, no object of any interest clinically has signal only on the lattice points. A uniform distribution of signal in the voxel will cause intravoxel dephasing. This results in a diminution of signal inside the voxel as described by Maudsley (1986). However, this signal is not really "lost" but is redistributed to neighboring voxels. A second and more pernicious source of localization error is caused by asymmetric distributions of signals inside the voxels. Both of these effects were recently described in detail (Wang *et al.*, 1991) along with a method of analyzing the data which eliminates both of these effects to some degree.

In most analyses of the Fourier-based CSI method, the signal from any pixel is taken to be concentrated at the center of each pixel and therefore sees only one discrete field strength induced by the gradient. Thus, the signal from each pixel in the phase encoding direction traces out cosine waves

$$s_T = \sum^{m} \int_{-\phi/2}^{\phi/2} \cos \phi \, d\phi \tag{14}$$

and the detected signal for each phase encoding step is merely a sum of the signals from each point:

$$S_j = \sum_{n}^{N} s_n \, e^{-i\gamma} G_j \tau r_n \tag{15}$$

Since the object does not consist of equally spaced points of signal but rather is a continuum of signal, each pixel is represented by an integral rather than a discrete point. Therefore, the detected signal becomes

$$S_j = \sum_n^N s_n \int e^{-i\gamma} G_j \tau r_n \, dx_i \tag{16}$$

In this case, different points in the pixel see a different field strength causing a phase dispersion of the signals and a cancellation of signal intensity as pointed out by Maudsley (1986). Thus, the cosine wave traced out by the single point is now multiplied by part of a sinc function, from $-\pi/4$ to $\pi/4$. Two problems now occur simultaneously. First, the data set has been truncated in k space. Second, the k space data have been modulated by the sinc function described. If one is to accurately reconstruct these data, it is necessary to correct for both of these effects. The example shown in Fig. 23 illustrates the problem. Here the object is considered to be continuous with a hole. The hole is aligned inside a pixel and one still finds 25% of the signal bleeding in from surrounding pixels. Thus, if the object were a brain with a tumor and the data were aligned so that a single pixel fit inside the tumor, the tumor spectrum would be contaminated by brain spectra. If the object is not aligned with the pixel or in other words, the voxel does not contain a uniform distribution of spins, then further localization errors occur.

All of these localization errors can be alleviated as long as information about the shape of the object is available from a high-resolution image.

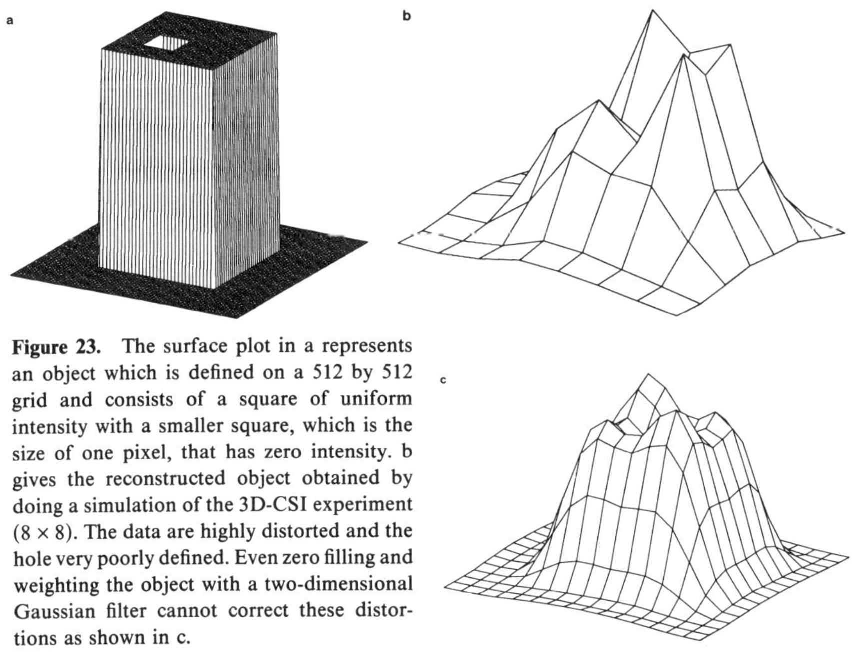

Figure 23. The surface plot in a represents an object which is defined on a 512 by 512 grid and consists of a square of uniform intensity with a smaller square, which is the size of one pixel, that has zero intensity. b gives the reconstructed object obtained by doing a simulation of the 3D-CSI experiment (8 × 8). The data are highly distorted and the hole very poorly defined. Even zero filling and weighting the object with a two-dimensional Gaussian filter cannot correct these distortions as shown in c.

This information is almost always available in order that one may determine the location of the interesting features (tumor) of the object (brain). This procedure is capable of correcting three errors:

1. The sinc point spread function resulting from truncation of the data set
2. Phase dispersion arising from linear gradients and continuous samples and to some degree
3. The arbitrary shape of the object

If each pixel is defined so that its spin density is uniform, then all we must correct is the phase dispersion which is easily done by dividing the data by a sinc function from $-\pi/4$ to $\pi/4$. The next step in correcting the data is to redefine the shape and size of each pixel so that the spectrum inside each is uniform. This is done with the guidance of a proton image.

The detected NMR signal after the jth phase encoding step is

$$S_j = \sum_{k=1}^{N} P_k f_{kj} \tag{17}$$

where P_k is the signal from each compartment in the absence of phase encoding and f_{kj} is the geometric phase factor:

$$f_{kj} = \int_k e^{-i\gamma \mathbf{G}_j \tau \mathbf{r}_k}\, d\mathbf{r}_k \tag{18}$$

Thus, f_{kj} is calculated from geometric factors:

1. The magnitude and direction of the jth phase encoding vector ($\gamma \mathbf{G} \tau$)
2. The size, shape, and location of the kth compartment

At the original lattice point of the pixel $r_k = 0$. Using the measured S_j's and the calculated f_{kj}'s the density of the signal for each pixel can be solved by using Eq. (18) or in matrix form

$$\mathbf{s} = \mathbf{F}\mathbf{p} \tag{19}$$

F, therefore, is an n by n matrix with elements f_{kj} and **p** can be found by inverting **F** and multiplying it by **s**. This technique does not require any knowledge about the spectrum inside each compartment. It does assume that the signal within each compartment is uniform. This hypothesis can be tested in principle by redefining the compartment and performing the entire procedure again. Of course, once the compartments become too small, the signal will become indistinguishable from the noise.

Having described the CSI methods based on Fourier encoding, it is also important to point out that there are other classes of encoding methods. As was pointed out earlier, the ISIS method is very similar to an encoding method based on Hadamard rather than Fourier transforms. The details of this method have been described by Bolinger (1989) and Bolinger and Leigh (1988). The Hadamard method as illustrated in Fig. 24 is a binary encoding

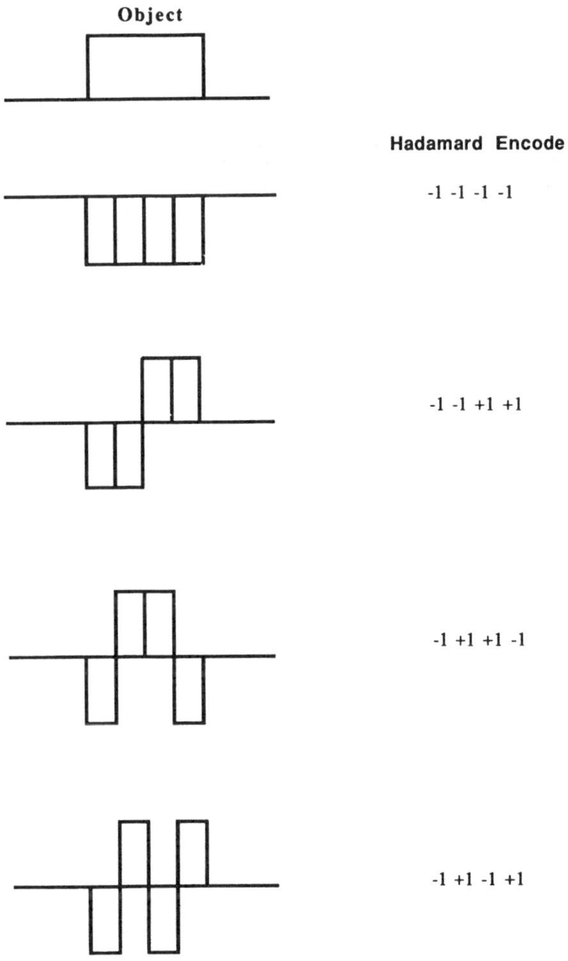

Figure 24. An illustration of a four-step, one-dimensional Hadamard encoding sequence. The object is shown in the top trace. Each of the four different spatially selective inversion steps is shown in the four lower traces. The designation of each step is given on the right. The addition of the four results, i.e., I + II + III + IV (where I is the result of the $-1-1-1-1$ inversion, II the result of the $-1-1+1+1$ inversion, etc.), gives -4 times the first slice of the object. The second slice is given by $-(1/4) \times (\mathrm{I} + \mathrm{II} - \mathrm{III} - \mathrm{IV})$; the third slice by $-(1/4) \times (\mathrm{I} - \mathrm{II} - \mathrm{III} + \mathrm{IV})$; and the fourth slice by $(1/4) \times (\mathrm{I} - \mathrm{II} + \mathrm{III} - \mathrm{IV})$.

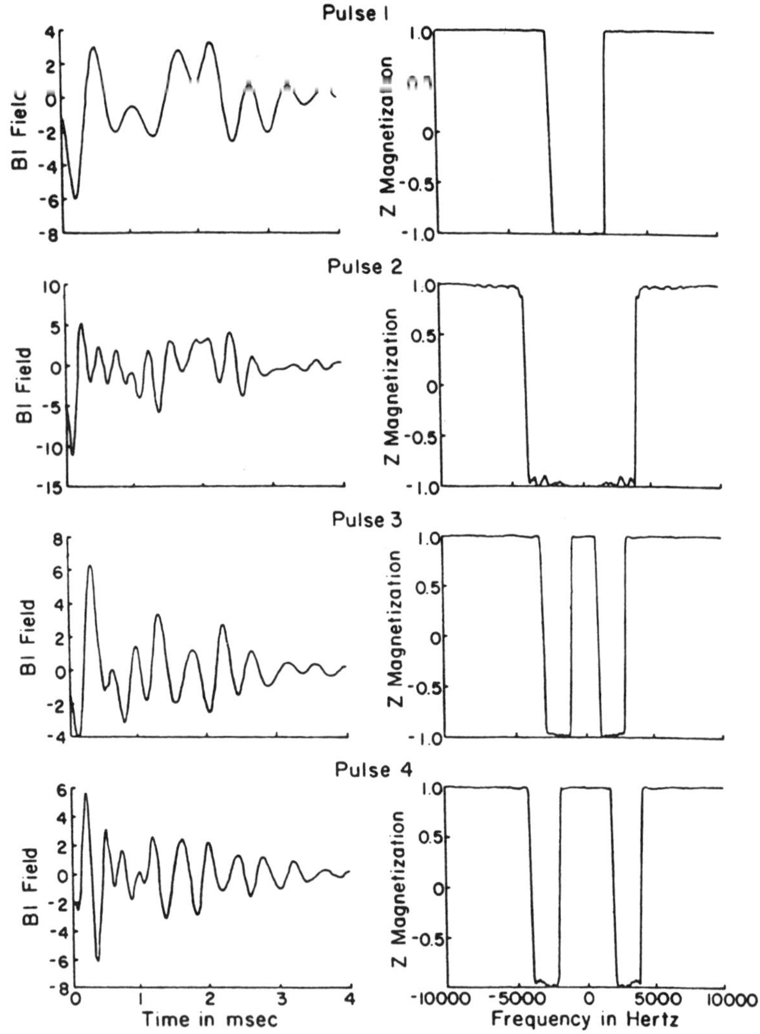

Figure 25. Examples of the pulses that can be employed in a four-step Hadamard encoding sequence. The pulses were synthesized by the methods indicated in the text, in order to obtain the Walsh functions corresponding to the fourth-order natural Hadamard matrix. The peak B_1 requirement for pulse 2 is only 1.5 times that of a sinc pulse with the same excitation bandwidth, 8 kHz. The excitation profiles shown here were obtained through a spin simulation program. These profiles have also been determined experimentally and are virtually identical to the simulations.

technique. In the 1D Hadamard method illustrated, four separate experiments are performed each of which selectively inverts the magnetization in different parts of the sample. The four separate slices can be determined by the appropriate addition and subtraction of the results of the experiments. Multiband inversion pulses can be calculated using the approach outlined by Shinnar and co-workers (Shinnar and Leigh, 1987; Eleff *et al.*, 1987; Shinnar *et al.*, 1989a, b). A set of these pulses suitable for a four-slice Hadamard encoding sequence together with their frequency profiles are shown in Fig. 25. The generalized pulse sequence for a 3D Hadamard sequence is shown in Fig. 26. The appropriate selective inversion pulses are applied in the presence of the appropriate gradients to encode the sample along the three orthogonal axes. For a 4 × 4 × 4 Hadamard sequence the four pulses shown in Fig. 25 are applied in the presence of the *x*, *y*, and *z* gradients. The encoded magnetization is then sampled with a sampling pulse or spin-echo sequence. The resulting 64 time-domain signals can be transformed by applying the appropriate 3D Hadamard transform followed by a Fourier transform of the spatially registered time-domain signals. The properties and limitations of this encoding method have been discussed in detail by Bolinger (1989). The major advantage is that this method is capable of producing well-localized multivoxel spectra even when the number of encoding steps is small (i.e., 4 × 4). Under these conditions the truncation artifacts present in the Fourier CSI method are very significant. Also, the

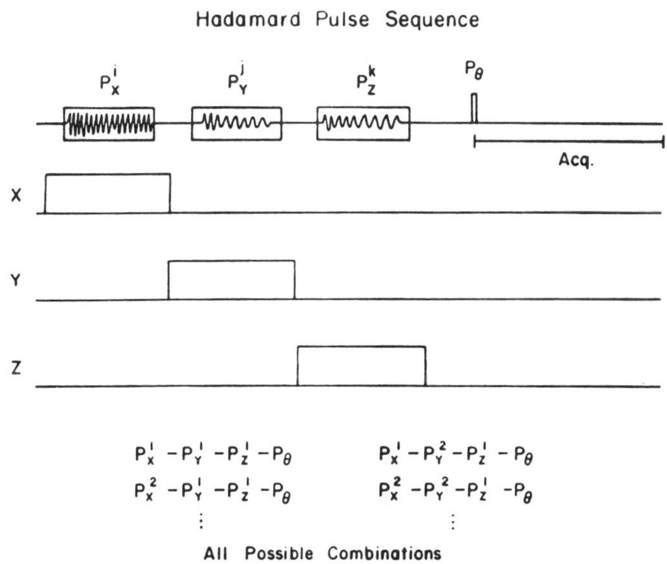

Figure 26. The pulse sequence for a four-dimensional (three spatial and one frequency) Hadamard encoding sequence.

Hadamard method can encode a field of view that is smaller than the object. The most significant problem is that of RF excitation bandwidth, since the Hadamard pulses are applied in the presence of a gradient. All of the Hadamard methods can be implemented with conformal, unequally sized, parallelepiped-shaped voxels.

Goelman and Leigh (1991) have developed a version of the Hadamard method which employs adiabatic pulses in combination with a surface coil. Since less power is necessary to excite the spins, this method may overcome some of the limitations described above.

5. SUMMARY AND FUTURE DEVELOPMENTS

We have described a variety of localization methods that are currently being employed in clinical NMR spectroscopy. The theoretical performance of some of these methods has been compared by Aue (1986), Galloway *et al.* (1987), Garwood *et al.* (1987), and Bolinger (1989). Rather than repeat the specific ratings of each of these methods, we have attempted to provide some indications of the general limitations of the various approaches employed to provide spatial localization. In this process we hope that we have provided some general insights into the technical and practical constraints that are currently present in these methods. In spite of all of the problems inherent in these methods, we feel that it is important to stress that there are many investigators who are actively applying these methods in the clinical setting (for reviews of the results of clinical studies, see Bottomley, 1989; Radda *et al.*, 1989; Weiner, 1990). The most popular single voxel method for ^{31}P is ISIS. The STEAM and PRESS methods are the most widely used methods for proton spectroscopy. All three of these methods provide excellent localization. They can all be combined with MR imaging sequences to give MR images of the volumes sampled.

The CSI methods based on Fourier encoding are gaining in popularity. These methods fulfill many of the criteria set out for the ideal localization technique. Moreover, there has been a great deal of progress made in solving many of the technical problems present in all of the localization methods. Examples include improvements in eddy current compensation methods and self-shielded gradient coil technology, better RF pulse generation and control, improvements in RF coil designs for spectroscopy, and better data processing algorithms.

In addition to these technical improvements, we wish to highlight several new techniques that have been reported recently, which may have an impact on localization in the clinical setting. Hu *et al.* (1988) have reported a method called SLIM (spectral localization by imaging) which offers interesting novel ways of combining high-resolution MRI spatial information with spectroscopic data. There have also been several reports dealing with nonstandard methods of sampling the spatial frequencies of an object. These methods involve the use of designer RF pulses in the presence of oscillating B_0 gradients to define sensitive volumes in space (see, e.g., Axel, 1990; Spielman *et al.*, 1991). Another interesting future development is the possible combination of cross-polarization methods such as INEPT with spatial localization methods to increase the spectral resolution and selectively edit *in vivo* spectra (Vogele, 1991).

Aue (1986) pointed out that the number of papers dealing with localization methods appeared to fit an exponential function. The author took this to indicate that the field was young and dynamic. Over the five years that have passed since his initial review, the number of papers dealing with localization methods has appeared to keep pace with the predicted function. The ultimate success in this area will be measured by the ease with which these techniques find their way into routine clinical use.

REFERENCES

Ackerman, J. J. H., Grove, T. H., Wong, G. G., Gadian, D. G., and Radda, G. K., 1980, *Nature* **283**:167.

Ahn, C. B., and Cho, Z. H., 1991, *Magn. Reson. Med.* **17**:149.

Allman, T., Holland, G. A., Lenkinski, R. E., and Charles, H. C., 1988, *Magn. Reson. Med.* **7**:88.

Aue, W. P., 1986, *Rev. Magn. Reson. Med.* **1**:21.

Aue, W. P., Mueller, S., Cross, T. A., and Seelig, J., 1984, *J. Magn. Reson.* **56**:350.

Axel, L., 1990, A Method for Rapid One-Dimensional Chemical Shift Imaging, RSNA Abstr., p. 133.

Bendall, M. R., and Gordon, R. E., 1983, *J. Magn. Reson.* **53**:365.

Blackledge, M. J., Rajagopalan, B., Oberhaensli, R. D., Bolas, N. M., Styles, P., and Radda, G. K., 1987, *Proc. Natl. Acad. Sci. USA* **84**:4283.

Bolinger, L., 1989, A Novel Technique for Nuclear Magnetic Resonance Spectroscopic Imaging Using Hadamard Transforms, Ph.D. dissertation, Department of Biochemistry and Biophysics, University of Pennsylvania.

Bolinger, L., and Leigh, J. S., 1988, *J. Magn. Reson.* **80**:162.

Bottomley, P. A., 1984, Selective Volume Method for Performing Localized NMR Spectroscopy, U.S. Patent 4,480,228.

Bottomley, P. A., 1989, *Radiology* **170**:1.

Bottomley, P. A., Foster, T. B., and Darrow, R. D., 1984, *J. Magn. Reson.* **59**:338.

Bottomley, P. A., Smith, L. S., Leue, W. M., and Charles, H. C., 1985, *J. Magn. Reson.* **64**:347.

Brown, T. R., Kincaid, B. M., and Ugurbil, K., 1982, *Proc. Natl. Acad. Sci. USA* **79**:3523.

Buchtal, S. D., Thoma, W. J., Taylor, J. S., Nelson, S. J., and Brown, T. R., 1989, *NMR Biomed.* **2**:298.

Buxton, R. B., Brady, T. J., Wismer, G. L., and Rosen, B. R., 1988, in *Magnetic Resonance Imaging*, 2nd ed., Vol. II (Partain C. L., Price, R. R., Patton, J. A., Kulkarni, M. V., and James, A. E., Jr., eds.), Saunders, Philadelphia.

Cooper, J. W., 1983, *The Minicomputer in the Laboratory*, Wiley, New York.

Cox, I. J., Coutts, G. A., Gadian, D. G., Ghosh, P., Sargentoni, J., and Young, I. R., 1991, *Magn. Reson. Med.* **17**:53.

Eleff, S. M., Subramanian, V. H., Shinnar, M., Renn, S., and Leigh, J. S., Jr., 1987, *J. Magn. Reson.* **72**:298.

Ernst, R. R., 1966, *Adv. Magn. Reson.* **2**:1.

Frahm, J., Merboldt, K. D., Hanicke, W., and Haase, A., 1985, *J. Magn. Reson.* **64**:81.

Frahm, J., Merboldt, K. D., and Hanicke, W., 1987, *J. Magn. Reson.* **72**:502.

Frahm, J., Bruhn, H., Gyngell, M. L., Merboldt, K. D., Hanicke, W., and Sauter, R., 1989a, *Magn. Reson. Med.* **9**:79.

Frahm, J., Bruhn, H., Gyngell, M. L., Merboldt, K. D., Hanicke, W., and Sauter, R., 1989b, *Magn. Reson. Med.* **11**:47.

Galloway, G. J., Brereton, I. M., Brooks, W. M., Field, J., Irving, M., Baddeley, H., and Doddrell, D. M., 1987, *Magn. Reson. Med.* **4**:393.

Garwood, M., Schleich, T., Matson, G. B., and Acosta, A., 1984, *J. Magn. Reson.* **60**:268.

Garwood, M., Schleich, T., Ross, B. D., Matson, G. B., and Winters, W. D., 1985, *J. Magn. Reson.* **65**:239.

Garwood, M., Schleich, T., and Bendall, M. R., 1987, *J. Magn. Reson.* **73**:191.

Goelman, G., and Leigh, J. S., Jr., 1991, *J. Magn. Reson.* **91**:93.

Gorenstein, D. G., 1984, in *Phosphorus-31 NMR: Principles and Applications* (Gorenstein, D. G., ed.), Academic Press, New York.

Granot, J., 1986, *J. Magn. Reson.* **70**:488.

Griffey, R. H., and Flamig, D. P., 1990, *J. Magn. Reson.* **88**:161.

Haase, A., 1986, *Magn. Reson. Med.* **3**:963.

Haase, A., and Frahm, J., 1985, *J. Magn. Reson.* **64**:94.

Haase, A., Frahm, J., Hanicke, W., and Matthei, D., 1985, *Phys. Med. Biol.* **30**:341.

Hafner, H.-P., Radu, E., and Seelig, J., 1990, *Magn. Reson. Med.* **15**:135.

Hahn, E. L., 1950, *Phys. Rev.* **80**:580.

Haselgrove, J. C., Subramanian, V. H., Leigh, J. S., Jr., Gyulai, L., and Chance, B., 1983, *Science* **220**:1170.

Hore, P. J., 1983, *J. Magn. Reson.* **55**:283.

Hoult, D. I., 1979, *J. Magn. Reson.* **33**:183.

Hu, X., Levin, D. N., Lauterbur, P. C., and Spraggins, T., 1988, *Magn. Reson. Med.* **8**:314.

Jensen, D. J., Narayana, P. A., and Delayre, J. L., 1987, *Med. Phys.* **14**:38.

Kimmich, R., and Hoepfel, D., 1987, *J. Magn. Reson.* **72**:379.

Kimmich, R., Rommel, E., and Knuttel, A., 9189, *J. Magn. Reson.* **81**:333.

Klose, U., 1990, *Magn. Reson. Med.* **14**:26.

Kreis, R., Farrow, N., and Ross, B. D., 1991, *NMR Biomed.* **4**:109.

Luyten, P. R., and den Hollander, J. A., 1986, *Magn. Reson. Imag.* **4**:237.

Luyten, P. R., Groen, J.P., Vermullen, J. W. A. H., and den Hollander, J. A., 1989, *Magn. Reson. Med.* **11**:1.

Maudsley, A. A., 1986, *J. Magn. Reson.* **68**:363.

Maudsley, A. A., Hilal, S. K., Perman, W. H., and Simon, H. E., 1983, *J. Magn. Reson.* **51**:147.

Merboldt, K. D., Chien, D., Bruhn, H., Gyngell, M. L., Hanicke, W., and Frahm, J., 1990, Soc. Magn. Reson. Med., Abstr. 9th Annu. Meet., Vol. 2, p. 1066.

Moonen, C. T. W., and van Zijl, P. C. M., 1990, *J. Magn. Reson.* **88**:28.

Moonen, C. T. W., Kienlin, M. V., van Zijl, P. C. M., Cohen, J., Gillen, J., Daly, P., and Wolf, G., 1989, *NMR Biomed.* **2**:201.

Mueller, A., Aue, W. P., and Seelig, J., 1985a, *J. Magn. Reson.* **65**:332.

Mueller, S., Aue, W. P., and Seelig, J., 1985b, *J. Magn. Reson.* **63**:87.

Muller, S., Hafner, H.-P., and Beckman, N., 1989, *NMR Biomed.* **4**:209.

Narayana, P. A., and Delayre, J. L., 1988, in *Magnetic Resonance Imaging*, 2nd ed., Vol. II (Partain, C. L., Price, R. R., Patton, J. A., Kulkarni, M. V., and James, A. E., Jr., eds.), Saunders, Philadelphia.

Ordidge, R. J., and Cresshull, I. D., 1986, *J. Magn. Reson.* **69**:151.

Ordidge, R. J., Connelly, A., and Lohman, J. A. B., 1986, *J. Magn. Reson.* **66**:283.
Ordidge, R. J., Mansfield, P., Lohman, J. A. B., and Prime, S. B., 1987, *Ann. N.Y. Acad. Sci.* **508**:376.
Ordidge, R. J., Bowley, R. M., and McHale, G., 1988, *Magn. Reson. Med.* **8**:323.
Pekar, J., Leigh, J. S., and Chance, B., 1985, *J. Magn. Reson.* **64**:115.
Plateau, P., Dumas, C., and Gueron, M., 1983, *J. Magn. Reson.* **54**:46.
Posse, S., Lazeyras, F., Herschkowitz, N., Aue, W. P., and Charles, H. C., 1990, Soc. Magn. Reson. Med., Abstr. 9th Annu. Meet., Vol. 2, p. 993.
Radda, G. K., Rajagopalan, B., and Taylor, D. J., 1989, *Magn. Reson. Q.* **5**:112.
Riddle, W. R., Gibbs, S. J., and Willcott, M. R., 1992, *Magn. Reson. Med.*, in press.
Rothman, D. L., Behar, K. L., Hetherington, H. P., den Hollander, J. A., Bendall, M. R., Petroff, O. A. C., and Shulman, R. G., 1985, *Proc. Natl. Acad. Sci. USA* **82**:1633.
Sauter, R., Mueller, S., and Weber, H., 1987, *J. Magn. Reson.* **75**:167.
Seeholzer, S. H., 1985, Carbon-13 Nuclear Magnetic Resonance Studies of Cardiac Metabolism, Ph.D. dissertation, Department of Biochemistry and Biophysics, University of Pennsylvania.
Shinnar, M., and Leigh, J. S., 1987, *J. Magn. Reson.* **75**:502.
Shinnar, M., Bolinger, L., and Leigh, J. S., 1989a, *Magn. Reson. Med.* **12**:88.
Shinnar, M., Bolinger, L., and Leigh, J. S., 1989b, *Magn. Reson. Med.* **12**:81.
Spielman, D., Meyer, C., Mackovski, A., and Enzmann, D., 1991, *Magn. Reson. Med.* **18**:269.
Styles, P., Scott, C. A., and Radda, G. K., 1985, *Magn. Reson. Med.* **2**:409.
Styles, P., Blackledge, M. J., Moonen, C. T. W., and Radda, G. K., 1987, *Ann. N.Y. Acad. Sci.* **508**:349.
Turner, C. J., and Garlick, P. B., 1984, *J. Magn. Reson.* **57**:221.
van Zijl, P. C. M., Moonen, C. T. W., and von Kienlin, M., 1989, Soc. Magn. Reson. Med., Abstr. 8th Annu. Meet., Vol. 2, p. 646.
Vogele, K., 1991, *In Vivo* Two-Dimensional Proton to Phosphorus Polarization Transfer Using Surface Coils, Ph.D. dissertation, Department of Biochemistry and Biophysics, University of Pennsylvania.
Wang, Z., Bolinger, L., Subramanian, H. V., and Leigh, J. S., 1991, *J. Magn. Reson.* **92**:64.
Wiener, M. W., 1990, *Administrative Radiology* p. 34.

2

Off-Resonance Rotating Frame Spin–Lattice Relaxation

Theory, and *in Vivo* MRS and MRI Applications

Thomas Schleich, G. Herbert Caines, and Jan M. Rydzewski

1. INTRODUCTION

The off-resonance rotating frame spin–lattice relaxation experiment represents a class of magnetic resonance techniques in which a low-power, continuous-wave radiofrequency (RF) field is applied off-resonance from a selected resonance. Application of such an irradiation field establishes nuclear spin polarization along an effective field inclined at an angle Θ away from the z axis of the rotating frame. In its original form, this approach was employed for the determination of the rotational correlation time, i.e., the rotational diffusion characteristics of a reorienting macromolecule (James *et al.*, 1978; Schleich *et al.*, 1989). More generally, the off-resonance rotating frame spin–lattice relaxation experiment can be thought of as a form of off-resonance saturation, in which a selected resonance is either

Thomas Schleich • Department of Chemistry and Biochemistry, Sinsheimer Laboratories, University of California, Santa Cruz, California 95064 and Department of Pharmaceutical Chemistry, School of Pharmacy, University of California, San Francisco, California 94143. **G. Herbert Caines and Jan M. Rydzewski** • Department of Chemistry and Biochemistry, Sinsheimer Laboratories, University of California, Santa Cruz, California 95064.

Biological Magnetic Resonance, Volume 11: In Vivo Spectroscopy, edited by Lawrence J. Berliner and Jacques Reuben. Plenum Press, New York, 1992.

completely or partially saturated. Viewed in these terms, the off-resonance rotating frame spin–lattice relaxation formalism, modified to include magnetization transfer phenomena, may be used to obtain chemical exchange kinetic and magnetic cross-relaxation information.

Various representations of the off-resonance rotating frame spin–lattice relaxation experiment displaying the reduction in signal intensity of a resonance by low-power off-resonance RF irradiation are schematically depicted in Fig. 1. Panel A displays the conventional representation of the experiment used for the determination of the rotational correlation time of a reorienting particle; panel B illustrates the conventional saturation transfer experiment; panels C and D incorporate magnetization transfer into the off-resonance rotating frame spin–lattice relaxation experiment.

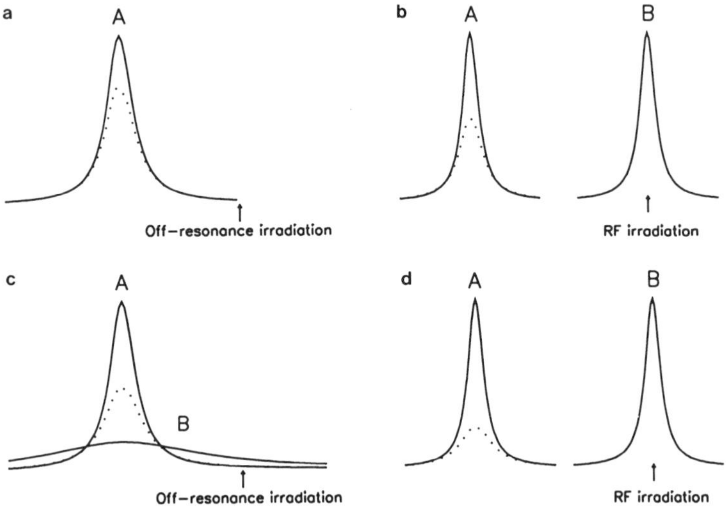

Figure 1. Schematic representation of the different forms of the off-resonance rotating frame spin–lattice relaxation experiment. (a) Conventional representation of the off-resonance rotating frame spin–lattice relaxation technique employed for the determination of the rotational correlation time. The far off-resonance limiting value of the intensity ratio equals unity. Magnetization transfer effects are assumed to be absent. (b) Conventional representation of the saturation transfer experiment employed for the determination of kinetic rate constants. The absence of off-resonance effects on resonance A is assumed. (c) The effect of exchange on the behavior of the off-resonance rotating frame spin–lattice relaxation experiment from a very broad spectral component (NMR "invisible") into an NMR "visible" pool of spins. The far off-resonance limiting value of the intensity ratio is less than unity. Off-resonance and magnetization transfer effects on resonance A are assumed. (d) The effect of exchange on the behavior of the off-resonance rotating frame spin–lattice relaxation experiment when both pools of spins are NMR "visible." Off-resonance and magnetization transfer effects are assumed. Resonances A and B are assumed to be in chemical exchange equilibrium. In each case the dotted line shape represents the response of the A resonance to the off-resonance RF irradiation.

Panel C illustrates the effect of exchange from a very broad spectral component, i.e., NMR "invisible" into a "mobile" and, hence, "visible" pool of spins, whereas panel D depicts the exchange case where both species are NMR "visible."

Applications of the formalism presented in this chapter include the *in vivo* determination of rotational diffusion characteristics of small as well as large molecules of biological interest (Section 5), the *in vivo* study of exchange between mobile and solidlike protein domains (Section 6), the detection of exchange between "mobile" and NMR "invisible" phosphorus metabolites (Section 6), the analysis of the saturation transfer experiment involving motionally restricted tissue metabolites (Section 6), the recently reported technique of magnetic cross-relaxation spectroscopy (Section 7) (Grad and Bryant, 1990; Grad *et al.*, 1990, 1991), and the magnetization transfer contrast (MTC)-enhanced proton magnetic resonance imaging experiment (Section 7) (Wolff and Balaban, 1989; Eng *et al.*, 1991).

2. THEORETICAL FORMALISM

The theoretical basis for the investigation of intermediate-time-scale isotropic molecular motions by off-resonance rotating frame spin–lattice relaxation was established by Jones (1966) and Blicharski (1972). The practical implementation of this technique for the study of isotropic and anisotropic protein reorientational motion has been reviewed by Schleich *et al.* (1989). Specific biophysical applications have included elucidation of protein reorientational motion and polypeptide chain dynamics characteristics in dilute and concentrated protein environments (James *et al.*, 1978; James, 1980; Morgan *et al.*, 1990a, b), nucleic acid relaxation dynamics characterization (James, 1984), and determination of phosphorus metabolite rotational diffusion behavior in lens tissue homogenates (Caines *et al.*, 1991).

The determination of the rotational correlation time for protein reorientation is made by an assessment of the ^{13}C rotating frame spin–lattice relaxation characteristics of the backbone ^{13}C peptide bond carbonyl resonances in the presence of an applied off-resonance RF field (James *et al.*, 1978, 1979; Schleich *et al.*, 1989; Morgan *et al.*, 1990a, b). Relaxation characteristics of the polypeptide backbone chain amino acid residue ^{13}C α-carbon resonances may also be used (Wilbur *et al.*, 1976). The rationale for these selections is based on the assumption that the polypeptide backbone motional dynamics mirror the overall rotational motion of the protein. This has been verified experimentally for the globular protein, carp muscle calcium binding protein (Bauer *et al.*, 1975). Protein motional correlation times determined by off-resonance rotating frame spin–lattice relaxation experiments are in excellent agreement with values determined by other techniques (James *et al.*, 1978; Schleich *et al.*, 1989). The study of nucleic acid chain dynamics and phosphorus metabolite rotational diffusion involves the assessment of ^{31}P NMR relaxation behavior.

The following sections describe the theoretical basis for $I = 1/2$ spin-lattice relaxation in the presence and absence of an off-resonance RF field for both isotropic and anisotropic reorientational motion arising from dipole–dipole (DD) interaction to protons (intra- and intermolecular) and chemical shift anisotropy (CSA). In addition, the relevant theory for the relaxation behavior of a quadrupolar ($I = 1$) nucleus in the presence and absence of an off-resonance RF field for isotropic reorientation is also presented to enable the study of deuterium *in vivo*-labeled metabolites in tissue (Rydzewski and Schleich, 1992).

2.1. Off-Resonance Rotating Frame Experiment

In a coordinate system rotating with angular velocity ω, the effective field $\mathbf{B}_e$ is given by

$$\mathbf{B}_e = \frac{\omega_e}{\gamma_I} = [B_0 - \omega/\gamma_I]\mathbf{k} + B_2\mathbf{i} \tag{1}$$

where ω_e is the angular precessional frequency about the effective field for an ensemble of $I = 1/2$ noninteracting spins with gyromagnetic ratio γ_I. B_0 is the static magnetic field strength, $\mathbf{i}$ and $\mathbf{k}$ are unit vectors, and B_2 is an RF field off-resonance by an amount Δ:

$$\Delta = 2\pi\nu_{\text{off}} = -\gamma_I\left(B_0 - \frac{\omega}{\gamma_I}\right) \tag{2}$$

where ν_{off} is the difference in frequency of the applied continuous-wave off-resonance RF field from the selected resonance. The effective field vector, $\mathbf{B}_e$, is inclined to the static magnetic field, $\mathbf{B}_0$, at an angle Θ defined by

$$\Theta = \text{atan}\left(\frac{\gamma_I B_2}{\Delta}\right) \tag{3}$$

The angular precessional frequency about the effective field vector is

$$\omega_e = \frac{\Delta}{\cos\Theta} \tag{4}$$

Assuming exponential relaxation of the magnetization aligned along $\mathbf{B}_e$ (M_e), defined by the time constant $T_{1\rho}^{\text{off}}$, the magnetization ratio, M_e/M_0, can be shown to be equal to $\cos\Theta(T_{1\rho}^{\text{off}}/T_1)$ in the high temperature limit [$kT \gg \gamma_I(h/2\pi)B_0$], with $\mathbf{B}_0 \gg \mathbf{B}_e$ (Jacquinot and Goldman, 1973). Experimentally, the projection of M_e on the z axis of the rotating frame is measured. Thus, the signal intensity ratio, $R = M_z/M_0$ [$=\cos\Theta(M_e/M_0)$] also equals $\cos^2\Theta(T_{1\rho}^{\text{off}}/T_1)$. This equation is valid under the condition that $(\gamma_I\beta_2)^2T_1T_2 \gg 1$. M_e is obtained in the presence of, whereas M_0 is obtained in the absence of, the off-resonance RF irradiation field. For $\Theta \leqslant 10°$ the $\cos^2\Theta$ term may be ignored.

Alternatively, the time constant $T_{1\rho}^{\text{off}}$ may be measured directly by use of the conventional inversion recovery pulse sequence modified to include off-resonance RF irradiation during the relaxation interval between the inversion and read pulses (Cornell and Pope, 1974).

The basis of the off-resonance rotating frame spin–lattice relaxation experiment is therefore the observation of the spin–lattice relaxation time in the presence of an applied off-resonance RF field. In the presence of such irradiation, a component of the sample nuclear magnetization is spin-locked by an effective field, $\mathbf{B}_e$. By contrast, for an on-resonance rotating frame spin–lattice relaxation experiment ($\Theta = 90°$) the sample nuclear magnetization is spin-locked by the B_2 field. Thus, for the off-resonance irradiation experiment the magnitude of the spectral density function at ω_e becomes important in addition to the Larmor frequencies of the interacting nuclei (*vide infra*).

2.2. Relaxation Expressions

2.2.1. Random Isotropic Reorientation

For random isotropic molecular reorientation involving two heteronuclear intramolecular dipolar coupled spin-1/2 nuclei, I and S, the following intramolecular dipole–dipole relaxation expressions for a particular spin S in the absence of cross-relaxation and under the far off-resonance condition, $|\Delta| \geqslant 5\gamma_S B_2$ and $\omega_e \ll \omega_0$ (off-resonance irradiation experiment only), may be written (James *et al.*, 1978; James, 1984):

$$\left(\frac{1}{T_1}\right)^{\text{SI}}_{\text{DD,rot}} = \sum_{i=1}^{N} (K_{\text{IS}})_i[J_0(\omega_{\text{I}_i} - \omega_{\text{S}}) + 3J_1(\omega_{\text{S}}) + 6J_2(\omega_{\text{I}_i} + \omega_{\text{S}})] \tag{5}$$

$$\left(\frac{1}{T_{1\rho}^{\text{off}}}\right)^{\text{SI}}_{\text{DD,rot}} = \left(\frac{1}{T_1}\right)^{\text{SI}}_{\text{DD,rot}} + (\sin^2\Theta)\sum_{i=1}^{N}(K_{\text{IS}})_i \times\left[2J_0(\omega_e) + \frac{3}{2}J_1(\omega_{\text{I}_i} + \omega_e) + \frac{3}{2}J_1(\omega_{\text{I}_i} - \omega_e)\right] \tag{6}$$

where $K_{IS} = (h/2\pi)^2\gamma_I^2\gamma_S^2/20r_{IS}^6$. $J_n(\omega)$ is the spectral density function in which motional dynamics information is embedded, h is Planck's constant, γ_I and γ_S are the gyromagnetic ratios for the I and S spins, respectively, ω_I and ω_S are the respective angular Larmor precessional frequencies, r is the distance between the I and S spins, and N is the number of I spins responsible for dipole–dipole IS relaxation. The summation over N is necessary in the event that more than one I spin is responsible for relaxation, and allows for the occurrence of different spectral densities for each of the interactions. For biological applications, such as those discussed in this chapter, I and S typically represent ^1H, and ^{13}C or ^{31}P, respectively.

For a homonuclear two-spin system with spin S = 1/2 (^1H or ^{31}P), the following expressions are applicable (James *et al.*, 1977):

$$\left(\frac{1}{T_1}\right)^{SS'}_{DD,rot} = \sum_{i=1}^{N} (K_{SS'})_i[J_1(\omega_{S_i'}) + 4J_2(2\omega_{S_i'})] \tag{7}$$

$$\left(\frac{1}{T_{1\rho}^{off}}\right)^{SS'}_{DD,rot} = \left(\frac{1}{T_1}\right)^{SS'}_{DD,rot} + \sum_{i=1}^{N} (K_{SS'})\left[\frac{3}{2}(\sin^2\Theta)J_0(\omega_e)\right] \tag{8}$$

where $K_{SS'} = 3\gamma_S^4(h/2\pi)^2/20r_{SS'}^6$.

In many instances the off-resonance rotating frame spin–lattice relaxation experiment is performed in water, a proton-containing solvent. Under such circumstances it is usually necessary to consider intermolecular contributions to the total dipole–dipole relaxation rate arising from solvent (water) translational motion. Such contributions are significant when the solvent molecules are near the nuclear spins under investigation. Translational contributions are especially significant for phosphorus metabolites, but appear to be slight for the protein backbone peptide carbonyl carbons (*vide infra*). The theoretical translational contribution to the dipole–dipole relaxation rate may be evaluated by using the simple model of Abragam (1961) which assumes Brownian diffusion of a particle through a homogeneous viscous fluid medium. For the heteronuclear case, where one of the spins is very dilute relative to the other, the following expression valid in the extreme narrowing limit ($\omega^2\tau^2 \ll 1$) was obtained by Caines *et al.* (1990):

$$\left(\frac{1}{T_1}\right)^{SH}_{DD,trans} = \left[\frac{h}{2\pi}\right]^2 \gamma_H^2\gamma_S^2 2\pi\left[2\frac{N}{15dD}\right] \tag{9}$$

where N is the solvent proton number density in the solution, d is the S–H distance of closest approach, and D is the average of the water and S species diffusion coefficients. For translation τ is defined to be $d^2/2D$.

Similarly for the off-resonance rotating frame spin–lattice relaxation rate the following expression was derived (Caines *et al.*, 1990):

$$\left(\frac{1}{T_{1\rho}^{\text{off}}}\right)_{\text{DD,trans}}^{\text{SH}} = \left(\frac{h}{2\pi}\right)^2 \gamma_{\text{H}}^2\gamma_{\text{S}}^2\pi(2+\sin^2\Theta)\left(2\frac{N'}{15dD}\right) \tag{10}$$

Since the angle Θ in the off-resonance experiment is small, $(1/T_{1\rho}^{\text{off}})_{\text{DD,trans}}^{\text{SH}} \approx (1/T_1)_{\text{DD,trans}}^{\text{SH}}$, as would be expected in the extreme narrowing limit.

CSA relaxation contributions assume significance at high magnetic field strengths for the protein peptide carbonyl carbon and the phosphoryl group of phosphorus metabolites. The relevant expressions are (James, 1984):

$$\left(\frac{1}{T_1}\right)_{\text{CSA}}^{\text{S}} = \frac{3}{20}(\gamma_{\text{S}}^2B_0^2\delta_{\text{z}}^2)\sum_{j=0}^{2} J_j(\omega_{\text{s}}) \tag{11}$$

$$\left(\frac{1}{T_{1\rho}^{\text{off}}}\right)_{\text{CSA}}^{\text{S}} = \left(\frac{1}{T_1}\right)_{\text{CSA}}^{\text{S}} + \frac{1}{10}(\gamma_{\text{S}}^2B_0^2\delta_{\text{z}}^2)(\sin^2\Theta)\sum_{j=0}^{2} J_j(\omega_{\text{e}}) \tag{12}$$

where δ_{z} is the anisotropy factor.

The total relaxation rate is obtained by summing the individual relaxation rates contributed by the different assumed mechanisms:

$$\left(\frac{1}{T_1}\right)_{\text{total}} = \left(\frac{1}{T_1}\right)_{\text{DD,rot}}^{\text{SI}} + \left(\frac{1}{T_1}\right)_{\text{DD,rot}}^{\text{SS}} + \left(\frac{1}{T_1}\right)_{\text{DD,trans}}^{\text{SS}} + \left(\frac{1}{T_1}\right)_{\text{CSA}}^{\text{S}} \tag{13}$$

$$\left(\frac{1}{T_{1\rho}^{\text{off}}}\right)_{\text{total}} = \left(\frac{1}{T_{1\rho}^{\text{off}}}\right)_{\text{DD,rot}}^{\text{SI}} + \left(\frac{1}{T_{1\rho}^{\text{off}}}\right)_{\text{DD,rot}}^{\text{SS}} + \left(\frac{1}{T_{1\rho}^{\text{off}}}\right)_{\text{DD,trans}}^{\text{SH}} + \left(\frac{1}{T_{1\rho}^{\text{off}}}\right)_{\text{CSA}}^{\text{S}} \tag{14}$$

For random isotropic reorientational motion of an $\text{I} = 1$ spin, the quadrupolar relaxation mechanism contribution is usually dominant. Assuming axial symmetry of the electric field tensor the following expressions are relevant (Blicharski, 1972; Rydzewski and Schleich, 1992):

$$\left(\frac{1}{T_1}\right)_{\text{Q}} = L[J_1(\omega) + 4J_2(2\omega)] \tag{15}$$

$$\left(\frac{1}{T_{1\rho}^{\text{off}}}\right)_{\text{Q}} = \left(\frac{1}{T_1}\right)_{\text{Q}} + L\left[\frac{3}{2}(\sin^2\Theta)J_0(\omega_{\text{e}})\right] \tag{16}$$

where $L = (3/20)\pi^2\chi^2$. χ is the quadrupole coupling or spectroscopic splitting constant in hertz. In some instances, where Θ is not small, it may be necessary to use the nontruncated expression for $(1/T_{1\rho}^{\text{off}})_{\text{Q}}$ (Rydzewski and Schleich, 1992).

2.2.2. Random Anisotropic Reorientation

Anisotropic molecular reorientation is encountered with particles whose shape deviates from sphericity, i.e., one represented by a hydrodynamic ellipsoid of revolution. The expressions appropriate for isotropic tumbling require modification to take into account differential motion about the major and minor axes of the ellipsoid. For an axially symmetric ellipsoid, with major semiaxis a, and minor semiaxis b, the following expressions for dipole–dipole relaxation are appropriate (James, 1984; Schleich *et al.*, 1989; Morgan *et al.*, 1990a; Woessner, 1962):

$$\left(\frac{1}{T_1}\right)_{\mathrm{DD}}^{\mathrm{SI}} = \sum_{i=1}^{N} (K_{\mathrm{IS}})_i [J_0'(\omega_{\mathrm{I}_i} - \omega_{\mathrm{S}}) + 3J_1'(\omega_{\mathrm{S}}) + 6J_2'(\omega_{\mathrm{I}_i} + \omega_{\mathrm{S}})] \tag{17}$$

$$\left(\frac{1}{T_{1\rho}^{\mathrm{off}}}\right)_{\mathrm{DD}}^{\mathrm{SI}} = \left(\frac{1}{T_1}\right)_{\mathrm{DD}}^{\mathrm{SI}} + \sum_{i=1}^{N} (K_{\mathrm{IS}})_i \times \left\{(\sin^2\Theta)\left[2J_0'(\omega_{\mathrm{e}}) + \frac{3}{2}J_1'(\omega_{\mathrm{I}_i} + \omega_0) + \frac{3}{2}J_1'(\omega_{I_i} - \omega_{\mathrm{e}})\right]\right\} \tag{18}$$

where K_{IS} is identical to the one used for isotropic tumbling, and $J_n'(\omega)$ is the spectral density function which differs from the isotropic case (*vide infra*).

For CSA relaxation, the following expressions (James, 1984; Schleich *et al.*, 1989; Morgan *et al.*, 1990a) pertain:

$$\left(\frac{1}{T_1}\right)_{\mathrm{CSA}}^{\mathrm{S}} = \frac{3}{20}\gamma_{\mathrm{S}}^2 B_0^2 \delta_z^2 \sum_{j=0}^{2} J_j'(\omega_{\mathrm{S}}) \tag{19}$$

$$\left(\frac{1}{T_{1\rho}^{\mathrm{off}}}\right)_{\mathrm{CSA}}^{\mathrm{S}} = \left(\frac{1}{T_1}\right)_{\mathrm{CSA}}^{\mathrm{S}} + \frac{1}{10}\gamma_{\mathrm{S}}^2 B_0^2 \delta_z^2 (\sin^2\Theta) \sum_{j=0}^{2} J_j'(\omega_{\mathrm{e}}) \tag{20}$$

where $J_j'(\omega)$ is the spectral density function which differs from the isotropic case (*vide infra*).

2.3. Spectral Density Functions

2.3.1. Random Isotropic Reorientation

The rapidity of reorientational motion responsible for magnetic relaxation is characterized by the rotational correlation time. For an isolated, i.e., at infinite dilution, uncharged spherical rigid rotor engaged in random isotropic reorientational motion, the rotational correlation time is given by

$$\tau_0 = \frac{V_{\text{hyd}}\eta_0}{k_{\text{B}}T} \tag{21}$$

where V_{hyd} is the hydrodynamic volume $[V_{\text{hyd}} = (M/N_{\text{A}})(\bar{v} + \delta_1 V_1^0) = (4/3)\pi r_0^3]$, M the anhydrous molecular weight, N_{A} Avogadro's number, $\bar{v}$ the macromolecular partial specific volume, δ_1 the hydration $(g_{\text{water}}/g_{\text{macromolecule}})$, V_1^0 the specific volume of pure solvent, r_0 the hydrodynamic radius of the spherical particle, η_0 the solvent viscosity (poise), k_{B} Boltzmann's constant, and T the absolute temperature. By contrast, the rotational relaxation time, τ_{r}, equals $3\tau_0$. The rotational diffusion coefficient, D_{r}, is given by

$$D_{\text{r}} = \frac{1}{6\tau_0} \tag{22}$$

Nonideal effects influence the rotational diffusion coefficient in a manner analogous to the translational diffusion coefficient [for a spherical rigid rotor $D_{\text{r}} = (6/8)D_{\text{t}}$] (Pessen and Kumosinski, 1985; Kumosinski and Pessen, 1989). The rotational correlation time at concentration c is given by:

$$\tau_{0,\text{c}} = \tau_0\left(1 + c\frac{d}{dc}\ln\gamma\right)^{-1} \tag{23}$$

where γ is the activity coefficient of the rotationally diffusing particle. Since excluded volume and hydrodynamic interaction terms as well as electrostatic effects arising from intrinsic charge and charge fluctuations can be significant, pH and ionic strength are important variables in rotational diffusion studies.

For uncharged spherical particles, Koenig (1980) has presented a semiempirical treatment relating the rotational correlation time to particle concentration, which implicitly includes excluded volume and hydrodynamic interaction contributions and, hence, interparticle distance. The original citation contains a typographical error. The correct form (Brooks, 1989) is:

$$\tau_{0,\text{c}} = \frac{\tau_0}{1 - [r_o/(D - r_0)]^3} \tag{24}$$

where r_0 is the hydrodynamic particle radius and D is the particle center-to-center distance (solution interparticle spacing). The interparticle spacing (D) by definition is equal to $\rho^{-1/3}$, where ρ is the number density of particles. The particle number density is obtained from (Brown *et al.*, 1975):

$$\rho = \frac{3c}{4\pi r^3 \rho_m} \tag{25}$$

where c is the protein concentration in g/cm^3 and ρ_m is the density of protein in g/cm^3. Plots showing the dependence of $\tau_{0,c}$ versus protein concentration using Eq. (24) are shown in Fig. 2. The simulations illustrate that the rotational correlation time becomes increasingly sensitive to increasing particle concentration and molecular weight.

The dependence of the isotropic rotational correlation time on protein concentration, i.e., water content, as modeled by Kimmich *et al.* (1990), is affected by the accessibility of the tumbling macromolecule to the volume of free water present in the system. Thus, macromolecular tumbling is only possible if sufficient solvent exists in the system. Using "free volume" theory adopted from translational diffusion formalism, Kimmich *et al.* (1990) developed the following expression for the dependence of $\tau_{0,c}$ on water content:

$$\tau_{0,c} = \tau_0 \exp\left[\gamma'(r-1)\left(\frac{1 - c_w + c_s}{c_w - c_s}\right)\right] \tag{26}$$

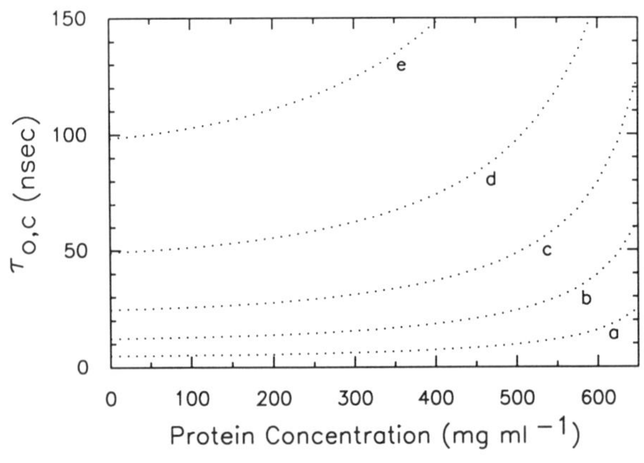

Figure 2. Theoretical dependence of $\tau_{0,c}$ on protein concentration for different molecular masses of an unchanged spherical rigid rotor [(a) 10 kDa; (b) 25 kDa; (c) 50 kDa; (d) 100 kDa; (e) 200 kDa] calculated using Eqs. (24) and (25) (see text). The following parameters were assumed: $\bar{v} = 0.72$ g/cm^3, $\delta = 0.5$ g H_2O/g protein, $\eta = 0.01$ poise, $T = 298.2$ K.

This expression is valid provided that $c_w > c_s$, where c_w and c_s are the mass fractions of total water and the maximum water of macromolecular hydration, respectively. γ' is a constant with a value between 0.5 and 1.0, r is the ratio of volumes of a circumscribing sphere (which represents the minimum volume necessary for rotational diffusion of a hydrated macromolecule) relative to that of a quasi-spherical reorienting particle [thus, the excess free volume is $(r-1)V_{hyd}$]. The mass fraction of free water, c_f, equals $c_w - c_s$. As the mass fraction of free water decreases, the rotational correlation time increases, and in the absence of free water $\tau_{0,c}$ goes to infinity. At the other extreme of infinite dilution, i.e., as $c_f \to 1$, $\tau_{0,c} \to \tau_0$. A critical water mass fraction, c_0, the point at which macromolecular tumbling commences, has also been defined by Kimmich and co-workers (Schauer *et al.*, 1988; Kimmich *et al.*, 1990). Theoretical arguments supported by experimental findings (Schauer *et al.*, 1988) indicate that this value is approximately 0.6.

In biological samples, nuclear spins are often distributed among different microenvironments, thus leading to a distribution of correlation times. Many studies assume for simplicity a single (average) rotational correlation time for the characterization of reorientational motion. We likewise adopt this simplifying assumption.

The spectral density function for dipolar relaxation is defined by

$$J_n(\omega) = \langle A(0)A(p)\rangle \frac{2\tau_0}{1+\omega^2\tau_0^2} \tag{27}$$

where $\langle A(0)A(p)\rangle$ is the plateau amplitude at time p, the time at which internal molecular motional averaging is completed, but before significant rotational motion has commenced (i.e., $p \ll \tau_0$), and τ_0 is the rotational correlation time for isotropic reorientational motion of a rigid uncharged spherical particle (Levy *et al.*, 1981; McCain and Markley, 1986; James *et al.*, 1978). Since a spectral intensity ratio, and, hence, a ratio of spectral densities, is measured in many implementations of the off-resonance rotating frame spin–lattice relaxation experiment (*vide ante*) the constant plateau amplitude factor of each spectral density contribution cancels out.

Correspondingly, the spectral density function for CSA-induced relaxation is (James, 1984; Schleich *et al.*, 1989):

$$J_j(\omega) = \langle A(0)A(p)\rangle \frac{2c_j\tau_j}{1+\omega^2\tau_j^2} \tag{28}$$

where $\tau_j = \tau_0$ for $j = 0, 1, 2$ (isotropic motion), with $c_0 = 1$, $c_1 = 0$, and $c_2 = \eta^2/3$ where η is the asymmetry parameter characterizing the CSA relaxation.

2.3.2. Random Anisotropic Reorientation

For an axially symmetric ellipsoid of revolution engaged in reorientational motion, $\tau_{\perp}$ is defined to be the correlation time for the motion of semiaxis a about semiaxis b, and $\tau_{\parallel}$ for the motion of semiaxis b about semiaxis a. The following relations hold for a particle in a homogeneous environment (Wilbur *et al.*, 1976; James, 1984; Woessner, 1962):

$$\tau_A = \tau_{\perp} \tag{29a}$$

$$\tau_B = 6\tau_{\perp}\tau_{\parallel}(\tau_{\perp} + 5\tau_{\parallel})^{-1} \tag{29b}$$

$$\tau_C = 3\tau_{\perp}\tau_{\parallel}(2\tau_{\perp} + \tau_{\parallel})^{-1} \tag{29c}$$

The spectral density function for dipolar relaxation, ignoring rapid internal motion (Wilbur *et al.*, 1976; James, 1984; Heatly, 1986; Schleich *et al.*, 1989), is given by:

$$J'_n(\omega) = A\frac{2\tau_A}{1+\omega^2\tau_A^2} + B\frac{2\tau_B}{1+\omega^2\tau_B^2} + C\frac{2\tau_C}{1+\omega^2\tau_C^2} \tag{30}$$

where $A = (1/4)(3\cos^2\alpha - 1)^2$, $B = (3/4)(\sin^2 2\alpha)$, and $C = (3/4)\sin^4\alpha$. The angle α is defined by the orientation of the I–S internuclear vector relative to the unique symmetry axis of the diffusing particle. The coefficients A, B, and C sum to unity at each value of α.

For CSA relaxation $J'_j(\omega)$ is identical to $J_j(\omega)$ in the isotropic case [see Eq. (28)], except that the correlation times $\tau_0 = \tau_A$, $\tau_1 = \tau_B$, $\tau_2 = \tau_C$ [see Eq. (29)] and the c_j coefficients are different (James, 1984; Schleich *et al.*, 1989):

$$c_0 = \frac{1}{4}[(3\cos^2\beta - 1) + \eta\sin^2\beta\cos 2\gamma]^2 \tag{31a}$$

$$c_1 = \frac{1}{3}\sin^2\beta[\cos^2\beta(3 - \eta\cos 2\gamma)^2 + \eta^2\sin^2 2\gamma] \tag{31b}$$

$$c_2 = \left[\sqrt{\frac{3}{4}}\sin^2\beta + \left(\frac{\eta}{2\sqrt{3}}\right)(1+\cos^2\beta)\cos 2\gamma\right]^2 + \left(\frac{\eta^2}{3}\right)\sin^2 2\gamma\cos^2\beta \tag{31c}$$

where β and γ are the Euler angles relating the rotation of the principal axes of the relevant chemical shift tensor (e.g., the peptide bond carbonyl carbon). The c_j coefficients sum to $1 + \eta^2/3$ for each combination of the Euler angles β and γ.

For the case of an axially symmetric ellipsoid of revolution of axial ratio less than 2, Kalk and Berendsen (1976) suggested that a reasonable estimate of τ_0 is obtained if the hydrodynamic spherical particle radius used in Eq. (21) is replaced by the geometric mean of the three axes ($r_0^3 = ab^2$).

2.3.2.1. Evaluation of Euler Angles β and γ and Internuclear Vector Angle α. The analysis of the CSA relaxation contribution to the spin–lattice relaxation of an aspherical particle, such as a protein, undergoing anisotropic tumbling is based upon the orientation of the carbonyl chemical shift tensor reported for the peptide bond of model dipeptides (Stark *et al.*, 1983; Oas *et al.*, 1987; Hartzell *et al.*, 1987). As noted above, the required parameters are the Euler angles β and γ and the angle α which relate the CSA principal axis system of the ith residue peptide bond carbonyl and the C–H_α internuclear vector, respectively, to the axially symmetric diffusion axis (D_z) (Hull and Sykes, 1975a, b; Morgan *et al.*, 1990a). The specific details of the procedure employed for the evaluation of the Euler angles for each amino acid residue present in the protein structure using X-ray structural information have been summarized by Morgan *et al.* (1990a).

2.3.2.2. Evaluation of δ_z and Asymmetry Parameter η. The evaluation of δ_z and η from principal values of the chemical shift tensor follows customary procedures (Hull and Sykes, 1975a). Principal values of the peptide bond carbonyl carbon chemical shift tensor for model dipeptide structures have been reported recently (Stark *et al.*, 1983; Oas *et al.*, 1987; Hartzell *et al.*, 1987). The assumption that each peptide bond, with index i, is characterized by the same value of the CSA tensor, is usually made. Corresponding literature citations of the phosphoryl group in phosphorus metabolites have been summarized (Seelig, 1978; Caines *et al.*, 1990).

2.4. Equivalence of Off-Resonance Rotating Frame Spin–Lattice Relaxation Formalism to the Bloch Equations for Off-Resonance Irradiation

Under appropriate conditions, the off-resonance rotating frame spin-lattice relaxation formalism leads to results equivalent to the steady-state solution of the Bloch equation for the z-magnetization. The necessary requirements were first noted by Bendel and James (1982).

In the limit of $\omega_e \to 0$ ($\omega_e^2\tau_0^2 \ll 1$), $J(\omega_e) = J(0)$, and as a result $1/T_{1\rho}^{\text{on}}$ converges to T_2 so that $1/T_{1\rho}^{\text{off}}$ can be expressed as

$$\frac{1}{T_{1\rho}^{\text{off}}} = \frac{\cos^2\Theta}{T_1} + \frac{\sin^2\Theta}{T_2} \tag{32}$$

Incorporating the relationship $M_z/M_0 = \cos^2\Theta T_{1\rho}^{\text{off}}/T_1$ [valid when $T_1T_2(\gamma B_2)^2 \gg 1$] into Eq. (32), recognizing that $(\tan\Theta)^2 = [\gamma B_2/(\omega - \omega_0)]^2$ [Eq. (3)], and multiplying the numerator and denominator by $[1 + (\omega - \omega_0)^2 T_2^2]$ leads to:

$$\left(\frac{M_z}{M_0}\right)_{\text{off-reson}} = \frac{1 + (\omega - \omega_0)^2 T_2^2}{1 + (\omega - \omega_0)^2 T_2^2 + T_1 T_2(\gamma B_2)^2[1 + 1/(\omega - \omega_0)^2 T_2^2]} \tag{33}$$

whereas the Bloch expression for the z-magnetization is:

$$\left(\frac{M_z}{M_0}\right)_{\text{Bloch}} = \frac{1 + (\omega - \omega_0)^2 T_2^2}{1 + (\omega - \omega_0)^2 T_2^2 + T_1 T_2(\gamma B_2)^2} \tag{34}$$

Equations (33) and (34) converge when

$$\frac{1}{(\omega - \omega_0)^2 T_2^2} \ll 1 \tag{35}$$

i.e., both z-magnetization formalisms converge under saturating conditions $[T_1T_2(\gamma B_2)^2 \gg 1]$, and when $(\omega - \omega_0) \gg 1/T_2$. However, when the offset frequency becomes close to the resonance of interest, the Bloch equation formalism deviates from off-resonance rotating frame spin-lattice relaxation formalism. As zero frequency offset is approached, $(M_z/M_0)_{\text{Bloch}}$ [Eq. (34)] approaches a finite spectral intensity ratio defined by $1/[1 + T_1T_2(\gamma B_2)^2]$, whereas $(M_z/M_0)_{\text{off-reson}}$ approaches zero.

Because Eq. (33) can be expressed as:

$$\left(\frac{M_z}{M_0}\right)_{\text{off-reson}} = \left[1 + \frac{T_1}{T_2}\left(\frac{\gamma B_2}{\omega - \omega_0}\right)^2\right]^{-1} \tag{36}$$

T_1/T_2 may be determined by the off-resonance rotating frame spin-lattice relaxation experiment. A similar experiment was proposed by Bain *et al.* (1981). The advantage of the off-resonance rotating frame spin-lattice relaxation experiment is that a broad range of B_2 field strengths and offset frequencies may be used to determine T_1/T_2, and if T_1 is known from an independent measurement, T_2 is readily evaluated, thereby avoiding the relatively stringent requirements imposed by the spin-echo-based Carr-Purcell-Meiboom-Gill experiment.

2.5. Computer Simulation of the Off-Resonance Rotating Frame Spin–Lattice Experiment

The behavior of the off-resonance rotating frame spin–lattice relaxation experiment for rigid hydrodynamic particles engaged in reorientational motion is conveniently explored by the use of simulation programs which incorporate the theoretical considerations outlined above. In this section we consider the off-resonance rotating frame spin–lattice relaxation behavior of the peptide chain backbone carbonyl carbon of hydrodynamically rigid proteins involved in isotropic or anisotropic reorientation, the phosphoryl group of phosphorus-containing metabolites engaged in isotropic reorientational motion, and the deuteron ($I = 1$) incorporated into a metabolite which is tumbling isotropically. The effect of intermolecular dipole–dipole relaxation, arising from solvent translational motion, is also considered.

The dependence of the spectral intensity ratio, $R = \cos^2 \Theta(T_{1\rho}^{\text{off}}/T_1)$, for a given resonance as a function of τ_0 may be simulated at a specified ν_{off} and B_2 field strength using computer programs incorporating the theoretical considerations outlined above. [Copies of these programs written in Borland TURBO Pascal (version 5.0) are available on request. Electronic mail requests may be sent to: yoti@aku.ucsc.edu or coyote@hydrogen.ucsc.edu.] For proteins the resonance intensity of peptide carbonyl carbons is usually assessed. Due to the relatively small differences in chemical shift of each of the peptide carbonyl carbons in the protein, an overall average resonance intensity ratio is experimentally determined. For the isotropic tumbling of a rigid spherical protein, each peptide bond carbonyl carbon of the *i*th residue is equivalent to all other residues in terms of its contribution to the resonance intensity in an off-resonance rotating frame spin-lattice relaxation experiment. This is true when it is assumed that local pepetide bond geometries and principal values of the pepetide bond carbonyl carbon chemical shift tensor are identical. By contrast, for the case of an anisotropically reorienting rigid particle, each carbonyl carbon contributes to the total resonance intensity in a manner that is explicitly dependent on (1) the relative orientation of the CSA principal axis system to the axially symmetric diffusion axis system (defined by the Euler angles β and γ), and (2) on the angle of the C–H internuclear vector to the unique diffusion axis of the anisotropic particle. The average value of the spectral intensity ratio for anisotropic tumbling is defined as follows:

$$\bar{R} = \frac{\sum_{i=1}^{N} \frac{T_{1\rho,i}^{\text{off}}}{T_{1,i}}}{N - 1} \tag{37}$$

where N is the number of amino acid residues, and therefore $N - 1$ is the number of peptide carbonyl carbons.

Polydispersity (molecular weight heterogeneity) effects on the spectral intensity ratio are incorporated by considering the weight fraction of each protein species j (Morgan *et al.*, 1990b) and is given by:

$$\bar{R} = \sum_j f_j \frac{T_{1\rho}^{\text{off}}}{T_1} \tag{38}$$

where f_j is the weight fraction of the jth protein species (this is equivalent to the mole fraction of the jth protein species with the assumption of a mean residue molecular weight).

The calculation of $\bar{R}$ for proteins takes into account intramolecular dipole–dipole interactions between the carbonyl carbon of the peptide bond and H_α (an average distance of 2.16 Å is usually assumed for r_{CH}), and a contribution from CSA relaxation of the peptide carbonyl carbon. Intramolecular dipole–dipole relaxation contributions from ^{13}C–^{14}N and intermolecular ^{13}C–^{1}H dipole–dipole interactions arising from solvent translational motion are negligible, and thus ignored.

For anisotropic protein tumbling, $\bar{R}$ is calculated using assumed values for the Euler angles β and γ, and the internuclear vector angle, α, or by calculating these values for each amino acid residue, i, from X-ray crystallographic data as described above. For a hydrodynamic ellipsoid of revolution, the dependence of $\bar{R}$ on axial ratio requires values for $\tau_\perp$ and $\tau_\parallel$ which are calculated using Perrin's equations (see Koenig, 1975). The value of $\tau_\perp$ for a prolate ellipsoid of revolution with a given axial ratio ($a/b > 1$) relative to the rotational correlation time for a sphere of equivalent volume should be calculated using corrected Perrin's equations provided by Koenig (1975). Most literature citations (see Koenig, 1975) of Perrin's equations for a prolate ellipsoid of revolution contain a typographical error.

In the case of phosphorus-containing metabolites (e.g., 2,3-bis-phosphoglycerate), ^{31}P NMR relaxation studies conducted in aqueous solution and in tissue (intact red blood cell) indicate that dipole–dipole and CSA relaxation mechanisms are dominant at high B_0 field strength (Christensen and Jacobsen, 1988) and are therefore assumed in the simulations. Intermolecular dipole–dipole (^{31}P–^{1}H) relaxation contributions arising from solvent translational motion are significant as noted by Caines *et al.* (1990) and are included as well. Dipolar relaxation contributions from phosphorus metabolite-bound Mg^{2+} may be ignored (Caines *et al.*, 1990). In many instances, paramagnetic ion contributions are minimal. A review of ^{31}P nuclear spin relaxation mechanisms has been published by McCain (1987), and the use of NMR for the investigation of protein–phosphorus interactions has been discussed by Brauer and Sykes (1987). Isotropic reorientation of the phosphorus metabolite is assumed.

For enriched metabolites containing an aliphatic ^{13}C carbon, the dominant relaxation mechanisms are intramolecular and intermolecular dipole-dipole (^{13}C–^{1}H), the latter arising from solvent translational motion in the vicinity of the ^{13}C-enriched carbon atom. A CSA relaxation contribution should be added in the case of a ^{13}C-enriched carbonyl carbon.

When applying the rotating frame spin–lattice relaxation formalism to a deuterated metabolite, only quadrupolar relaxation need be considered since this is the dominant relaxation mechanism for a nuclear spin greater than one-half. This simplifies the problem because the only parameter required for the spectral density functions is the quadrupolar coupling constant. Because this constant cancels in the spectral intensity ratio expression, it is not necessary that the value be known. Like the phosphorus-containing metabolites, isotropic reorientation is assumed.

It should be noted that the $1/T_{1\rho}^{\text{off}}$ quadrupolar relaxation expression [Eq. (16)] is derived by assuming axial symmetry of the electric field tensor, i.e., the asymmetry parameter (η) is zero. This assumption is made so that quantum mechanically the Hamiltonian can be transformed to the doubly rotating frame which precesses about B_e while in the singly rotating frame. The impact of this assumption has not been thoroughly investigated. However, because the off-resonance technique uses the ratio of $T_{1\rho}^{\text{off}}$ to T_1, it can be tentatively assumed that the omission of the asymmetry parameter in the derivation of both expressions has minimal consequences.

Because of the narrow chemical shift range for ^{2}H and the occurrence of relatively broad resonances, the possibility of spectral overlap is great. For a metabolite labeled with nonequivalent deuterium nuclei, a unique quadrupolar coupling constant cannot be associated with the overlapping resonances. Thus, an average spectral intensity ratio is defined in a manner analogous to CSA relaxation of an anisotropically reorienting particle [see Eq. (37)]. The term $N-1$ in this case is the number of deuterium nuclei.

If the resonances do not overlap, the rotating frame spin–lattice relaxation experiment can be used to obtain quadrupolar coupling constants. This is accomplished by performing a T_1 measurement on the resonance of interest, and determining a rotational correlation time from a subsequent off-resonance experiment. These values are then inserted into the expression for the spin–lattice relaxation rate constant where the quadrupolar coupling constant is the only unknown parameter.

2.5.1. Isotropic Reorientation

2.5.1.1. Proteins. The spectral intensity ratio, R, is typically measured in an off-resonance rotating frame spin–lattice relaxation experiment, and

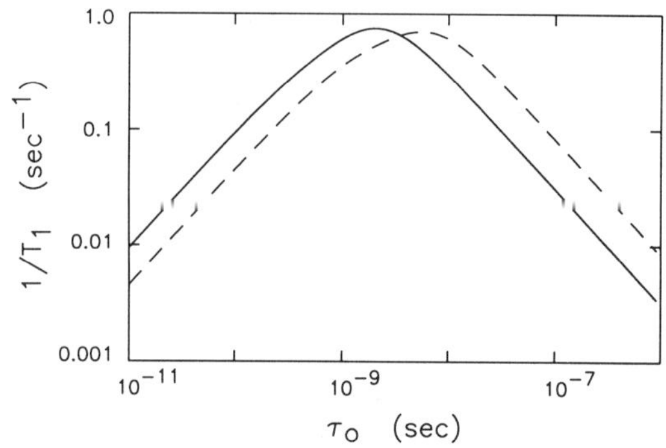

Figure 3. Theoretical dependence of the ^{13}C spin–lattice relaxation rate constant ($1/T_1$) on the correlation time for random isotropic reorientation (τ_0) at 2.35 T (- - -) and 7.05 T (——). The computed curves assume dipolar relaxation by a proton 2.16 Å from the ^{13}C and CSA relaxation contributions. Values of 81.8 ppm and −0.82 were used for δ_z and η, respectively (see Section 2.3.2.2). Reproduced, with permission, from Schleich *et al.* (1989).

is represented by the ratio of the off-resonance spin–lattice relaxation time in the rotating frame ($T_{1\rho}^{\text{off}}$) to the spin–lattice relaxation time (T_1) (see Section 2.1). The theoretical dependencies of these magnetic relaxation times on the correlation time for isotropic reorientation of the peptide carbonyl carbon in a rigid tumbling protein are shown in Figs. 3 and 4. As shown in Fig. 3, the characteristic maximum for $1/T_1$ occurs when $\omega_C^2\tau_0^2 = 1$, and is therefore dependent on the static magnetic field strength, B_0, moving to shorter τ_0 values with increasing field strength. In Fig. 4, $1/T_{1\rho}^{\text{off}}$ is plotted versus τ_0 at three values of ν_{off} (at constant B_2 field strength); at values of

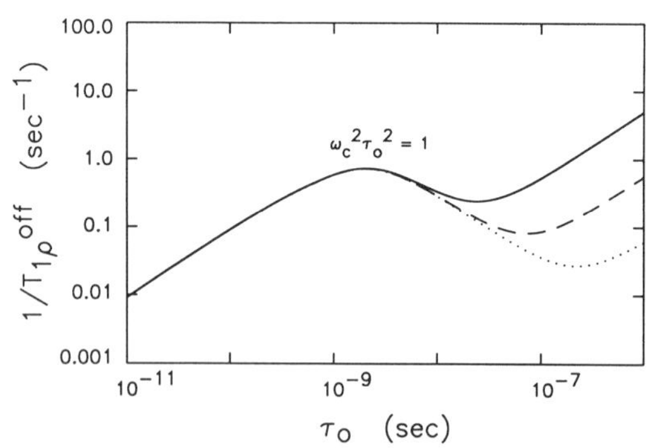

Figure 4. Theoretical dependence of the ^{13}C off-resonance rotating frame spin–lattice relaxation rate constant ($1/T_{1\rho}^{\text{off}}$) on the correlation time for random isotropic reorientation (τ_0). The computed curves assume dipolar and CSA relaxation contributions at 7.05 T as described in Fig. 3. An off-resonance field of 0.5 G applied 5 (——), 15 (– – –), and 45 kHz (· · · · ·) off-resonance was assumed. Adapted from Schleich *et al.* (1989).

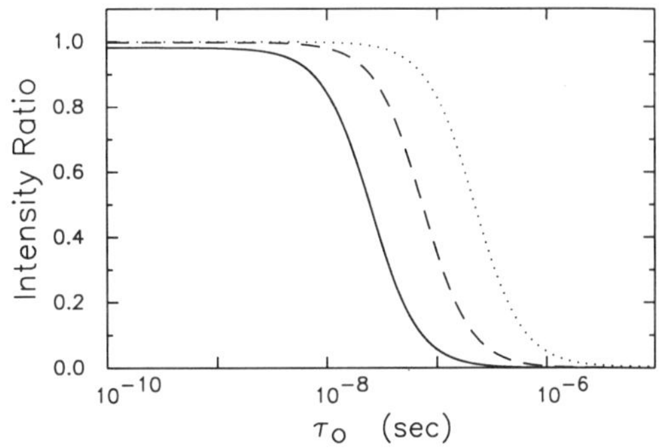

Figure 5. Simulation of the ^{13}C intensity ratio ($R = T_{1\rho}^{\text{off}}/T_1$) versus τ_0 at 7.05 T. The computed curves assume dipolar and CSA relaxation mechanism contributions using the parameters listed in Fig. 3. An off-resonance field of 0.5 G applied 5 (——), 15 (– – –), and 45 kHz (· · · · ·) off-resonance was assumed. Adapted from Schleich *et al.* (1989).

$\omega_C^2\tau_0^2 \leqslant 1$, $1/T_{1\rho}^{\text{off}}$ is independent of ν_{off}, whereas for τ_0 values such that $\omega_C^2\tau_0^2 \geqslant 1$ the dependence of $1/T_{1\rho}^{\text{off}}$ on ν_{off} becomes appreciable. A minimum occurs at a value of τ_0 corresponding to an R value of 0.5, i.e., the inflection point in a plot of R versus τ_0 (see Fig. 5). Both simulations take into account dipole-dipole and CSA relaxation mechanism contributions. In Fig. 5 the intensity ratio (R) is plotted against τ_0 for a monodisperse protein preparation at various values of ν_{off} (B_2 is kept constant). The effect of increasing ν_{off} at constant B_2 is to shift the inflection point of the R versus τ_0 curve to longer values of τ_0. The window of motional sensitivity for the parameters adopted in this simulation (and for most other combinations of parameters) is about 1 to 1000 nsec. Analogous plots of R versus τ_0 where, by contrast, ν_{off} is held fixed, and a variable B_2 field strength is used are shown in Fig. 6. A decrease in B_2 shifts the inflection point of the R versus τ_0 curve to larger τ_0 values. The consequence of altering either ν_{off} or B_2 is to change both the angle that the effective field makes with the z axis (Θ) as well as the precessional frequency (ω_e) of the nuclear spins about B_e. The intensity ratio can also be plotted versus ω_e (at constant B_2), i.e., represented in the form of a frequency-dependent or dispersion curve. Representative plots

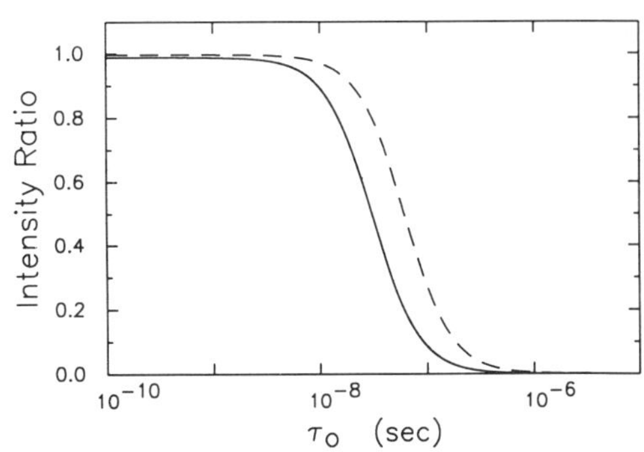

Figure 6. Simulation of the ^{13}C intensity ratio ($R = T_{1\rho}^{\text{off}}/T_1$) versus τ_0 at 7.05 T. The computed curves assume dipolar and CSA relaxation mechanism contributions using the parameters listed in Fig. 3. An off-resonance field of 0.4 (——) and 0.2 G (– – –) was applied 5 kHz off-resonance. Adapted from Schleich *et al.* (1989).

for several different τ_0 values are shown in Fig. 7. The dispersion representation of an intensity ratio plot has the advantage that the effects of polydispersity and chemical exchange (*vide infra*) may be readily incorporated into the formalism. The frequency-dependent form of the experiment incorporates the results of a number of individual off-resonance rotating frame experiments performed at different values of ν_{off}, but at constant B_2, in contrast to a measurement under a single set of experimental parameters.

The above simulations assume intramolecular dipole–dipole relaxation of the peptide bond carbonyl carbon and the neighboring proton of the amino acid residue α-carbon. An additional source of intramolecular dipole–dipole relaxation of the carbonyl carbon arises from the amide proton of the peptide bond. Although the majority of amide protons are likely to participate in chemical exchange with the aqueous solvent, and thus limit this interaction, some protein-buried environments may exist which are not readily accessible to solvent, thereby hindering proton exchange reactions. To assess the contribution of the amide proton to the overall relaxation characteristics of an isotropically tumbling protein, simulations were performed by Caines and Schleich (unpublished) which assumed the complete absence of amide proton exchange. The results of such simulations which include a neighboring amide proton located 2.0 Å from the peptide carbonyl carbon, in addition to the α-carbon proton,

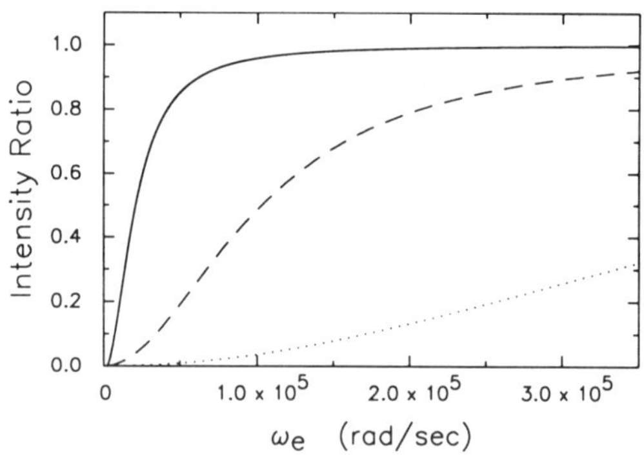

Figure 7. Simulation of the ^{13}C intensity ratio ($R = T_{1\rho}^{\text{off}}/T_1$) versus the angular precessional frequency of the effective field (ω_e) produced by a radiofrequency field applied off-resonance at three different values of τ_0: 20 (——), 100 (– – –), and 500 (· · · · ·) nsec. The computed curves assume an off-resonance field strength of 0.4 G and a B_0 strength of 7.05 T. Dipolar and CSA relaxation mechanism contributions were assumed using the parameters listed in Fig. 3. In aqueous solution at 20°C, assuming a protein partial specific volume of 0.73 cm^3/g, and an average hydration value of 0.51 g H_2O/g protein (Squire and Himmel, 1979), these τ_0 values for isotropic reorientation represent molecular masses of 39.2, 196.1, and 980.0 kDa, respectively. Reproduced, with permission, from Schleich *et al.* (1989).

reveal that this additional source of dipole–dipole relaxation has a negligible effect (ca. 1%) at 2.35 and 7.05 T on the computed intensity ratio dispersion data. This error represents an upper limit since many of the peptide bond amide protons are in exchange with the aqueous solvent. Consequently, amide proton interactions can be generally omitted in the evaluation of τ_0 from peptide bond carbonyl carbon intensity ratio dispersion curves. This conclusion is supported by experimental determinations of proteins τ_0 values in ^1H_2O and ^2H_2O, which, when corrected for viscosity differences, are indistinguishable within the experimental error of the measurement (Caines and Schleich, unpublished).

The theoretical effect of intermolecular dipole–dipole spin–lattice relaxation of the peptide bond carbonyl carbon situated on the surface of an isotropically tumbling protein arising from aqueous solvent translational diffusional motion was examined by Caines and Schleich (unpublished) as a function of protein molcular weight. The theoretical translational contribution is evaluated using Eqs. (9) and (10). The solvent proton density of pure water (6.75×10^{22} protons/cm^3) is assumed, and the distance of carbonyl carbon–water proton closest approach is taken to be 2.0 Å. An average of the water and protein translational diffusion coefficients is calculated from the value of pure water (Simpson and Carr, 1958) and the Stokes–Einstein equations for translational and rotational diffusion, for different assumed isotropic values of τ_0. Since only approximately 15% of the peptide bond carbonyl carbons are located on the protein surface and thus accessible to solvent translational diffusion effects, the amount of intermolecular dipole–dipole spin–lattice relaxation is weighted in the simulations by 15%, the average percentage of side-chain carbonyl carbons (carboxyl plus amide side chains) present in a typical globular protein. The effect of including an intermolecular dipole–dipole spin–lattice relaxation contribution at 7.05 T is to shift the intensity ratio dispersion curves to shorter rotational correlation times by an amount that varies directly with increasing rotational correlation time. At a τ_0 value of 10 nsec the error is 0.4% whereas at 100 nsec the error increases to 3.4%. This estimated error is within the usually encountered experimental error of the measurement, and thus, aqueous proton solvent translational diffusion does not appear to be a complication in the determination of native globular protein rotational correlation times. A possible exception is at very high molecular weight, in which case it is possible to make a correction by inclusion of a solvent-derived intermolecular dipole–dipole relaxation contribution in the analysis. This conclusion is supported by experimental determinations of protein τ_0 values in ^1H_2O and ^2H_2O (*vide ante*).

If monodispersity can be realistically assumed, the rotational correlation time may be determined by a single measurement of the intensity ratio at a known value of ν_{off} and B_2, and evaluated from a theoretically generated plot of R versus τ_0. By contrast, for a polydisperse system it is possible to obtain different τ_0 values when either ν_{off} or B_2 is altered. This is due to the fact that the motional window of the experiment is shifted as each of these two parameters is varied, as illustrated in Figs. 5 and 6. In this case, the offset frequency irradiation or dispersive variation of the off-resonance rotating frame spin–lattice relaxation experiment (at constant B_2) yields by application of nonlinear regression techniques an effective value of τ_0 ($\tau_{0,\text{eff}}$) which is described more fully below. Thus, the restrictive assumption of monodispersity of the reorienting macromolecular population is not necessary.

Rotational correlation time distributions arising from polydispersity (i.e., a distribution of molecular sizes) may be treated statistically to yield different moments of the mean to achieve a single overall description of the average correlation time. The different moments of the mean that define the *average* correlation time are expressed by the following equation (Morgan *et al.*, 1990b):

$$\bar{\tau} = \frac{\sum_j f_j \tau_j^a}{\sum_j f_j \tau_j^{a-1}} \tag{39}$$

where τ_j is the rotational correlation time of the jth species, f_j is the corresponding weight fraction, and a is a constant. When $a = 0, 1, 2, 3, \ldots,$ the number, weight, z and $z + 1$ average correlation time, respectively, are obtained. Using a heuristic approach, Morgan *et al.* (1990b) determined that the appropriate value for a in the off-resonance rotating frame spin-lattice relaxation experiment was 0.56, which places the average correlation time ($\tau_{0,\text{eff}}$) approximately midway between the number and weight averages. This permits reconciliation of $\tau_{0,\text{eff}}$ values with either known or assumed molecular weight distributions.

The rotational diffusion information obtained from an off-resonance rotating frame spin–lattice relaxation experiment is critically dependent on the model adopted for the analysis of the relaxation data. The model utilized for the evaluation of τ_0 from off-resonance rotating frame spin–lattice relaxation experimental data assumes that all of the carbonyl carbons contributing to the resonance signal are part of the rigid polypeptide backbone, and thus their magnetic relaxation characteristics reflect isotropic reorientation characteristics of the tumbling protein. Globular proteins, on average, contain about 15% amino acid residues which possess side-chain carboxyl or amide residues. Since the carbonyl carbons of these more rapidly reorienting residues ($\tau_0 \approx 0.1$ nsec) contribute to the overall carbonyl resonance signal, the effect of potential internal reorientational motions needs to be considered. Spectral simulations (Schleich *et al.*, 1989) reveal that the result of internal motion is to decrease the isotropic $\tau_{0,\mathrm{eff}}$ value. The percentage decrease is dependent on both the correlation time for side-chain motion as well as the overall rotational correlation time.

The ^{13}C α-carbon resonance relaxation behavior is also sensitive to molecular motion. In this case, the CSA relaxation mechanism contribution is negligible. The dominant dipolar relaxation mechanism is efficient at the α-carbon due to the directly bonded proton at a vibrationally averaged distance of 1.11 Å.

2.5.1.2. Metabolites. The utility of the off-resonance rotating frame spin–lattice relaxation experiment for the elucidation of phosphorus metabolite rotational diffusion characteristics is illustrated in Fig. 8. This figure compares the simulated signal intensity ratio dispersion curves, R versus ω_e, for different values of the rotational correlation time. The family of curves depicted by the solid lines represents an isotropically tumbling phosphorus metabolite assuming both intramolecular dipole–dipole and CSA relaxation mechanism contributions, whereas the dotted lines include an additional dipole–dipole relaxation contribution from solvent translational motion. Both simulations employed parameters characteristic of a phosphomonoester (see figure legend). With increasing correlation time, the attainment of the maximum value of the spectral intensity ratio (at constant off-resonance irradiation field strength) requires larger values of

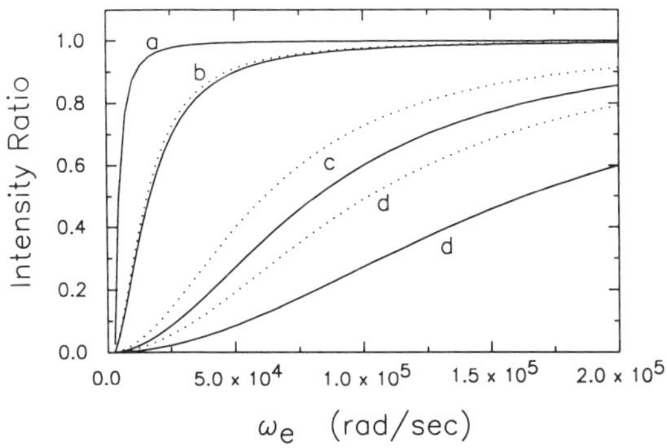

Figure 8. Simulated ^{31}P intensity ratio frequency (dispersion) curves as a function of isotropic rotational correlation time [(a) 1.0 nsec; (b) 10.0 nsec; (c) 50.0 nsec; (d) 100.0 nsec] for glycerol-3-phosphate (G-3-P), a representative tissue phosphomonoester. A B_0 field of 7.05 T, an off-resonance B_2 field of 0.25 G, dipole–dipole relaxation by two protons located 2.6 Å from the ^{31}P, and CSA relaxation (−68.33 ppm for δ_z and −0.67 for η) were assumed (solid lines). The dotted lines indicate the effect of an additional intermolecular dipole–dipole relaxation contribution from solvent (water) translational motion at 20°C to the G-3-P simulated intensity ratio dispersion curves (see text for details). Reproduced, with permission, from Caines *et al.* (1990).

ν_{off} (or ω_e). The simulations indicate that a wide range of rotational correlation times from less than 1 nsec to at least 1000 nsec may be investigated. Changing the off-resonance irradiation field strength alters the window of rotational motion sensitivity slightly. The theoretical contribution of solvent translational motion is assessed by the use of Eqs. (9) and (10); a value of 2.09 Å is used for the distance of closest solvent approach (McCain and Markley, 1986) and the proton density, N, is set to the pure solvent value. The average water and PME translational diffusion coefficient is calculated using the PME translational diffusion coefficient, estimated with the Stokes–Einstein equation assuming the viscosity value and self-diffusion coefficient of pure water at 20°C. For a particle with a rotational correlation time of 10 nsec, a $(1/T_1)^{\text{PH}}_{\text{DD,trans}}$ value of $\sim 2.4 \times 10^{-2}\ \text{sec}^{-1}$ may be calculated for PME, a value equal to 0.6 times the intramolecular relaxation contribution, and 0.2 times the chemical shift anisotropy contribution. As shown by the family of dotted lines in Fig. 8, the effect of including solvent translational motion produces the same general trend as was observed for protein carbonyl carbons. The effect is negligible at small τ_0 values ($\leqslant 10$ nsec), but assumes increasing significance as τ_0 increases. Unlike the protein carbonyl carbons, the effect cannot be ignored. The effect of solvent translational motion is to shift the R versus ω_e curves to a smaller τ_0 value. Even for large isotropically tumbling particles (e.g., $\tau_0 = 100$ nsec, $\bar{D}_{20,\text{w}} = 1.05 \times 10^{-5}$ cm/sec), $\omega^2\tau^2$ equals 1.1×10^{-3}, thereby fulfilling the assumption of "extreme narrowing" used in the derivation of Eq. (9). Thus, a solvent translation dipole–dipole relaxation contribution would also be significant for protein-bound ligand.

The rotational diffusion characteristics of metabolites containing an enriched ^{13}C carbon may also be investigated by use of the off-resonance rotating frame spin-lattice relaxation technique. The choice of relaxation mechanism contributions depends on the type of carbon monitored. The appropriate mechanism for aliphatic carbons is intramolecular dipole–dipole relaxation whereas for carbonyl carbons both intramolecular dipole–dipole and CSA relaxation contribute. In both cases an intermolecular dipole–dipole relaxation contribution arising from solvent translational motion would have to be included if a protonated solvent (e.g., ^1H_2O) was employed in the measurements. Extension to other spin-1/2 nuclei is straightforward.

Deuterium-labeled metabolites are also attractive for study since ^{2}H quadrupolar relaxation is exquisitely sensitive to molecular motion. Depicted in Fig. 9 are intensity ratio versus ω_e dispersion plots for a deuterium-labeled metabolite as a function of τ_0. Simulations indicate that solvent (^1H_2O) translational motion-mediated intermolecular dipole–dipole relaxation mechanism under various assumed conditions are negligible, as expected. Thus, intermolecular contributions may be ignored.

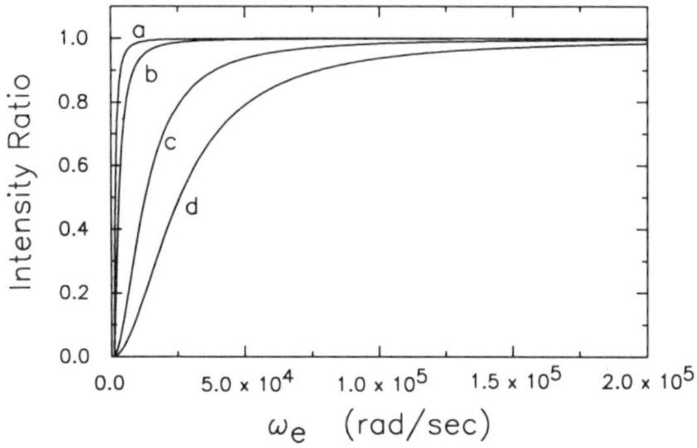

Figure 9. Simulated ^{2}H intensity ratio frequency (dispersion) curves as a function of isotropic rotational correlation time [(a) 1.0 nsec; (b) 10.0 nsec; (c) 50.0 nsec; (d) 100.0 nsec] for a deuterated metabolite. A B_0 field of 7.05 T and an off-resonance B_2 field of 0.25 G were assumed.

Due to the smaller gyromagnetic ratio, the ^{2}H dispersion curves (see Fig. 9) rise faster at the same τ_0 value than the corrresponding ^{31}P curves for glycerol-3-phosphate (Fig. 8) because different relaxation mechanisms are operable. This rapid rise can be reduced by using higher B_2 field strengths as discussed in Section 2.5.1.1. This would make the faster tumbling species easier to analyze since a wider low-power RF offset frequency range would be used.

2.5.2. Anisotropic Reorientation: Proteins

Axially symmetric prolate ($a/b > 1$) or oblate ($a/b < 1$) ellipsoids of revolution are utilized for the modeling of hydrodynamic behavior of globular proteins which deviate from spherical shape. Rotation about the long axis of a prolate ellipsoid occurs more easily than about the center of mass of a sphere of equivalent volume; however, rotation about the short axis encounters considerably more friction. By contrast, oblate ellipsoids experience approximately the same friction about either axis, but both rotational motion modes experience more friction than a sphere of equivalent volume. These qualitative considerations form the basis for understanding the magnetic relaxation behavior of anisotropically tumbling rigid particles.

The axially symmetric prolate ellipsoid of revolution is considered first since the greatest difference between $\tau_\perp$ and $\tau_\parallel$ occurs for this shape. The angle α describes the orientation of the C–H inernuclear vector r_{CH} with respect to the principal diffusion axis D_z ($= D_\parallel$). The spectral density function for dipolar relaxation is dependent on α by way of the coefficients A, B, and C [see Eq. (30)]. Orientations which place r_{CH} parallel to D_z ($\alpha = 0$ or 180°) correspond to minima in the relaxation rate, whereas a maximum effect occurs when r_{CH} is perpendicular to D_z ($\alpha = 90°$). For CSA relaxation the orientation of the peptide carbonyl CSA with respect to the diffusion axis system assumes critical importance. If the CSA tensor of all peptide carbonyl carbons is assumed to be similar to that found in the model peptide bond, then the CSA tensor has the orientation relative to the atoms of the peptide bond as shown in Fig. 10a. The orientation of the CSA tensor relative to the diffusion axis for axially symmetric ellipsoids is described by two Euler angles, β and γ, as shown in Fig. 10b. The spectral density functions are influenced by angles β and γ through the coefficients c_0, c_1, and c_2 [see Eq. (31)]. The peptide carbonyl CSA tensor is schematically depicted in terms of an oblate ellipsoid with semilaxes $|\delta_z| \geqslant |\delta_x| \gg |\delta_y|$. The orientation of the δ_z axis to D_z is described by the angle β. Orientations which place D_z perpendicular to the $\delta_z\delta_x$ plane ($\beta = 90$ or $270°$ and $\gamma = 90$ or $270°$) give maxima in the relaxation rate, since δ_z and δ_x are the components of the CSA tensor along which the largest changes in chemical shift occur. Furthermore, for $\beta = 90$ or $270°$ the relaxation rate varies strongly with γ and sustains a minimum value when D_z is rotated into the $\delta_z\delta_x$ plane at $\gamma = 0$ or $180°$. When $\beta = 0$ or $180°$, D_z is in the $\delta_z\delta_x$ plane, and such values of β have a weak dependence on γ.

Similar correlation times about both $D_{\parallel}$ and $D_{\perp}$ (i.e., $\tau_{\parallel}$ or $\tau_{\perp}$) occur for oblate ellipsoids. This similarity of τ values renders the orientation effects less pronounced than for prolate ellipsoids.

The dependence of the spin–lattice relaxation rate and the off-resonance rotating frame spin–lattice relaxation rate on the axial ratio of an ellipsoid of revolution relative to that for an isotropically tumbling sphere of equivalent volume for various values of α, β, and γ has been considered by Schleich *et al.* (1989). The intensity ratio versus axial ratio plot (Fig. 11) displays behavior similar to the corresponding $1/T_{1\rho}^{\text{off}}$ plots presented by Schleich *et al.* (1989) and is in accord with the relaxation rate behavior described above.

Deviations from spherical shape, i.e., isotropic tumbling, for both prolate and oblate ellipsoids of revolution are characterized by reduced values of R at fixed ω_e with increasing (prolate) or decreasing (oblate) axial ratio, as shown in Figs. 12 and 13.

Intensity ratio ($\bar{R}$) versus ω_e curves have been calculated for lysozyme and concanavalin A using the theoretical formalism described above and the X-ray crystal structure coordinates. In each case, the hydrodynamically derived axial ratio was used. The theoretical dispersion curves for these proteins are shown in Fig. 14. For reference, the corresponding isotropic dispersion curve is also given. For both proteins, or more generally for anisotropic molecular reorientation, the dispersion curve is shifted to a smaller $\bar{R}$ value at a given ω_e. For lysozyme the two dispersion curves are

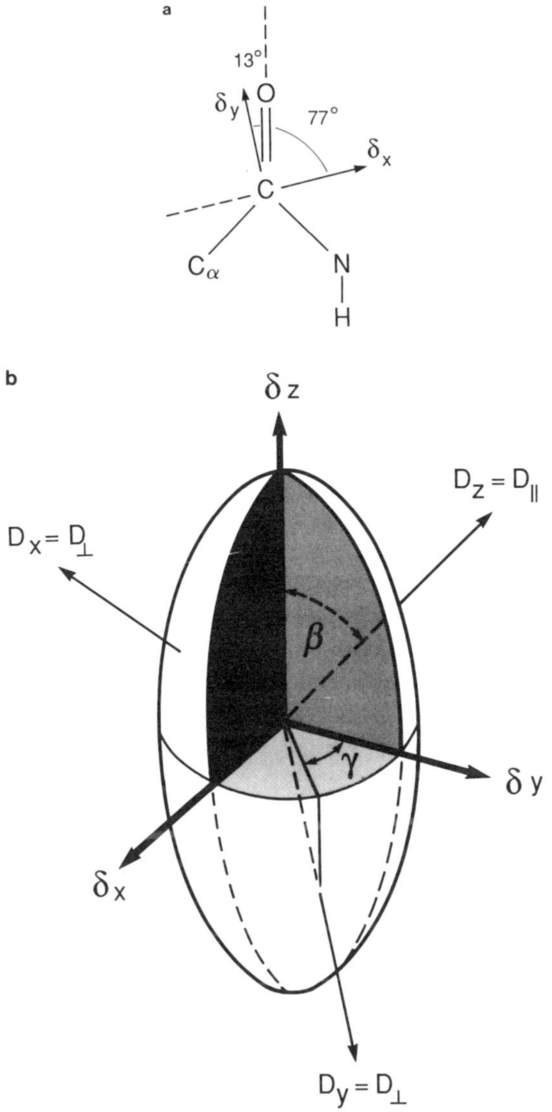

Figure 10. (a) Orientation of the peptide ^{13}C chemical shift tensor of glycerol·glycine·HCl·H_2O. The α-carbon, carbonyl carbon, and amide nitrogen atoms are virtually coplanar; δ_x and δ_y are also in this plane, and form an angle of 77 and 13°, respectively, with the C–O bond. The direction of δ_z (not shown) is perpendicular to this plane. Adopted from Stark *et al.* (1983). (b) The chemical shielding ellipsoid for the peptide bond carbonyl carbon indicating the relationship of the chemical shielding tensor principal axis system to the diffusion axis coordinate system. The Euler angles β and γ relate the CSA principal axis system to the axially symmetric diffusion axis system. Reproduced, with permission, from Schleich *et al.* (1989).

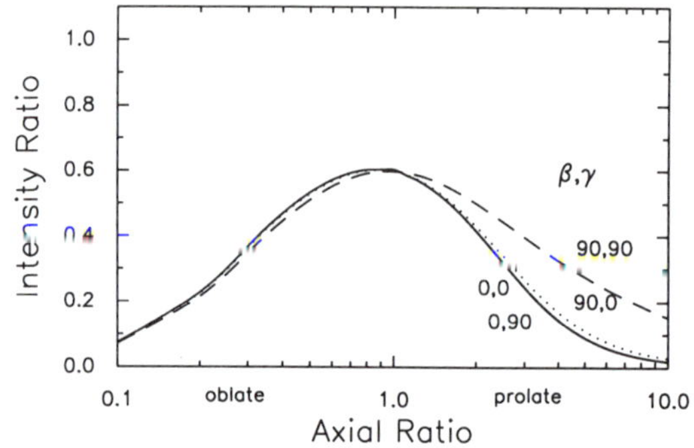

Figure 11. Simulation of the ^{13}C intensity ratio for an anisotropically tumbling particle ($T^{off}_{1\rho,aniso}/T_{1,aniso}$) versus axial ratio for different β and γ Euler angles. The computed curves assume a B_0 field strength of 7.05 T, an off-resonance irradiation field of 0.5 G applied 5 kHz off-resonance, dipolar and CSA relaxation contributions as described in Fig. 3, a C–H internuclear vector angle α of 0°, and a τ_0 value of 20 nsec for isotropic reorientation of the equivalent volume sphere. Reproduced, with permission, from Schleich *et al.* (1989).

fairly similar, whereas for concanavalin A, a protein with an axial ratio of ~2, the deviation between the isotropic and anisotropic cases is more significant. Analysis of $\bar{R}$ (at fixed ω_e) versus residue number for different proteins of known structure indicates, as expected, that the individual R values are independent of secondary structure type. The spectral intensity ratio is solely dependent on the orientation of the internuclear dipole vector and CSA principal axis system relative to the particle diffusion axis (Morgan *et al.*, 1990a), rather than the location of the amino acid residue in the protein.

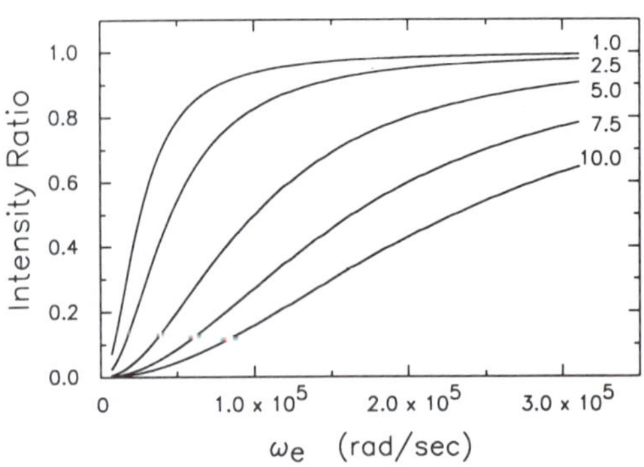

Figure 12. Simulation of the ^{13}C intensity ratio for an anisotropically tumbling particle ($T^{off}_{1\rho,aniso}/T_{1,aniso}$) versus ω_e at different axial ratios for prolate ellipsoids. The computed curves assume a B_0 field strength of 7.05 T, an off-resonance irradiation field of 0.5 G applied 5 kHz off-resonance, dipolar and CSA relaxation contributions as described in Fig. 3, a C–H internuclear vector angle α of 0°, β and γ Euler angles of 0°, and a τ_0 value of 20 nsec for isotropic reorientation of the equivalent volume sphere. Reproduced, with permission, from Schleich *et al.* (1989).

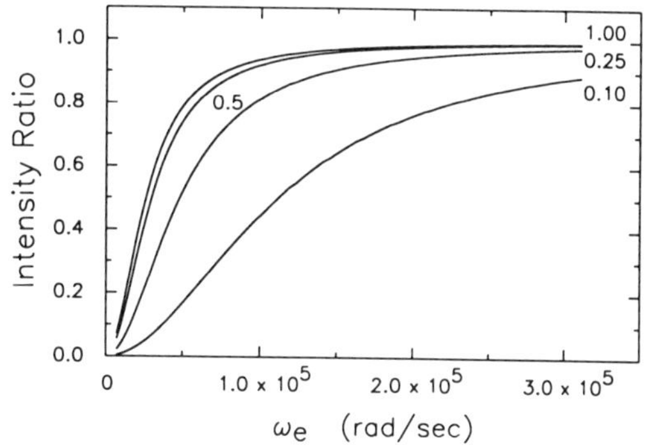

Figure 13. Simulation of the ^{13}C intensity ratio for an anisotropically tumbling particle ($T^{\text{off}}_{1\rho,\text{aniso}}/T_{1,\text{aniso}}$) versus ω_e at different axial ratios for oblate ellipsoids. The computed curves assume a B_0 field strength of 7.05 T, an off-resonance irradiation field of 0.5 G applied 5 kHz off-resonance, dipolar and CSA relaxation contributions are described in Fig. 3, a C-H internuclear vector angle α of 0°, β and γ Euler angles of 0°, and a τ_0 value of 20 nsec for isotropic reorientation of the equivalent volume sphere. Reproduced, with permission, from Schleich *et al.* (1989).

For an anisotropically reorienting particle, the spectral density functions employed for the calculation of the spectral intensity ratio require three correlation times (τ_A, τ_B, τ_C). By contrast, only a single correlation time is required for an isotropically tumbling particle since there is no unique diffusion axis. If the observed quantities for an anisotropically

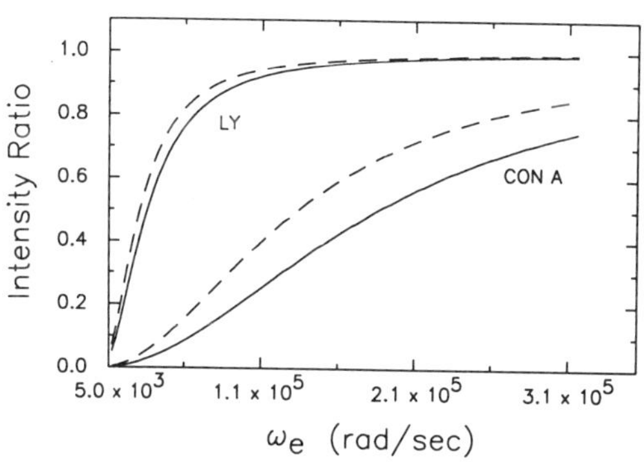

Figure 14. Simulation of the ^{13}C intensity ratio for an anisotropically tumbling particle ($T^{\text{off}}_{1\rho,\text{aniso}}/T_{1,\text{aniso}}$) (——) versus ω_e for the protein lysozyme (LY) and concanavalin A (CON A). Corresponding curves for assumed isotropic (axial ratio = 1.0) reorientation are also presented (– – –). The computed curves assume a B_0 field strength of 7.05 T, an off-resonance irradiation field of 0.5 G applied 5 kHz off-resonance, dipolar and CSA relaxation contributions as described in Fig. 3, a C-H internuclear vector angle α, and β and γ Euler angles calculated for each protein amino acid residue peptide bond present in the protein from the X-ray structure (see Section 2.3.2.1); τ_0 for isotropic reorientation of the equivalent volume sphere and axial ratios, of 20 nsec, 1.67; 100 nsec, 2.13, respectively, for lysozyme and concanavalin A. Reproduced, with permission, from Schleich *et al.* (1989).

tumbling protein are fitted using an assumed isotropic model, then an "effective" correlation time, $\tau_{0,\text{eff}}$, is obtained representing a weighted value of τ_A, τ_B, and τ_C.

A direct method of evaluating the error associated with the application of an isotropic reorientation model to an anisotropically tumbling particle is afforded by a plot (Fig. 15) of $\tau_{0,\text{eff}}/\tau_0$, i.e., the ratio of the "effective" rotational correlation time to the correlation time for a sphere of equivalent volume (Morgan *et al.*, 1990a). The polynomial describing the dependence of $\tau_{0,\text{eff}}/\tau_0$ on axial ratio (a/b) is:

$$\frac{\tau_{0,\text{eff}}}{\tau_0} = 1.012 + 0.282x + 2.773x^2 + 0.244x^3 + 0.720x^4 - 0.446x^5 \quad (40)$$

where $x = \log(a/b)$ $[0.1 \leqslant a/b \leqslant 10.0]$. This polynomial expression may be used to estimate the error introduced by the application of the isotropic analysis model to an anisotropically tumbling protein such as lysozyme, which has been shown to be hydrodynamically equivalent to a prolate ellipsoid of axial ratio 1.67 (Dubin *et al.*, 1971). The application of an isotropic reorientation model for data analysis of lysozyme results in an overestimation of the correlation time by 21%. The error associated with an assumed isotropic analysis increases significantly with departure from spherical shape.

The α-carbon relaxation behavior is similar to the carbonyl carbon with the exception that CSA relaxation is not considered. Therefore, the conclusions arrived at in the above treatment of the carbonyl carbon are pertinent to the α-carbon with the exclusion of CSA relaxation mechanism contributions.

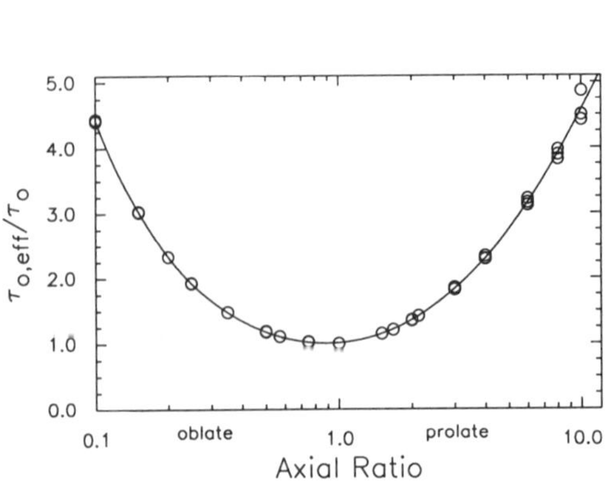

Figure 15. The dependence of the ratio of the effective rotational correlation time ($\tau_{0,\text{eff}}$) to the assumed isotropic correlation time of an equivalent volume sphere (τ_0) on axial ratio for the proteins lysozyme, concanavalin A, and myoglobin. The solid line represents the fifth-order polynomial fit to the data. See Section 2.5.2 for details of the $\tau_{0,\text{eff}}/\tau_0$ calculation and the numerical values of the coefficients of the polynomial. Reproduced, with permission, from Schleich *et al.* (1989).

3. EXPERIMENTAL PROCEDURES

3.1. Pulse Schematic and Block Diagram

The timing diagram of the off-resonance rotating frame spin-lattice relaxation pulse sequence is shown in Fig. 16. The off-resonance RF radiation is applied at a specific low-power B_2 field strength and offset frequency for a period of at least 3–5 T_1 to establish nuclear spin polarization and subsequent precession about the effective field. This is followed by a nonselective 90° read pulse which samples the z-magnetization in the rotating frame and results in a free induction decay (FID). The procedure is repeated at different RF offsets relative to the frequency of the signal of interest. Application of an intervening homospoil pulse prior to the read pulse has been found to be unnecessary in most circumstances. Quadrature phase detection and phase cycling are used in sampling the FID to eliminate coherent noise. The diagram in Fig. 17 shows essential instrumental components for the implementation of the off-resonance rotating frame spin-lattice relaxation experiment. The X-nucleus broadband decoupler (f_1) synthesizes the low-power RF which is applied under continuous-wave mode. To eliminate residual power interference effects from the broadband decoupler when gated off, a crossed diode circuit is used in line between the decoupler and a low band filter, and the preamplifier (probe interface module). The low band filter removes harmonic frequencies and signals which are outside the range of interest. In addition, a wattmeter can be used to monitor the forward and reflected power from the broadband decoupler. The power coupler functions as a means of protecting the broadband decoupler amplifier from the high-power observe amplifier which utilizes the same route to the probe. The spectrometer frequency (f_0) and

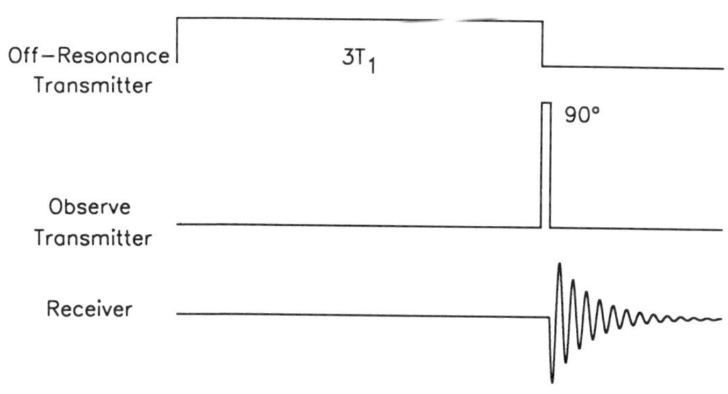

Figure 16. The pulse sequence for the off-resonance rotating frame spin-lattice relaxation experiment. Reproduced, with permission, from Schleich *et al.* (1989).

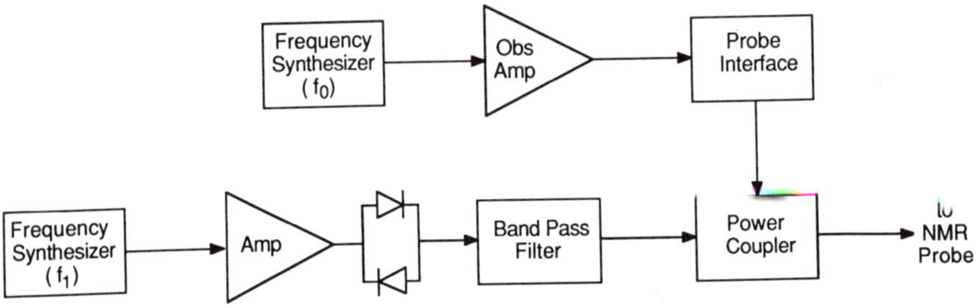

Figure 17. Instrumentation block diagram for the off-resonance rotating frame spin–lattice relaxation experiment. Frequency synthesizer f_1 provides the low-power off-resonance RF irradiation field. Crossed diodes and a band pass filter block interfering background signals. The power coupler provides the means of protecting the off-resonance frequency amplifier from the high-power RF pulses used for signal detection simultaneous with the routing of the off-resonance RF to the probe. Reproduced, with permission, from Schleich *et al.* (1989).

the decoupler frequency (f_1) are independent sources and lack phase coherence. This arrangement between the two synthesizers produces certain demands when the off-resonance field calibration is performed.

3.2. Off-Resonance Field Calibration

Suitable calibration standards for ^{13}C and ^{31}P studies are 2% w/v aqueous solutions of 1-^{13}C (99 atom%) sodium acetate or disodium phosphate, respectively, which contain ~35 μM $MnCl_2$ (to reduce the ^{13}C or ^{31}P T_1) and NaCl to appropriately adjust the dielectric load. For ^2H studies, 5% w/v ^2H (99 atom%) acetone in glycerol may be used over a range of temperatures whereas glycerol-*d*5 (99 atom%) is adequate only at room temperature, due to excessive line broadening at lower temperatures. These solutions provide adequate signal-to-noise with the acquisition of a single FID. The B_2 field strength, typically 0.2 to 2.0 G, is calibrated at a given RF power level by determining the 180° pulse width from an on-resonance X-band decoupler experiment. The receiver (f_0) and decoupler (f_1) frequencies are set to the resonance frequency of the nuclide in the standard sample, which is determined from a one-pulse experiment. A number of X-band pulse widths (~20 to 32), varying from 100 μsec to 2.0 msec, are used in the calibration experiment. Following a delay period of at least 5 T_1 and a continuous-wave X-band pulse, the receiver is gated on and the FID collected. Only one FID is collected for each of the various pulse durations. In the event that the receiver and decoupler lack phase coherence, only the magnitude, or power, spectrum at each pulse width should be used in determining the 180° pulse width. The nutational frequency is equal to twice the inverse of the 180° pulse width. The B_2 field strength is calculated by relating the nutational frequency (ν) to the gyromagnetic ratio [$B_2 = 2\pi\nu/\gamma_I = \pi/(t_{180°}\gamma_I)$]. Calibration of the proton decoupler field used in ^1H off-resonance irradiation experiments is accomplished using a technique described by Bax (1983). A solution of 1.5 M sucrose in ^2H_2O, adjusted to an appropriate dielectric load with NaCl, is employed as a calibration standard. The decoupled calibration pulse sequence is comprised of a pulse delay (to allow for ^{13}C relaxation), an off-resonance 90° carbon pulse, a specific interpulse delay (equal to $1/2J_{HC}$), and a continuous-wave on-resonance proton pulse (varying from 1 to 150 μsec), followed by acquisition of the FID. The pulse sequence is repeated until sufficient ^{13}C signal is accumulated (~40 transients with 1.5 M sucrose). The specific interpulse delay time is determined from a measurement of the J_{CH} coupling constant of the C-1 sucrose carbon with its directly bonded protons. The measured value of J_{CH} may have a 5% deviation and still yield satisfactory results (Bax, 1983). The proton pulse which produces a null in the C-1 carbon spectrum is the proton 90° pulse width. The B_2 pulse strength is calculated from the 180° nutational frequency as described above. Any changes in the probe tune and match of the biological sample compared to the calibration sample should be noted since this will directly affect the quality factor (Q) of the coil and therefore the B_2 calibration. If a large difference in the tune and match is observed, the dielectric load of the standard should be adjusted, or a more suitable standard obtained.

3.3. Experimental Procedure

A series of computer-generated intensity ratio versus offset frequency dispersion curves (R versus ν_{off}) as a function of B_2 field strength and rotational correlation time can be very useful in choosing the optimum B_2 power level and offset frequencies for an off-resonance rotating frame spin–lattice relaxation experiment. In addition, these simulations provide insight into the range of molecular reorientational motions which can be studied under the selected conditions. They are also useful in estimating how large ν_{off} should be for the reference spectrum. The majority of biological samples will require signal averaging due to low concentration or low natural abundance. To average changes which might occur during the course of a ^{13}C protein solution experiment, a small number of transients (e.g., 20 to 100) should be collected at each ν_{off} cycling through all selected values of ν_{off}, including the reference spectrum, until the total number of desired transients is obtained. In ^2H-labeled and ^{31}P metabolite experiments the impact of slow changes in metabolite levels with time can be minimized by collecting consecutively the desired total number of transients (e.g., 100) at the particular offset frequency and the corresponding reference spectrum by alternating between the two in blocks of 20 accumulations. In ^1H

off-resonance experiments the water proton signal is usually very strong and only a small number of transients (~16) need be acquired. Additionally, a randomly selected order of frequency offsets should be used. Reference spectra are collected in the presence of a very large ν_{off} value such that no off resonance signal attenuation effects will occur. This is done to assure that the effects of RF heating will be approximately the same for all spectra. The reference spectrum yields the Boltzmann thermal equilibrium distribution magnetization (M_0). At higher B_2 field strengths such as 1 to 2 G, care must be taken so that adequate temperature regulation is utilized to avoid excessive RF heating. It may be necessary to insert an additional delay between acquisitions during which no RF is applied. The experimental intensity ratio (R), as a function of offset frequency, is the ratio of the measured intensity in the presence of an off-resonance irradiation field to a reference intensity obtained with very far off-resonance irradiation. For ^{13}C spectra the carbonyl intensity is obtained by integration of the protein carbonyl resonance envelope. Line-shape deconvolution is generally required for ^{31}P spectra in order to provide the relevant signal intensities since overlapping resonances are frequently encountered. ^{2}H spectra may be analyzed by either method depending upon the spectral characteristics. In ^{1}H spectra the intensity of the water resonance is obtained by signal integration.

4. ANALYSIS OF EXPERIMENTAL OFF-RESONANCE ROTATING FRAME SPIN–LATTICE RELAXATION DATA

Quantitative analysis of the R versus ω_e dispersion curve is achieved by nonlinear least-squares regression with R and ω_e the respective dependent and independent variables. Nonlinear regression is performed by incorporating the appropriate (for the specific case under consideration), off-resonance rotating frame spin–lattice relaxation formalism (described above) into a program such as NONLINWOOD (Daniel and Woods, 1980). The effective (average) rotational correlation time, $\tau_{0,\text{eff}}$, and the intensity ratio very far off-resonance, $R(\omega_e \to \infty)$, are fit to the experimental data using a theoretical model which assumes a single correlation time for random isotropic tumbling. In those instances where the dispersion curves plateau at R (large ω_e) values less than unity, saturation transfer effects are operant (*vide ante*). In this situation, the effective rotational correlation time and the rate constant, k_A, for exchange from site A → B can be fit to the experimental data using the modified off-resonance formalism (see Section 6). The resulting value of the effective rotational correlation time, $\tau_{0,\text{eff}}$, is insensitive to initial estimate. Necessary parameters are obtained experimentally or estimated theoretically. Other models may be employed which recognize anisotropic motion, or polydispersity arising from discrete mixtures or from assumed models of association.

5. APPLICATIONS

5.1. Protein Rotational Diffusion

5.1.1. Aqueous Solution Studies

The off-resonance rotating frame spin–lattice relaxation technique has been applied to a variety of dilute aqueous protein solutions for the purpose of determining protein rotational correlation times. As noted by James *et al.* (1978), values of the protein rotational correlation time determined for six monodisperse protein preparations, varying in axial ratio from 1.2 to at least 3.5, and spanning the molecular mass range from 14.6 to 150 kDa, obtained by assuming isotropic reorientational motion, were in excellent agreement with those evaluated by the use of well-established rotational diffusion techniques.

Morgan *et al.* (1990b) demonstrated that polydisperse protein systems may be analyzed if a model describing the molecular weight distribution is known or a reasonable one can be assumed. Several different protein systems were investigated by these workers. Fig. 18 shows the intensity ratio

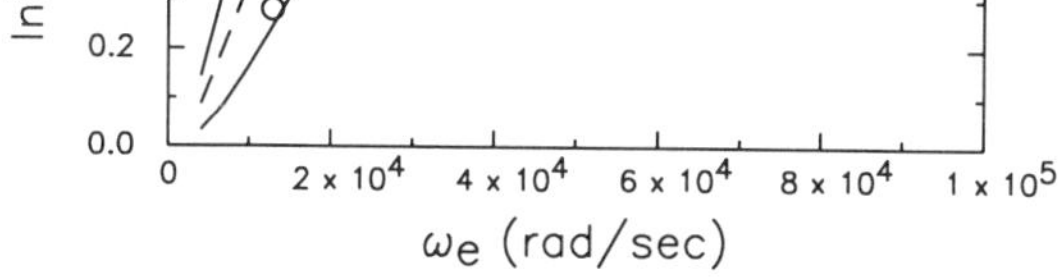

Figure 18. Carbonyl ^{13}C intensity ratio (R) versus ω_e plots for 100% bovine serum albumin (BSA) ($\tau_{0,\text{eff}} = 22.1$ nsec) (○), 100% lysozyme ($\tau_{0,\text{eff}} = 10.1$ nsec) (□), and a synthetic mixture of the two proteins (48% by weight BSA) (●). The solid lines for the two protein preparations represent the theoretical line calculated for isotropic tumbling with the indicated best-fit $\tau_{0,\text{eff}}$ value, whereas for the mixture the dashed line represents the calculated curve for the best-fit value of 45% (by weight) of BSA in the mixture, and the assumed $\tau_{0,\text{eff}}$ values for the constituent pure proteins. The calculations assume a B_0 field strength of 7.05 T and an off-resonance irradiation field of 0.4 G. The total protein concentration is 100 mg/ml. Reproduced, with permission, from Morgan *et al.* (1990a).

frequency-dependent (dispersion) curves (R versus ω_e) for monodisperse lysozyme and bovine serum albumin (BSA) preparations under conditions favoring the monomeric state of lysozyme, and synthetic binary mixture. Nonlinear regression analyses of the dispersion curves obtained for the pure protein preparation solutions yielded $\tau_{0,\text{eff}}$ values of 10.1 ± 0.5 and 22.1 ± 0.8 nsec at 35°C, respectively, for the lysozyme and BSA solutions. For the binary synthetic protein mixture a weight fraction of 0.45 ± 0.05 for BSA and a $\tau_{0,\text{eff}}$ value of 14.4 ± 0.7 nsec were obtained. The latter is in excellent agreement with that calculated with equation (39), which predicts a value of 14.8 nsec.

In another study, the pH-induced association of lysozyme (Sophianopoulos and van Holde, 1964), which occurs at pH values greater than 5.5, was also investigated by Morgan *et al.* (1990b). The results at pH 4.4 and 8.0 are shown in Fig. 19. At pH 4.4 lysozyme was observed to have a $\tau_{0,\text{eff}}$ value of 11.6 ± 0.5 nsec, whereas at pH 8.0, $\tau_{0,\text{eff}}$ increased to 21.2 ± 0.6 nsec, indicating aggregation, and thus most likely a polydisperse protein mixture. Assuming a model of indefinite self-association (isodesmic) in the nonlinear regression analysis of the pH 8.0 intensity ratio dispersion curve yielded an apparent association constant of $66 \pm 9\ \text{M}^{-1}$. This value is somewhat lower than that obtained by rigorous thermodynamic-based methods at lower protein concentration (Wills *et al.*, 1980). For comparison, the simulated dispersion curve for a dimer of lysozyme ($\tau_0 = 23.2$ nsec) is included in Fig. 19. As shown, this trace is a fair representation of the experimental data obtained at pH 8.0.

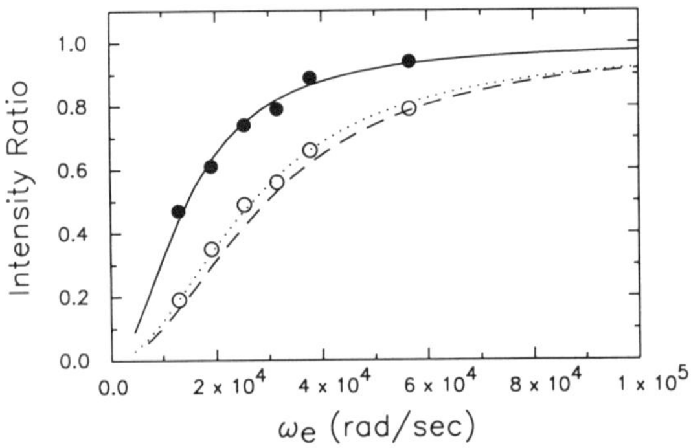

Figure 19. Carbonyl ^{13}C intensity ratio (R) versus ω_e plots for lysozyme at pH values of 4.4 (●) and 8.0 (○) at 15°C. The solid line represents the best-fit $\tau_{0,\text{eff}}$ value of 11.6 nsec at pH 4.4; the dotted line represents a $\tau_{0,\text{eff}}$ of 21.2 nsec at pH 8.0. The dashed line represents the simulated dispersion curve for lysozyme dimer ($\tau_0 = 23.2$ nsec) assuming isotropic reorientation. Assuming an indefinite self-association constant (isodesmic) model in the analysis of R versus ω_e yielded an apparent association constant of $66 \pm 9\ \text{M}^{-1}$. Reproduced, with permission, from Morgan *et al.* (1990a).

Lysozyme exhibits a reversible phase separation under appropriate solvent conditions (100 mg/ml, pH 5.5, and 0.5 M NaCl) upon cooling to below a critical temperature of ~5°C (Tanaka *et al.*, 1977). Intensity ratio dispersion curves for this system at temperatures above and below the phase transition temperature are shown in Fig. 20. At 35°C a $\tau_{0,\mathrm{eff}}$ value of 7.7 ± 0.4 nsec was obtained by nonlinear regression, whereas at 1°C the value of this parameter increased to 53.7 ± 6.8 nsec, assuming an isotropic reorientation tumbling model. The increase in $\tau_{0,\mathrm{eff}}$ to a value much larger than that predicted theoretically (2.7 times) indicates protein association. Assuming an isodesmic association model for the 1°C data, as employed in the above example, yields an apparent association constant of $178 \pm 60\ \mathrm{M}^{-1}$. The theoretically calculated intensity ratio dispersion trace for a lysozyme dimer (see Fig. 20) is a substantially poorer fit of the experimental data than the isodesmic association model.

The association–dissociation behavior of γ-crystallin, a well-characterized major mammalian eye lens protein, was studied by Stevens *et al.* (1992) using a combination of spectroscopic techniques. Off-resonance rotating frame spin–lattice relaxation concentration-dependent studies, such as those in Fig. 21, indicate that γ-crystallin reversibly associates. Additional studies (not shown here) indicate that the properties of these aggregates are sensitive to the exact solution conditions (pH, ionic strength, and temperature) employed. The maximum size of the aggregate determined in this study does not exceed 300 kDa. Interpretation of NMR dispersion studies

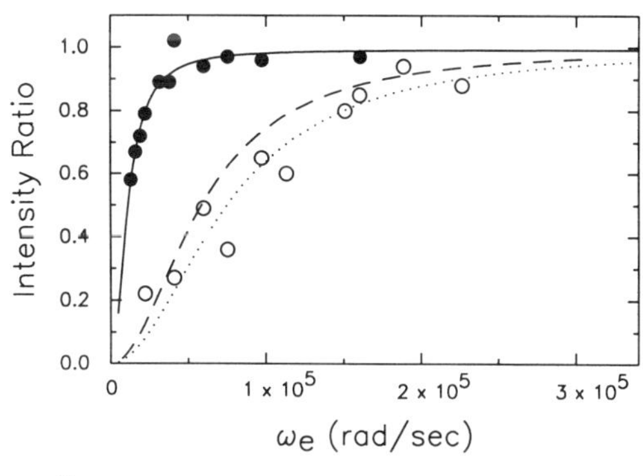

Figure 20. Carbonyl ^{13}C intensity ratio (R) versus ω_e plots for lysozyme in 0.5 M NaCl (pH 5.5) at 35°C (●) and 1°C (○). The latter temperature induces a phase transition resulting in mimicking the "cold cataract" condition of juvenile mammalian eye lenses and γ-crystallin solutions. The solid line represents the best fit assuming a model of isotropic reorientation with $\tau_{0,\mathrm{eff}} = 7.7$ nsec at 35°C; the dotted line represents a $\tau_{0,\mathrm{eff}}$ of 53.7 nsec at 1°C, using the same isotropic monodisperse model. The dashed line represents the simulated dispersion curve for a lysozyme dimer at 1°C ($\tau_0 = 41.6$ nsec) assuming isotropic reorientation. A nonlinear regression analysis yielded an apparent association constant of $178 \pm 60\ \mathrm{M}^{-1}$. Reproduced, with permission, from Morgan *et al.* (1990a).

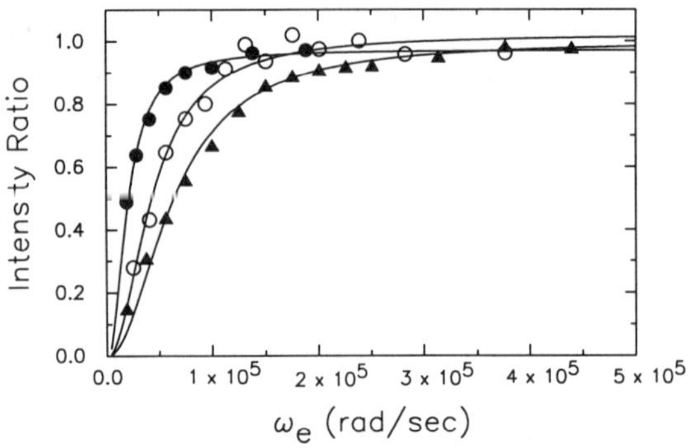

Figure 21. Carbonyl ^{13}C intensity ratio (R) versus ω_e plots for γ-crystallin at different concentrations in 20 mM sodium phosphate buffer, pH 6.9, at 22°C. In addition, the buffer contains 120 mM KCl, 1 mM EDTA, 0.2 mM DTT, and 0.005% NaN_3. The solid lines represent the best fit assuming a model of isotropic reorientation with $\tau_{0,\text{eff}}$ values of 40, 60, and 108 nsec, at 80 (●), 160 (○), and 240 mg/ml (▲), respectively. Taken from Stevens *et al.* (1992).

(NMRD) of γ-crystallin (Koenig *et al.*, 1990), discussed below (see Section 5.1.4), results in a somewhat different conclusion regarding the solution conformational and interaction behavior of this protein.

These studies serve to illustrate the power as well as the limitations of the off-resonance rotating frame spin–lattice relaxation technique for the analysis of protein reorientational motion in aqueous solution. If a model describing the molecular weight distribution is known, or a reasonable one is assumed, the analysis is relatively straightforward. Thus, the derived rotational hydrodynamic information is dependent on the model employed to interpret the relaxation data. Application of the off-resonance rotating frame spin–lattice relaxation formalism, modified to include the effects of polydispersity, should prove useful for the study and characterization of protein rotational dynamics in complex biological systems. Specific applications are illustrated below.

5.1.2. Intact Red Blood Cell

The ^{13}C off-resonance rotating frame spin–lattice relaxation technique has been applied by James *et al.* (1979) to the study of hemoglobin S (HbS) aggregation in oxygenated sickle cells by assessment of protein rotational diffusion characteristics. For comparison, hemoglobin rotational diffusion was also examined in normal adult red blood cells. A value of 95 nsec was obtained for τ_0 of hemoglobin in normal erythrocytes, whereas a significantly larger value (130 nsec) was obtained for HbS in sickle cells. In such comparative studies it is important to ensure that comparable hematocrit values are used. Because the molecular properties of hemoglobin A and hemoglobin

S are similar, except for the tendency of enhanced aggregation, the increased rotational correlation time is therefore indicative of hemoglobin S aggregation in the intact red blood cell.

5.1.3. Eye Lens Tissue Homogenates

To explore the motional dynamics behavior of eye lens proteins under conditions approximating those found *in vivo*, Morgan *et al.* (1989) applied the ^{13}C off-resonance rotating frame spin-lattice relaxation technique to lens tissue cortical and nuclear homogenates. The frequency dependence of the peptide bond carbonyl carbon intensity ratio of lens proteins present in native calf nuclear homogenate (42% protein) at 35°C revealed the presence of a polydisperse mobile protein fraction with a $\tau_{0,\mathrm{eff}}$ value of 57 nsec. This value for $\tau_{0,\mathrm{eff}}$ is consistent with the average value calculated from the known water-soluble nuclear lens protein polydispersity assuming a cytoplasmic viscosity three times that of pure water. At a temperature below the cold cataract phase transition temperature (~16°C), $\tau_{0,\mathrm{eff}}$ *decreased* to 43 nsec (theoretically an *increase* of 2.7 times would be expected for a temperature decrease from 35 to 1°C), indicating a selective removal of a specific lens protein fraction under cold cataract conditions to an NMR "invisible" fraction in this type of experiment. Intensity ratio dispersion curves for nuclear homogenates are shown in Fig. 22. Similar experiments (see Fig. 23) performed on native cortical homogenate (25% protein) gave $\tau_{0,\mathrm{eff}}$ values of 34 and 80 nsec at 35 and 1°C, respectively. Both values could be reconciled with the known lens cortex soluble protein polydispersity using an assumed cytoplasmic viscosity 1.5 times that of pure water at the same temperature. In combination with the application of solid-state NMR techniques (Morgan *et al.*, 1989; Rydzewski *et al.*, 1992), these studies

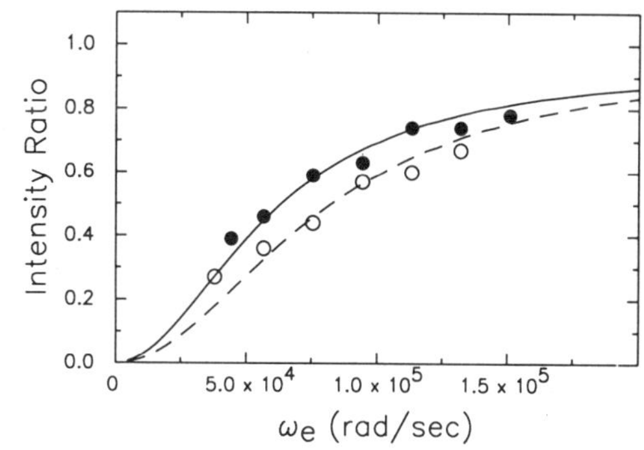

Figure 22. Carbonyl ^{13}C intensity ratio (R) versus ω_e plots for calf lens nuclear homogenate proteins at 35°C (○) and 1°C (●). The latter temperature induces a phase transition resulting in a "cold cataract" characterized by intense light scattering (opacification). The solid and dashed lines represent best-fit lines obtained by nonlinear regression assuming a model of monodisperse isotropic reorientation with $\tau_{0,\mathrm{eff}}$ values of 43 and 57 nsec, respectively. Reproduced, with permission, from Morgan *et al.* (1989).

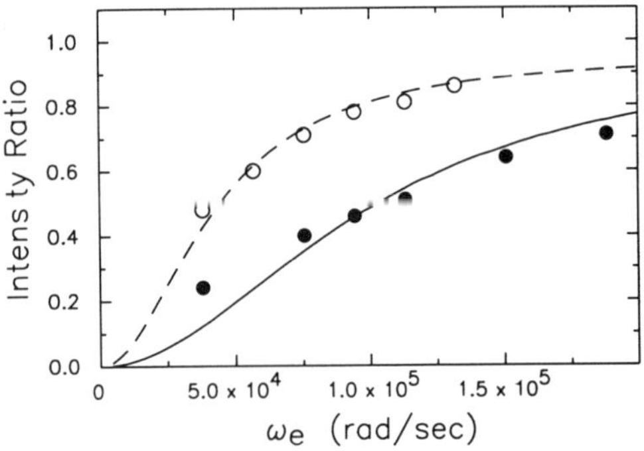

Figure 23. Carbonyl ^{13}C intensity ratio (R) versus ω_e plots for calf lens cortical homogenate proteins at 35°C (○) and 1°C (●). The solid and dashed lines represent best-fit lines obtained by nonlinear regression assuming a model of monodisperse isotropic reorientation with $\tau_{0,\text{eff}}$ values of 34 and 80 nsec, respectively. Reproduced, with permission, from Morgan *et al.* (1989).

established the presence of both mobile and solidlike protein phases in calf lens nuclear homogenate, whereas for the native cortical homogenate, within the detection limits of NMR, the protein phase is mobile. Application to the intact calf lens gives a $\tau_{0,\text{eff}}$ value intermediate between that obtained for nuclear and cortical homogenates.

5.1.4. Relationship to NMRD

NMRD is a relaxometry technique in which the water proton longitudinal relaxation rate constant of an aqueous protein solution or a tissue is measured as a function of the static magnetic field strength (Hallenga and Koenig, 1976; Koenig, 1980; Koenig *et al.*, 1984; Koenig and Brown, 1987). The resulting NMR curve is usually characterized heuristically by the Cole–Cole expression originally employed to describe dielectric relaxation dispersion data and, thus, has no *a priori* validity. The Larmor precessional frequency at the inflection point of the dispersive portion of the curve has been related to the rotational relaxation time, τ_R $(=3\tau_0)$. Thus, the NMRD curve is related phenomenologically to the rotational Brownian motion of the macromolecular solute, which appeared to arise because of a hydrodynamic coupling of the solute to the solvent (Hallenga and Koenig, 1976; Koenig, 1980; Koenig *et al.*, 1984). More recent investigations suggest this interpretation to be incorrect, and instead emphasize the importance of cross-relaxation between protein and solvent protons (Koenig *et al.*, 1978; Grad *et al.*, 1990, 1991; Lester and Bryant, 1990; Zhong *et al.*, 1990).

Relative to the off-resonance rotating frame spin-lattice relaxation technique, NMRD offers the advantage of a large amount of data over a broad range of Larmor precessional frequencies on dilute aqueous protein solutions (or tissue preparations) in a relatively short time. Limitations of NMRD stem from the phenomenological character of the analysis. These include the inability to rigorously take into account effects arising from polydispersity and anisotropic molecular reorientation. Both of these effects are formally included in the off-resonance rotating frame spin–lattice relaxation experiment.

The pH-induced dimerization of lysozyme has been investigated by both NMRD (Raeymaekers *et al.*, 1989) and the ^{13}C off-resonance rotating frame spin–lattice relaxation technique (Morgan *et al.*, 1990b). Both studies were performed at identical ionic strength, but at somewhat different pH values, temperature, and total protein concentrations. The NMRD technique gave τ_0 values of 14 and 30 nsec at 4°C for monomer (pH 4) and dimer (pH 9), respectively. The corresponding values corrected to a temperature of 15°C are 9.8 and 21 nsec, respectively. The corresponding values for monomer (pH 4.4) and dimer (pH 8.0) obtained by ^{13}C off-resonance rotating frame spin–lattice relaxation at 15°C were 11.6 ± 0.5 and 21.2 ± 0.6 nsec, respectively. These two techniques demonstrate excellent agreement in the derived rotational correlation times. Less agreement was obtained in the value of the association constant for the aggregation reaction. Using a two-state model, Raeymaekers *et al.* (1989) obtained a value of 208 M^{-1} (pH 7.0, 4°C) whereas Morgan *et al.* (1990b) employing an isodesmic model obtained a value for the association constant equal to 66 M^{-1} (pH 8.0, 15°C).

The rotational diffusion characteristics of hemoglobin in both normal human and sickle red blood cells have been studied by NMRD (Lindstrom and Koenig, 1974; Lindstrom *et al.*, 1976; Koenig and Brown, 1987). Intracellular values for τ_0 of 180 and 360 nsec (mean corpuscular hemoglobin concentration 5.0–5.2 mM) were obtained for oxy normal (HbA) and sickle cell (HbS) hemoglobin, respectively. These values are ~2 and 3 times larger than the corresponding values determined by ^{13}C off-resonance rotating frame spin–lattice relaxation (James *et al.*, 1979).

The concentration-dependent solution behavior of γ_{II}-crystallin has been investigated by NMRD (Koenig *et al.*, 1990) in a manner analogous to the off-resonance rotating frame spin–lattice relaxation studies described above. Interpretation of the NMRD results, neglecting the complicating effects of polydispersity, yields approximate molecular masses of 100, 320, and ≥2500 kDa at 9.5, 16.0 and 24.4% (v/v), respectively. Corresponding estimates of molecular mass derived by analysis of off-resonance rotating frame spin–lattice relaxation dispersion curves (Stevens *et al.*, 1992) are 80, 120, and 210 kDa, at respective concentrations comparable to those employed by Koenig *et al.* (1990). It is apparent that the NMRD technique provides estimates of molecular size that are higher than those afforded by the off-resonance rotating frame spin–lattice relaxation technique. This disparity in size estimation becomes more pronounced at higher protein concentrations. A similar disparity in size estimation by these two techniques was also noted for hemoglobin in the red blood cell (*vide ante*). Koenig *et al.* (1990) also concluded from concentration-dependent NMRD measurements that γ_{II}-crystallin monomers, at concentrations in excess of 16% (v/v), "unfold and reassemble into an extended, interconnected configuration." That is, intramolecular hydrogen bonding is disrupted and replaced by intermolecular hydrogen bonding. By use of Fourier transform infrared spectroscopy, a technique exquisitely sensitive to amide bond secondary structure environment (Surewicz and Mantsch, 1988), Stevens *et al.* (1992) were unable to find evidence in support of the concentration-dependent conformational change of γ_{II}-crystallin proposed by Koenig *et al.* (1990).

The NMRD relaxometry technique has also been applied to bovine cortical and nuclear lens homogenates (Beaulieu *et al.*, 1988, 1989). Below 19% (v/v) the NMRD curves resemble those elaborated by globular proteins in dilute solution, whereas at concentrations above this point the resulting curves are characteristic of heat-denatured (polymerized) egg albumin and tissue samples (Koenig *et al.*, 1984; Koenig and Brown, 1987). The change in NMRD curve behavior was interpreted by Koenig's group in terms of a concentration-dependent change in three-dimensional organization, characterized as an "extended, highly open, network of protein chains" (Beaulieu *et al.*, 1989). Other NMR studies (Morgan *et al.*, 1989; Rydzewski *et al.*, 1992), employing a combination of off-resonance rotating frame spin–lattice relaxation and solid-state NMR (cross-polarization, protein dipolar decoupling) techniques, have demonstrated the occurrence of both mobile and solidlike protein components in mammalian lens homogenates whose relative proportions are sensitive to water content (protein concentration). The presence of ^{14}N quadrupolar peaks observed in the water proton NMRD profiles of lens homogenates (Beaulieu *et al.*, 1988, 1989) are also dependent on concentration in a manner similar to the concentration-dependent amount of solidlike lens protein detected by the use of solid-state NMR techniques (Rydzewski *et al.*, 1992). Lester and Bryant (1990) have argued that the presence of ^{14}N quadrupolar peaks is a reflection of magnetic coupling between water protons and the protons of a semirigid protein system. Koenig's analysis (Koenig, 1988), by contrast, does not take into account the occurrence of solidlike protein components. It is likely that the change in NMRD curve behavior with increasing protein concentration is related to the occurrence of solidlike protein components in tissue and not to a change in protein conformation. Such a conclusion is in fact supported by the recent investigations of Lester and Bryant (1990). In this context, it is interesting to note that the maximum in the theoretically calculated Rayleigh ratio, a direct measure of scattered light intensity, using a hard sphere random fluctuation approach (Bettelheim and Siew, 1983), occurs at a volume fraction of 0.13 for spherical particles 100 nm or greater in diameter, a value not too different from that observed for the onset of the concentration-dependent effects described above. The maximum in the Rayleigh ratio–particle volume fraction curve corresponds to the transition between uncorrelated and correlated particle fluctuations in defining scattered light intensity. In addition, the polymerized egg albumin NMRD results of Koenig and Brown (1987) and corresponding proton relaxation studies of Grad *et al.* (1990, 1991) support the suggestion of the importance of solidlike protein components in defining solvent relaxation characteristics, as do the concentration-independent Fourier transform infrared results obtained for γ-crystallin conformation by Stevens *et al.* (1992). In turn, this suggests that the heuristic description in terms of the Cole–Cole expression for dilute protein solutions may be inappropriate for those systems containing solidlike protein components, such as exemplified by concentrated protein solutions and tissues. If this suggestion survives further scrutiny by experimental and theoretical experiments, of the sort described recently by Lester and Bryant (1990) and Kimmich *et al.* (1990), then the interpretation of the proton relaxation behavior of tissues cannot be considered simply in terms of dilute solution protein relaxation behavior, a point made independently by Zhong *et al.* (1990).

5.2. Phosphorus Metabolite Rotational Diffusion: Lens Tissue Homogenates

The rotational diffusion behavior of phosphorus metabolites in calf lens cortical and nuclear homogenates was investigated by ^{31}P off-resonance rotating frame spin–lattice relaxation as a means of assessing the occurrence and extent of phosphorus metabolite–protein interactions (Caines *et al.*, 1990) in a tissuelike environment. Effective rotational correlation times ($\tau_{0,\text{eff}}$) for the major phosphorus metabolites present in cortical and nuclear bovine calf lens homogenates were derived from nonlinear least-squares analysis of spectral intensity ratio dispersion plots with the assumption of isotropic reorientational motion. Intramolecular dipole–dipole (^{31}P–^{1}H, ^{31}P–^{31}P), chemical shift anisotropy, and solvent (water) translational dipole–dipole (^{31}P–^{1}H) relaxation contributions were assumed in the analysis (see Section 2.2). In those cases where the limiting value of the spectral intensity ratio fails to reach unity at large offset frequency, a modified formalism (see Section 6) incorporating chemical exchange-mediated saturation transfer between two sites should be used. Intensity ratio dispersion plots for selected phosphorus metabolites present in cortical and nuclear lens tissue homogenates are shown in Figs. 24, 25, and 26. Values of the derived phosphorus metabolite $\tau_{0,\text{eff}}$ values are shown in bar graph form in Fig. 27. As shown in the figures, the phosphorus metabolite $\tau_{0,\text{eff}}$ values varied considerably and in one case (PME) evidence was obtained for binding and subsequent exchange with solidlike protein domains (*vide infra*). The diversity in $\tau_{0,\text{eff}}$ values for lenticular phosphorus metabolites was interpreted by Caines *et al.* (1990) in terms of differential binding to more slowly tumbling macromolecular species, specifically the lens crystallin proteins.

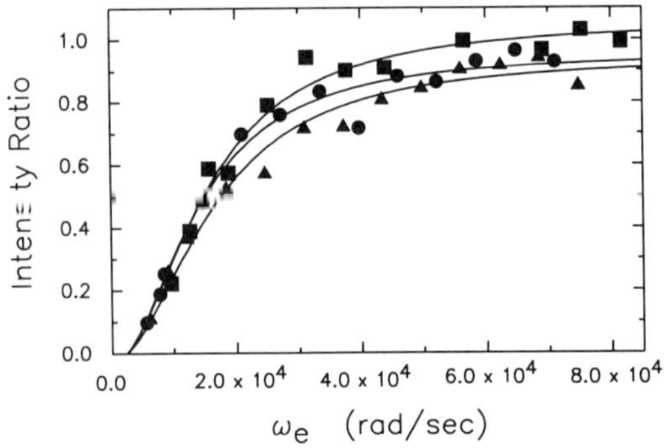

Figure 24. ^{31}P spectral intensity ratio dispersion curves for the α- (▲), β- (■), and γ- (●) ATP resonances present in calf lens cortical homogenate at 10°C. The solid lines represent the best fit obtained by nonlinear regression assuming isotropic reorientation, with $\tau_{0,\text{eff}}$ values of 12.2 ± 1.0 (α-ATP), 11.3 ± 0.8 (β-ATP), 10.3 ± 0.9 (γ-ATP) nsec, respectively. Intra- and intermolecular dipole–dipole and CSA relaxation mechanism contributions were assumed. Reproduced, with permission, from Caines *et al.* (1990).

A fast-exchange model (T_1 time scale), between free and bound forms, was employed by these workers to obtain the fraction of free (unbound) metabolite. The following expressions are relevant:

$$\frac{1}{T_1} = \frac{\Theta_{\text{free}}}{T_{1,\text{free}}} + \frac{\Theta_{\text{bound}}}{T_{1,\text{bound}}} \tag{41}$$

$$\frac{1}{T_{1\rho}^{\text{off}}} = \frac{\Theta_{\text{free}}}{T_{1\rho,\text{free}}^{\text{off}}} + \frac{\Theta_{\text{bound}}}{T_{1\rho,\text{bound}}^{\text{off}}} \tag{42}$$

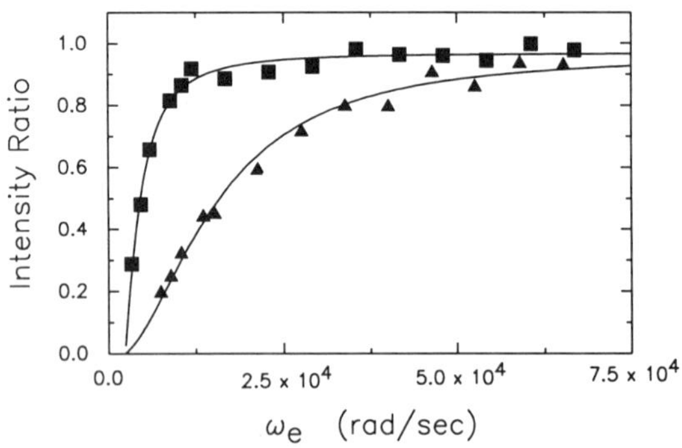

Figure 25. ^{31}P spectral intensity ratio dispersion curves for the GPC (■) and P_i (▲) resonances present in calf lens cortical homogenate at 10°C. The solid lines represent the best fit obtained by nonlinear regression assuming isotropic reorientation, with $\tau_{0,\text{eff}}$ values of 1.6 ± 0.2 (GPC) and 10.6 ± 0.5 (P_i) nsec, respectively. Intra- and intermolecular dipole–dipole and CSA relaxation mechanism contributions are assumed. Reproduced, with permission, from Caines *et al.* (1990).

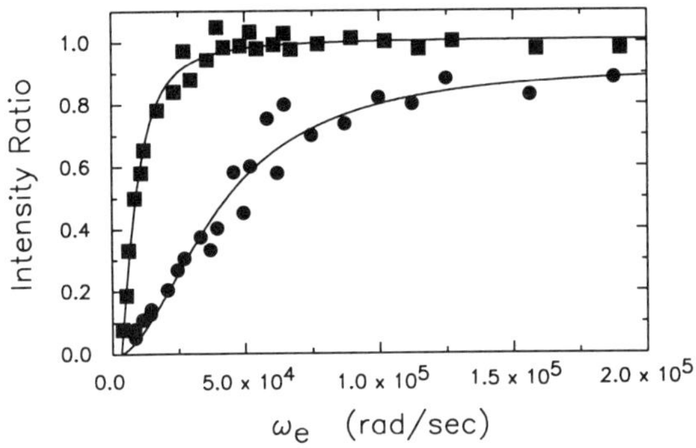

Figure 26. [13]P spectral intensity ratio dispersion curves for the GPC (■) and PME (●) resonances present in calf lens nuclear homogenate at 18°C. The solid lines represent the best fit obtained by nonlinear regression assuming isotropic reorientation. Intra- and intermolecular dipole–dipole and CSA relaxation mechanism contributions were assumed. The results of the fits were $\tau_{0,\mathrm{eff}} = 25.5 \pm 1.4$ nsec and $k_A = (7.6 \pm 2.6) \times 10^{-3}\ \mathrm{sec}^{-1}$ for PME, and $\tau_{0,\mathrm{eff}} = 3.7 \pm 0.2$ nsec and $R(\omega_e \to \infty) = 1.01 \pm 0.01$ for GPC. Reproduced, with permission, from Caines *et al.* (1990).

where the fractional amounts of a particular metabolite are represented by Θ_{free} and Θ_{bound}, and sum to unity.

The results are summarized in Fig. 28, and demonstrate that different degrees of fractional binding are dependent on tissue type (cortex versus nucleus) and temperature. As noted in Fig. 28, GPC is virtually unbound under all conditions, whereas PME and inorganic phosphate are observed to be partially bound under some conditions. By contrast, ATP exhibits partial binding under all conditions examined.

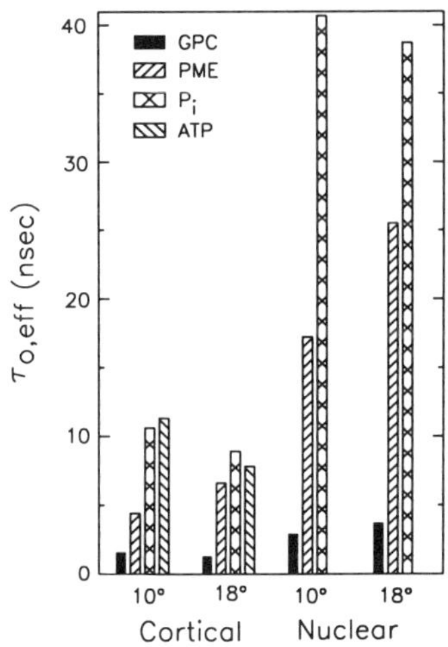

Figure 27. Summary of phosphorus metabolite $\tau_{0,\mathrm{eff}}$ values for cortical and nuclear calf lens homogenates at 10 and 18°C obtained by the ^{31}P off-resonance rotating frame spin–lattice relaxation method. Data from Caines *et al.* (1990).

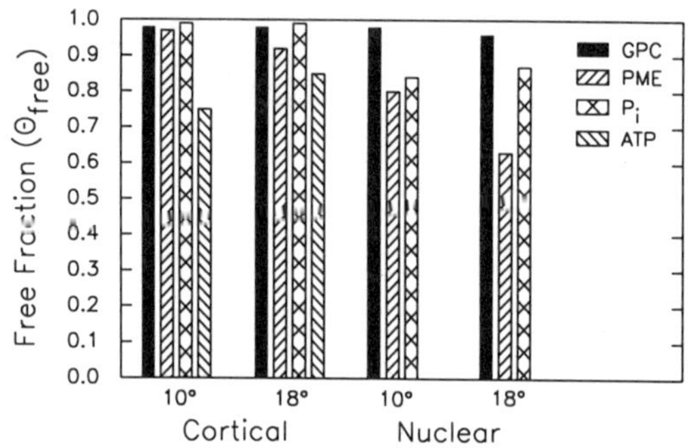

Figure 28. Fraction of free (Θ_{free}) phosphorus metabolite in cortical and nuclear calf lens homogenates at 10 and 18°C obtained by the ^{31}P off-resonance rotating frame spin–lattice relaxation method. A two-site fast-exchange model was assumed in the analysis. See text for details. Data from Caines *et al.* (1990).

These experiments serve to illustrate the utility of the off-resonance rotating frame spin–lattice relaxation experiment. Extension to other tissues is indeed possible, as well as performing the experiment on a spatially localized basis by inclusion of a spatially selective pulse sequence in place of the hard 90° pulse which follows the preparation pulse (see Fig. 16).

6. OFF-RESONANCE ROTATING FRAME SPIN–LATTICE RELAXATION: APPLICATION TO SATURATION TRANSFER

6.1. Introduction

The extension of off-resonance rotating frame spin–lattice relaxation theory to include the effects of chemical exchange-mediated saturation transfer affords knowledge about the occurrence of magnetization exchange kinetic processes in addition to the usually obtained rotational diffusion information. Excellent reviews of the application of magnetization transfer methodology to biological systems have been presented by Brown (1980), Alger and Shulman (1984), and Brindle (1988). Retaining features similar to the classical on-resonance saturation transfer experiment of Forsen and Hoffman (1963a, b), analytical expressions were derived by Caines and Schleich (1991) describing the time-dependent and steady-state behavior of the exchanging nuclear magnetization experiencing off-resonance field-induced partial or complete saturation. Applications of this formalism include the study of phosphorus exchange between two chemically different species, exchange between free and protein-bound metabolites, and exchange between mobile and relatively immobile, solidlike protein domains. Moreover, the extended formalism easily permits an evaluation of conventional saturation transfer experiments of slowly tumbling or motionally restricted metabolites when nonselective, off-resonance, partial saturation becomes appreciable. Extension to cross-relaxation phenomena is straightforward (see Section 7).

6.2. Theory

Saturation transfer off-resonance rotating frame spin–lattice relaxation formalism was developed by Caines and Schleich (1991) to study the situation of chemical exchange-mediated transfer of nuclear magnetization between two sites (A ⇔ B) in which either partial or complete saturation arises because of off-resonance irradiation. The time-dependent behavior of the chemical exchange-coupled longitudinal magnetization at sites A and B in the rotating frame is described by a set of nonhomogeneous linear differential equations:

$$\frac{d}{dt} M_z^A = \frac{M_{ze}^A}{T_{1\rho,A}^{off}} - \frac{M_z^A}{\tau_{1A}} + \frac{M_z^B}{\tau_B} \tag{43}$$

$$\frac{d}{dt} M_z^B = \frac{M_{ze}^B}{T_{1\rho,B}^{off}} - \frac{M_z^B}{\tau_{1B}} + \frac{M_z^A}{\tau_A} \tag{44}$$

where

$$\frac{1}{\tau_{1A}} = \frac{1}{T_{1\rho,A}^{off}} + \frac{1}{\tau_A} \tag{45}$$

$$\frac{1}{\tau_{1B}} = \frac{1}{T_{1\rho,B}^{off}} + \frac{1}{\tau_B} \tag{46}$$

The forward (k_A) and reverse (k_B) first-order rate constants are equal to $1/\tau_A$ and $1/\tau_B$, respectively. At equilibrium, in the absence of RF irradiation, the principle of detailed balance requires that $k_A M_0^A = k_B M_0^B$. The mole fractions of the spins at site A (f_A) and site B (f_B) are related to the rate constants by the expressions:

$$f_A = \frac{\tau_A}{\tau_A + \tau_B} = \frac{k_B}{k_A + k_B} \tag{47}$$

$$f_B = \frac{\tau_B}{\tau_A + \tau_B} = \frac{k_A}{k_A + k_B} \tag{48}$$

where $f_A + f_B = 1$. The thermal equilibrium macroscopic magnetizations at A and B established in the static magnetic field are therefore:

$$M_0^A = f_A M_0^T \tag{49}$$

$$M_0^B = f_B M_0^T \tag{50}$$

where M_0^T $(= M_0^A + M_0^B)$ is the total magnetization. The theoretical equilibrium z-magnetization at sites A and B in the presence of RF irradiation employed at a particular RF field strength (B_2) and frequency offset (ν_{off}) in the absence of exchange is given by:

$$M_{ze}^A = M_0^A \cos^2 \Theta_A \frac{T_{1\rho,A}^{off}}{T_{1,A}} \tag{51}$$

$$M_{ze}^B = M_0^B \cos^2 \Theta_B \frac{T_{1\rho,B}^{off}}{T_{1,B}} \tag{52}$$

where $\Theta = \tan^{-1}(\gamma B_2/\Delta)$ (see Section 2.1) and assumed expressions for $T_{1\rho}^{off}$ and T_1 have been given above (see Section 2.2). With the application of continuous-wave, low-power, off-resonance RF irradiation, the exchanging macroscopic magnetization at sites A and B will reach a partially saturated steady-state given by:

$$M_z^A(\infty) = \frac{\dfrac{1}{\tau_{1B}}\left(\dfrac{M_{ze}^A}{T_{1\rho,A}^{off}}\right) + \dfrac{1}{\tau_B}\left(\dfrac{M_{ze}^B}{T_{1\rho,B}^{off}}\right)}{D} \tag{53}$$

$$M_z^B(\infty) = \frac{\dfrac{1}{\tau_{1A}}\left(\dfrac{M_{ze}^B}{T_{1\rho,B}^{off}}\right) + \dfrac{1}{\tau_A}\left(\dfrac{M_{ze}^A}{T_{1\rho,A}^{off}}\right)}{D} \tag{54}$$

where $D = 1/\tau_{1A}\tau_{1B} - 1/\tau_A\tau_B$. With complete saturation of the signal at site B, the magnetization M_z^B is eliminated, and thus the time dependence of the z-magnetization of A under off-resonance RF irradiation is given by

$$\frac{d}{dt} M_z^A = \frac{M_{ze}^A}{T_{1\rho,A}^{off}} - \frac{M_z^A}{\tau_{1A}} \tag{55}$$

The steady-state intensity of the z-magnetization of A with continued saturation of B is given by

$$M_z^A(t \to \infty) = M_{ze}^A(1 - k_A\tau_{1A}) \tag{56}$$

Equation (56) is also equal to $M_{ze}^{A}[1 + k_A T_{1\rho,A}^{off}]^{-1}$. When the off-resonance effects at site A are negligible (i.e., $T_{1\rho,A}^{off} = T_{1A}$ and $M_{ze}^{A} = M_0^{A}$), the off-resonance irradiation saturation transfer formalism reduces to the expression for classical saturation transfer.

6.3. Applications

The behavior of chemical exchange-modified off-resonance rotating frame spin–lattice relaxation formalism may be readily investigated by computer simulation. The following sections consider the application of this formalism to the exchange of free and G-actin-bound adenosine triphosphate (ATP), creatine kinase-catalyzed phosphocreatine (PCr), and ATP γ-phosphate exchange, and phosphorus metabolite exchange between two different relaxation environments representing mobile and relatively immobile (solidlike) pools of metabolites present in bovine calf nuclear lens tissue homogenates (Caines *et al.*, 1990; Caines and Schleich, 1991).

6.3.1. ATP and G-actin Exchange

The reversible binding of the ligand ATP to G-actin (Brauer and Sykes, 1981; Brauer and Sykes, 1987)

$$\text{ATP} + \text{G-actin} \Leftrightarrow \text{ATP–G-actin}$$

can be represented by a two-site ($A \Leftrightarrow B$) exchange process, with pseudo-first-order rate constants, $k_A = k_f$ [G-actin] and $k_B = k_r$, where k_f and k_r are the forward and reverse rate constants, respectively. The theoretical dependence of the ^{31}P spectral intensity ratio versus resonance frequency offsct (R versus ν_{off}) for free ligand (site A, $\tau_0 = 0.83$ nsec) and G-actin-bound ligand (site B, $\tau_0 = 60$ nsec) is shown in Figs. 29–31 as a function of pseudo-first-order association (k_A) and dissociation (k_B) rate constants for three assumed equilibrium conditions: $k_A/k_B = 0.1$, 1.0, and 10.0. The steady-state spectral intensity ratios for ligand at site A ($R^A = M_z^A(\infty)/M_0^A$) and for ligand at site B were calculated according to Eqs. (53) and (54) at the applied off-resonance RF, ν_{RF}, and off-resonance irradiation field strength, B_2. The frequency offsets are defined as: $\nu_{off,A} = \nu_{RF} - \nu_A$ and $\nu_{off,B} = \nu_{RF} - \nu_B$.

In Figs. 29–31 the solid lines characterize the theoretical behavior of free and bound ligand in the *absence* of exchange, i.e., k_A and $k_B \to 0$. The free ATP dispersion curves (Figs. 29A–31A) attain the limiting value $R^A = 1$ at an offset of ~5 kHz, whereas the bound ATP curves (Figs. 29B–31B) at this offset show substantial off-resonance saturation with a reduced value

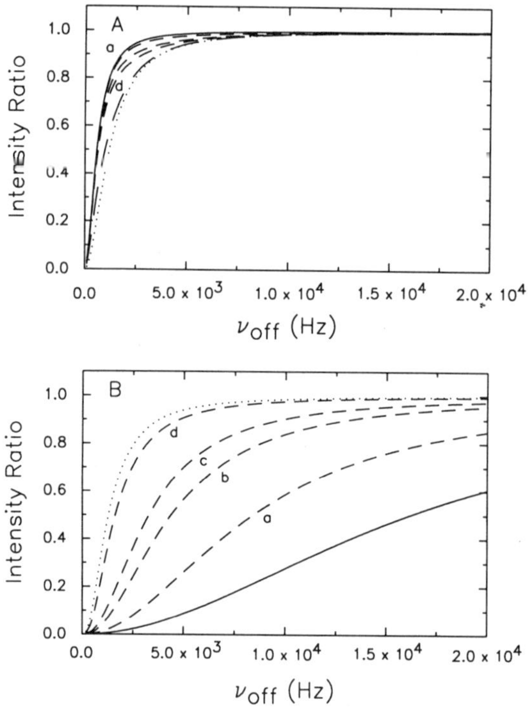

Figure 29. Computer-simulated $[^{31}P]\gamma$-ATP intensity ratio dispersion curves as a function of rate constant for two-site exchange between free and G-actin-bound ATP assuming $\nu_A \neq \nu_B$. (A) R^A versus $\nu_{off,A}$ for free ATP (site A). (B) R^B versus $\nu_{off,B}$ for bound ATP (site B). For the assumed equilibrium condition $k_A/k_B = 0.1$, the rate constants k_A and k_B were varied: (a) 0.01 sec^{-1}, 0.1 sec^{-1}; (b) 0.05 sec^{-1}, 0.5 sec^{-1}; (c) 0.1 sec^{-1}, 1.0 sec^{-1}; (d) 1.0 sec^{-1}, 10.0 sec^{-1}. The solid lines represent the theoretical behavior without considering exchange effects, the dotted lines represent the limit of fast exchange, and the dashed lines represent slow to intermediate exchange (a–d). The simulations assume an off-resonance RF field strength of 0.25 G, a B_0 field strength of 7.05 T, isotropic tumbling with τ_0 values of 0.83 nsec (free) and 60 nsec (bound), and chemical shift values: $\nu_A = -607.5$ Hz (free) and $\nu_B = -692.6$ Hz (bound). Taken from Caines and Schleich (1991).

of R^B, a consequence of the longer rotational correlation time for bound ATP. The limiting value of $R^B = 1$ is attained for bound ATP at offsets of ~50 kHz.

The effects of magnetization transfer under slow to intermediate rates of chemical exchange are shown by the series of dashed curves in Figs. 29–31. Values of the rate constants employed are increased while maintaining the specified equilibrium condition, $K_{assoc} = k_A/k_B$. It is apparent that changes in the rate constants at constant K_{assoc} affect the R versus ν_{off} curves differently as the equilibrium condition ($K_{assoc} = k_A/k_B$) and hence the equilibrium populations (f_A and f_B) are varied. When the system contains

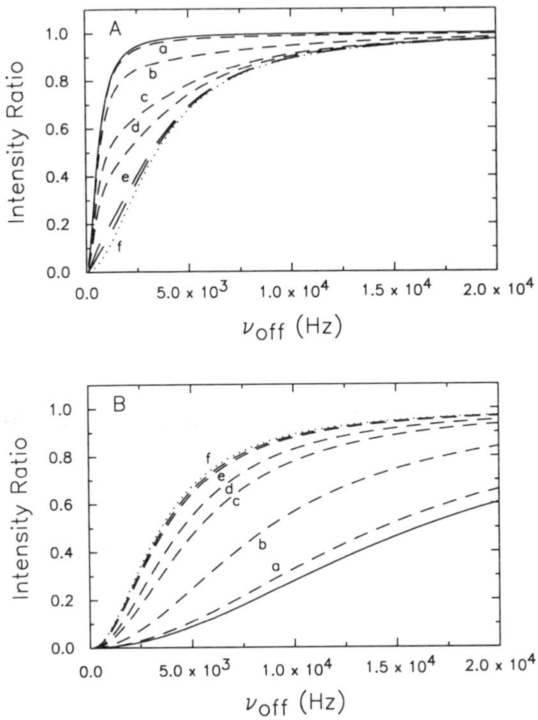

Figure 30. Computer-simulated [^{31}P]γ-ATP intensity ratio dispersion curves as a function of rate constant for two-site exchange between free and G-actin-bound ATP assuming $\nu_A \neq \nu_B$. (A) R^A versus $\nu_{off,A}$ for free ATP (site A). (B) R^B versus $\nu_{off,B}$ for bound ATP (site B). For the assumed equilibrium condition $k_A/k_B = 1.0$, the rate constants k_A and k_B were varied: (a) 0.01 sec^{-1}, 0.01 sec^{-1}; (b) 0.1 sec^{-1}, 0.1 sec^{-1}; (c) 0.5 sec^{-1}, 0.5 sec^{-1}; (d) 1.0 sec^{-1}, 1.0 sec^{-1}; (e) 5.0 sec^{-1}, 5.0 sec^{-1}; (f) 10 sec^{-1}, 10 sec^{-1}. The solid lines represent the theoretical behavior without considering exchange effects, the dotted lines represent the limit of fast exchange, and the dashed lines represent slow to intermediate exchange (a–f). Other parameters are listed in Fig. 29. Taken from Caines and Schleich (1991).

mostly free ATP, as in Fig. 29A where $k_A/k_B = 0.1$ and $f_A = 0.91$, the variation in the dispersion curves as a function of rate constant at the particular value of K_{assoc} are more significant for G-actin-bound ATP (see Fig. 29B). When the system contains mostly G-actin-bound ATP (Fig. 31A, B) where $k_A/k_B = 10$ and $f_A = 0.09$, the opposite behavior is observed, and the variation is now more significant for the free ATP. In the fast-exchange limit, on the spin–lattice relaxation time scale, the dotted curves of the free and G-actin-bound ATP converge, as expected. Thus, under fast-exchange conditions the free ATP spectrum reports on the off-resonance dispersion behavior of the bound ATP (see the proton Z-spectrum discussion in Section 7.2).

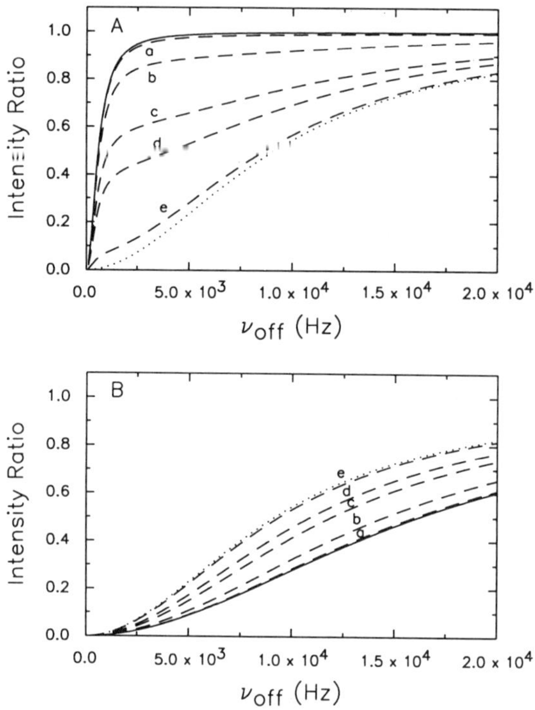

Figure 31. Computer-simulated $[^{31}P]\gamma$-ATP intensity ratio dispersion curves as a function of rate constant for two-site exchange between free and G-actin-bound ATP assuming $\nu_A \neq \nu_B$. (A) R^A vesus $\nu_{\text{off,A}}$ for free ATP (site A). (B) R^B versus $\nu_{\text{off,B}}$ for bound ATP (site B). For the assumed equilibrium condition $k_A/k_B = 10.0$, the rate constants k_A and k_B were varied: (a) 0.01 sec^{-1}, 0.001 sec^{-1}; (b) 0.1 sec^{-1}, 0.01 sec^{-1}; (c) 0.5 sec^{-1}, 0.05 sec^{-1}; (d) 1.0 sec^{-1}, 0.1 sec^{-1}; (e) 10.0 sec^{-1}, 1.0 sec^{-1}. The solid lines represent the theoretical behavior without considering exchange effects, the dotted lines represent the limit of fast exchange, and the dashed lines represent slow to intermediate exchange (a–e). Other parameters are listed in Fig. 29. Taken from Caines and Schleich (1991).

The theoretical intensity ratio dispersion curves for both free and G-actin-bound ATP are quite sensitive to the assumed rate constant values when the equilibrium association constant (k_A/k_B) is unity (see Fig. 30). These observations suggest that a quantitative determination of experimental rate constants is possible in those instances where the fraction of ATP is approximately equally distributed between the free and G-actin-bound sites, a situation that can be tailored experimentally. Use of the exchange modified off-resonance rotating frame spin–lattice relaxation formalism in the determination of exchange rate constants and rotational correlation times would require the simultaneous analysis of intensity ratio dispersion curves of both free and G-actin-bound ATP. When the equilibrium composition is dominated by either the free or bound species as in Figs. 29 and 31, it becomes difficult to obtain quantitative estimates of the kinetic parameters, since the variation in the intensity ratio dispersion curves of the predominate species is small. In addition, the signal-to-noise ratio would be relatively low for the dilute component, which would impose dynamic range limitations. The use of larger off-resonance RF field strengths which produce greater dispersions in the intensity ratio of free and G-actin-bound ATP could improve the opportunity to extract chemical exchange information, especially for the conditions assumed in Figs. 29 and 31.

To study the behavior of ligand exchange between two sites with identical chemical shift values ($\nu_A = \nu_B$), calculations were made according to $R^{AB} = [M_z^A(\infty) + M_z^B(\infty)]/M_0^T$, employing the same parameter values as in the example simulations of Figs. 29–31. The intensity ratio dispersion curves (R^{AB} versus ν_{off}) as a function of the pseudo-first-order rate constant are presented in Fig. 32A, B, C for the three equilibrium conditions. In Fig. 32B the assumed equilibrium association constant is unity, $f_A = 0.50$, and ATP is distributed equally between the free and G-actin-bound sites. The solid curve in this figure, for example, represents the behavior without exchange between sites A and B. The sudden rise to $R^{AB} = 0.5$ at an offset of 5 kHz is dominated by the more rapidly tumbling free ATP, whereas the slower tumbling G-actin-bound ATP remains largely saturated. When the equilibrium is shifted toward free ATP as in Fig. 32A, where $k_A/k_B = 0.1$ and $f_A = 0.91$, the solid curve rises sharply to $R^{AB} \sim 0.9$ at an offset of 5 kHz. By comparison, the rise to $R^{AB} \sim 0.1$ in Fig. 32C at this offset is due to the small fraction of free ATP where $k_A/k_B = 10.0$ and $f_A = 0.09$. With larger offsets, each curve attains the limiting value $R^{AB} = 1$. The magnetization transfer simulations of slow to intermediate exchange rate are represented by the dashed curves and the behavior under fast-exchange conditions is given by the dotted curves. The unique features of the intensity ratio dispersion curves shown in Fig. 32 demonstrate the potential for obtaining useful chemical exchange information from systems which exhibit overlapping chemical shifts. In principle, the formalism could also be applied in the study of two-compartment exchange if the rotational mobility of the exchanging molecules is suitably different.

6.3.2. Phosphocreatine and γ-ATP Exchange

The reversible creatine kinase-catalyzed phosphoryl group transfer between phosphocreatine (PCr) and the γ-phosphate of ATP (γ-ATP),

$$PCr + ADP + H^+ \Leftrightarrow ATP + Cr$$

is represented by a two-site ($A \Leftrightarrow B$) exchange process, PCr $\Leftrightarrow$ ATP, with pseudo-first-order rate constants, $k_A = k_f[ADP][H^+]$ and $k_B = k_r[Cr]$,

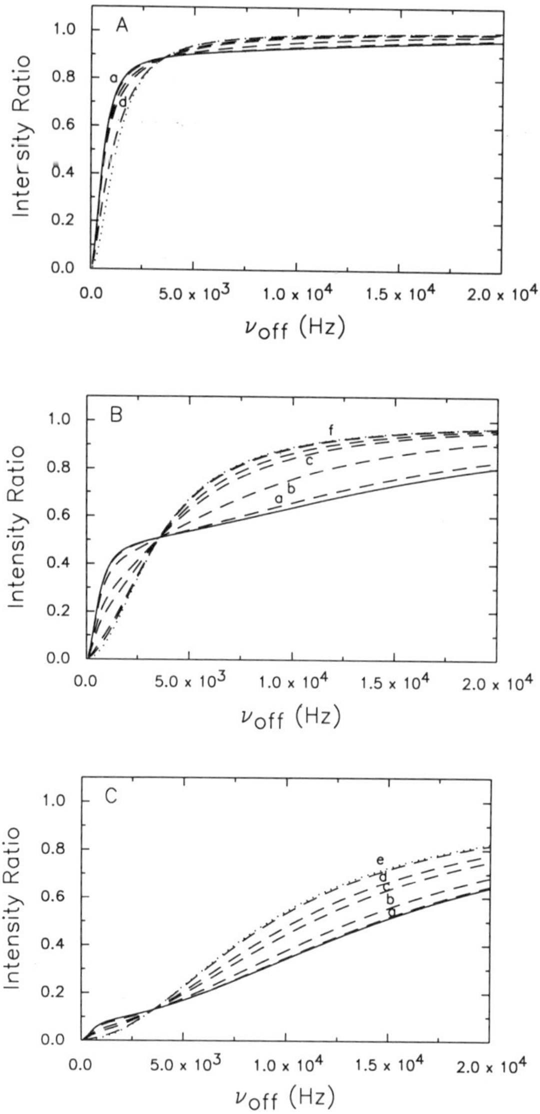

Figure 32. Computer-simulated $[^{31}P]\gamma$-ATP intensity ratio dispersion curves as a function of rate constant for two-site exchange between free and G-actin-bound ATP assuming $\nu_A = \nu_B$. (A) $k_A/k_B = 0.1$ using k_A and k_B of Fig. 29. (B) $k_A/k_B = 1.0$ using k_A and k_B of Fig. 30. (C) $k_A/k_B = 10.0$ using k_A and k_B of Fig. 31. The solid lines represent the theoretical behavior without considering exchange effects, the dotted lines represent the limit of fast exchange, and the dashed lines represent slow to intermediate exchange (a–f). For the chemical shifts, ν_A and ν_B, a value of -650 Hz was assumed. Other parameters are listed in Fig. 29. Taken from Caines and Schleich (1991).

where k_f and k_r are the forward and reverse rate constants, respectively (Eldar and Degani, 1989). Assuming rotational correlation time values representative of tissue metabolites, Caines *et al.* (1990) calculated theoretical spectral intensity ratio dispersion curves for PCr (site A, $\tau_0 = 1.2$ nsec) and γ-ATP (site B, $\tau_0 = 12.0$ nsec) as a function of rate constant using the experimental equilibrium condition, $k_A/k_B = 0.88$ at 16°C (Eldar and Degani, 1989).

The simulations employ the two-site exchange steady-state magnetization transfer equations used in the ATP–G-actin example above. The curves in Fig. 33A, B were calculated with a B_2 field strength of 0.25 G whereas in Fig. 34A, B the value of B_2 is increased to 1.0 G. The behavior of the

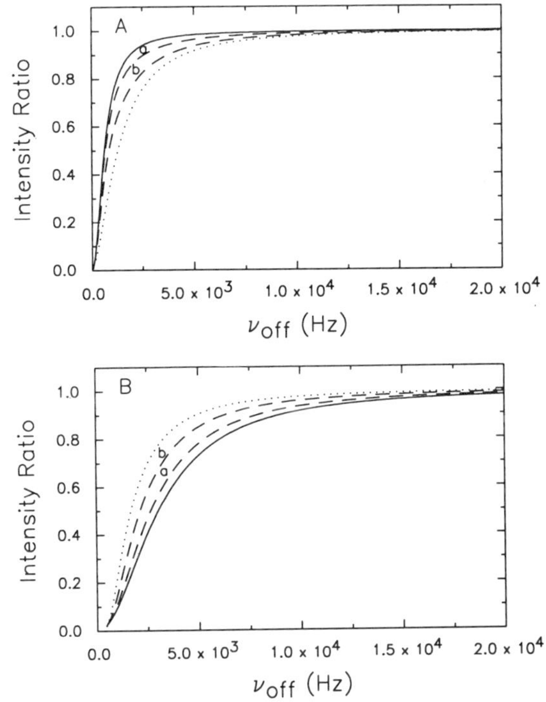

Figure 33. Computer-simulated ^{31}P intensity ratio dispersion curves as a function of rate constant for two-site exchange between PCr and γ-ATP. (A) R^A versus $\nu_{off,A}$ for PCr (site A). (B) R^B versus $\nu_{off,B}$ for γ-ATP (site B). The simulations assume an off-resonance RF field strength of 0.25 G, a B_0 field strength of 7.05 T, isotropic tumbling with τ_0 values of 1.2 nsec (PCr) and 12.0 nsec (γ-ATP), and chemical shift values: $\nu_A = -351.15$ Hz (PCr) and $\nu_B = -752.12$ Hz (γ-ATP). The solid lines represent the theoretical behavior without considering exchange effects, the dotted lines represent the limit of fast exchange, and the dashed lines represent slow to intermediate exchange (a, b) assuming the equilibrium condition $k_A/k_B = 0.88$ and the rate constant pairs (k_A, k_B): (a) 0.07 sec^{-1}, 0.08 sec^{-1}; (b) 0.38 sec^{-1}, 0.434 sec^{-1}. Taken from Caines and Schleich (1991).

steady-state intensity ratio for PCr (R^A) and γ-ATP (R^B) in the absence of exchange, i.e., k_A and $k_B \to 0$, is denoted by the solid lines. With the fast-exchange simulation, represented by the dotted lines, the R^A and R^B dispersion curves at the same B_2 field strength are identical when corrected for the chemical shift difference between PCr and γ-ATP. The simulations of the slow to intermediate-exchange cases, given by the dashed lines, assume the rate constant values given in the legend of Fig. 33. The variations in the dispersion curves as a function of rate constant are more dramatic in Fig. 34 compared with those in Fig. 33 due to the larger employed B_2 field strength. Increasing the intensity of the B_2 field thus highlights the effects of chemical exchange saturation transfer by increasing the off-resonance dispersion, and renders the determination of kinetic information more accessible.

^{31}P NMR magnetization transfer studies are widely used to measure enzyme-catalyzed metabolite chemical exchange rates (Brown, 1980; Alger and Shulman, 1984; Brindle, 1988). A potential difficulty with the application of the classical on-resonance saturation transfer technique, which assumes the extreme motional narrowing condition for the investigation of motionally restricted *in vivo* metabolites engaged in chemical exchange, is the effect of nonselective off-resonance partial saturation at the site under investigation (e.g., site A) which occurs during complete on-resonance saturation of the chemically exchange-coupled site (e.g., site B). Although minimal RF power irradiation is typically employed, off-resonance partial saturation effects sometimes cannot be avoided and in fact were observed in the monitored NMR signals of the PCr–ATP phosphate exchange reaction in heart and skeletal muscle (Bittl *et al.*, 1987). Therefore, application of the chemical exchange-modified off-resonance rotating frame spin–lattice relaxation formalism can be a feasible alternative for the study of *in vivo* metabolites by saturation transfer methodologies.

6.3.3. Phosphorus Metabolite Exchange between NMR "Visible" and "Invisible" Sites

The exchange of metabolite between a free (A) and a bound (B) immobilized pool of metabolites was investigated by simulating the intensity ratio (R^A) versus ω_e for a PME as a function of the association rate constant (k_A). With complete saturation of B by off-resonance irradiation, the steady-state intensity ratio of A [$R^A = M_z^A(\infty)/M_0^A$] is calculated assuming a short (Fig. 35A) and long (Fig. 35B) PME rotational correlation time according to Eqs. (51) and (56), and $M_0^A = 1$. The B_2 field (0.25 G) was varied over a 32-kHz range. Clearly shown by the group of curves in Fig. 35A, B, the effect of exchange (increase in k_A) is to reduce the limiting value of R^A [$R(\omega_e \to \infty)$] obtained far off-resonance, i.e., at large ω_e, or ν_{off} values.

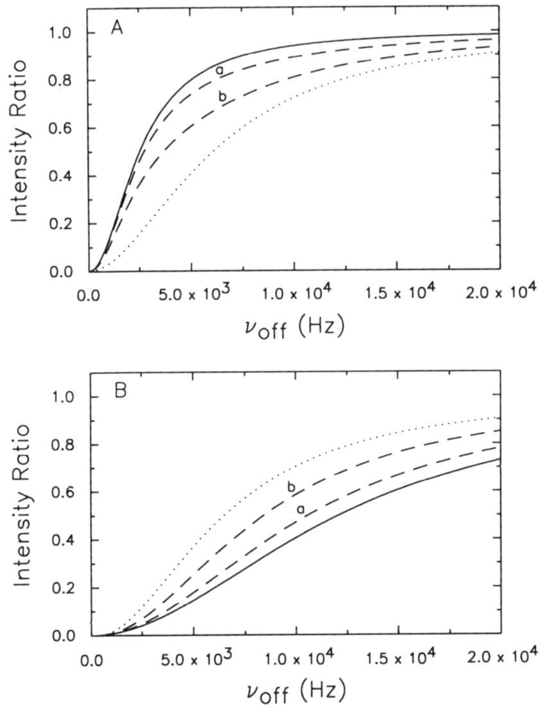

Figure 34. Computer-simulated ^{31}P intensity ratio dispersion curves as a function of rate constant for two-site exchange between PCr and γ-ATP. (A) R^{A} versus $\nu_{\text{off,A}}$ for PCr (site A). (B) R^{B} versus $\nu_{\text{off,B}}$ for γ-ATP (site B). The solid lines represent the theoretical behavior without considering exchange effects, the dotted lines represent the limit of fast exchange, and the dashed lines represent slow to intermediate exchange (a, b). The simulations assume an off-resonance RF field strength of 1.0 G. Other parameters are listed in Fig. 33. Taken from Caines and Schleich (1991).

Additionally, from a comparison of Fig. 35A, B, the effect of an increasing rotational correlation time is to decrease the steepness of approach to the limiting plateau value, $R(\omega_e \rightarrow \infty)$. The plateau region is where $T_{1\rho,A}^{\text{off}} = T_1^{A}$, $M_{ze}^{A} = M_0^{A}$, and where the conventional saturation transfer theory applies (Caines and Schleich, 1991).

The application of the chemical exchange-modified off-resonance rotating frame spin-lattice relaxation formalism to off-resonance saturation data obtained for two phosphorus metabolites [L-α-glycerolphosphate (G-3-P) and L-α-glycerolphosphoryl-choline (GPC)] present in nuclear eye lens tissue homogenates of bovine calves (Caines *et al.*, 1990) is shown in Fig. 26. The limiting value of $R(\omega_e \rightarrow \infty)$ for GPC approaches unity, whereas for PME the limiting value of $R(\omega_e \rightarrow \infty)$ is ~0.9. This observation thus

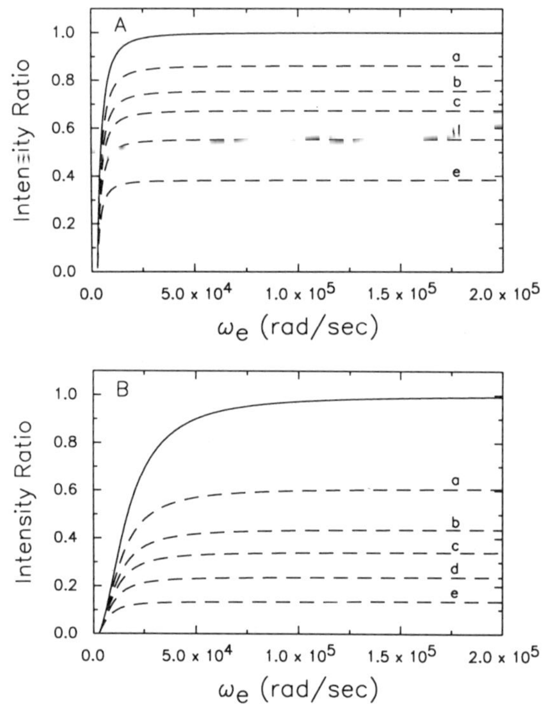

Figure 35. Computer-simulated ^{31}P intensity ratio curves of G-3-P as a function of forward exchange rate constant (k_A) for two-site exchange between a "mobile" (site A) and a saturated "immobile" (site B) species. The simulations assume an off-resonance RF field strength of 0.25 G, a B_0 field strength of 7.05 T, dipole–dipole relaxation by two protons located 2.6 Å from the ^{31}P, and CSA relaxation ($\delta_z = -58.33$ ppm and $\eta = -0.67$). Two cases of isotropic rotational correlation time were assumed for site A: (A) $\tau_0 = 1.0$ nsec and (B) $\tau_0 = 10.0$ nsec. Values for k_A are: (a) 0.1 sec^{-1}; (b) 0.2 sec^{-1}; (c) 0.3 sec^{-1}; (d) 0.5 sec^{-1}; (e) 1.0 sec^{-1}. The solid lines represent the theoretical behavior without considering exchange effects, whereas the dashed lines represent slow to intermediate exchange (a–e). Taken from Caines and Schleich (1991).

indicates exchange with an immobilized and, hence, NMR "invisible," metabolite pool. Nonlinear least-squares analysis of the entire off-resonance PME data set, using Eqs. (51) and (56) with assumed relaxation mechanisms (see Section 2.2.1), results in a determination of both $\tau_{0,\text{eff}}$ and k_A; for GPC values of $\tau_{0,\text{eff}}$ and $R(\omega_e \rightarrow \infty)$ are determined (see Fig. 26 legend). The computer simulations and results of the off-resonance irradiation experiments of the eye lens metabolites suggest that the chemical exchange-modified off-resonance rotating frame spin–lattice relaxation formalism is potentially important in the characterization of motional dynamics and exchange kinetics in diverse biochemical systems.

7. MAGNETIZATION TRANSFER CONTRAST-ENHANCED PROTON MAGNETIC RESONANCE IMAGING

7.1. Background

The proton longitudinal magnetic relaxation characteristics of water in heterogeneous biological systems, exemplified in this chapter by aqueous protein solutions and tissue, have been extensively studied (Edzes and Samulski, 1977, 1978; Koenig *et al.*, 1978, 1984; Bryant and Shirley, 1980; Fung and McGaughy, 1980; Shirley and Bryant, 1982; Gore *et al.*, 1989; Zhong *et al.*, 1989, 1990; Wolff and Balaban, 1989; Eng *et al.*, 1990; Grad *et al.*, 1990, 1991; Grad and Bryant, 1990; Lester and Bryant, 1990). These studies have generally employed a two-phase proton system, i.e., macromolecular protons and bulk water protons, as an approximate model for proton relaxation behavior in heterogeneous biological systems. Each proton phase is assumed to have a uniform spin temperature, the result of efficient spin diffusion and cross-relaxation processes (macromolecular proton phase) and rapid chemical exchange and translational diffusion processes (bulk water proton phase). A prominent hallmark of proton relaxation behavior in heterogeneous biological systems is the occurrence of cross-relaxation between macromolecular protons and the bulk water protons, thereby significantly affecting bulk water proton spin–lattice relaxation.

Recently, Wolff and Balaban (1989) introduced a novel magnetic resonance technique capable of detecting, and hence, probing, the interaction of macromolecular protons with those present in the bulk water pool of tissue. This technique, reminiscent of the selective hydration inversion technique of Edzes and Samulski (1978), is based on the occurrence of magnetization transfer, and has important implications for contrast enhancement in a proton magnetic resonance imaging experiment. It involves the application of a low-power RF irradiation field off-resonance to the mobile bulk water proton resonance, capable of selectively saturating the protons associated with the more slowly reorienting macromolecular species. Assessment of the bulk water proton magnetization by the application of a hard pulse completes the experiment. In principle, the resulting decrease in bulk water proton magnetization can arise from off-resonance effects of the sort described in Section 2, as well as by the transfer of magnetization, the consequence of dipole–dipole-mediated cross-relaxation, and chemical exchange processes. The experiment, in practice, is performed in a manner minimizing off-resonance RF irradiation effects [$B_2 = 0.1$ G, $\nu_{\text{off}} = 5$–10 kHz (from the bulk water ^{1}H resonance)]. The following sections discuss the theoretical and experimental considerations dictating the behavior of ^{1}H magnetization transfer spectroscopy and imaging experiments of aqueous protein solutions and tissues, as well as consider applications.

7.2. Theory

An aqueous macromolecular solution or a tissue may be modeled by a pair of proton spin baths, designated A and B, coupled magnetically by way of intermolecular dipole–dipole interactions (Edzes and Samulski, 1978, Koenig *et al.*, 1978; Shirley and Bryant, 1982; Grad and Bryant, 1990). *Neglecting chemical exchange effects*, but including the presence of an RF field, the system is described by the following set of six coupled Bloch equations:

$$\frac{d}{dt} M_z^A = -(R_T + R_A)(M_z^A - M_{z0}^A) + \frac{R_T}{f}(M_z^B - M_{z0}^B) + \omega_{1A} M_y^A \tag{57a}$$

$$\frac{d}{dt} M_z^B = -\left(\frac{R_T}{f} + R_B\right)(M_z^B - M_{z0}^B) + R_T(M_z^A - M_{z0}^A) + \omega_{1B} M_y^B \tag{57b}$$

$$\frac{d}{dt} M_x^A = -\left(\frac{M_x^A}{T_{2A}}\right) - 2\pi\Delta_A M_y^A \tag{57c}$$

$$\frac{d}{dt} M_x^B = -\left(\frac{M_x^B}{T_{2B}}\right) - 2\pi\Delta_B M_y^B \tag{57d}$$

$$\frac{d}{dt} M_y^A = -\frac{M_y^A}{T_{2A}} + 2\pi\Delta_A M_x^A - \omega_{1A} M_z^A \tag{57e}$$

$$\frac{d}{dt} M_y^B = -\frac{M_y^B}{T_{2B}} + 2\pi\Delta_B M_x^B - \omega_{1B} M_z^B \tag{57f}$$

where $R_{A,B}$ and R_T are the rate constants for longitudinal relaxation of the A or B spins in the absence of cross-relaxation, and the cross-relaxation rate constant for *magnetization transfer from the A to B spin systems*, respectively. By contrast, for magnetization transfer from the B to the A spin system the rate constant is equal to R_T/f, where f represents the ratio of the number of B spins to the number of A spins. $\Delta_{A,B}$ are the offsets of the RF irradiation field from the A or B spins, $\omega_{1A,B}$ are the B_2 field strengths of the preparation RF irradiation field of the A or B spins, respectively, and $M_{z0}^{A,B}$ is the thermal equilibrium longitudinal magnetization of the A or B spins in the absence of the B_2 irradiation field. The other symbols have their usual meaning.

The steady-state solution of Eqs. (57a–f) is given by (Caines *et al.*, 1991):

$$R^A = \frac{M_z^A}{M_{z0}^A} = \frac{\left\{R_A\left[\left(\frac{R_T}{f} + R_B\right) + \frac{\omega_{1B}^2 T_{2B}}{1 + 4\pi^2\Delta_B^2 T_{2B}^2}\right] + R_B R_T\right\}}{D} \tag{58a}$$

$$R^{\mathrm{B}} = \frac{M_{\mathrm{z}}^{\mathrm{B}}}{M_{\mathrm{z0}}^{\mathrm{B}}} = \frac{\left\{ R_{\mathrm{B}} \left[(R_{\mathrm{T}} + R_{\mathrm{A}}) + \dfrac{\omega_{1\mathrm{A}}^2 T_{2\mathrm{A}}}{1 + 4\pi^2 \Delta_{\mathrm{A}}^2 T_{2\mathrm{A}}^2} \right] + \dfrac{R_{\mathrm{A}} R_{\mathrm{T}}}{f} \right\}}{D} \tag{58b}$$

where

$$D = \left[(R_{\mathrm{T}} + R_{\mathrm{A}}) + \frac{\omega_{1\mathrm{A}}^2 T_{2\mathrm{A}}}{1 + 4\pi^2 \Delta_{\mathrm{A}}^2 T_{2\mathrm{A}}^2} \right] \left[\frac{R_{\mathrm{T}}}{f} + R_{\mathrm{B}} + \frac{\omega_{1\mathrm{B}}^2 T_{2\mathrm{B}}}{1 + 4\pi^2 \Delta_{\mathrm{B}}^2 T_{2\mathrm{B}}^2} \right] - \frac{R_{\mathrm{T}}^2}{f} \tag{58c}$$

In the event of negligible off-resonance effects, as would be the case if spin bath B were partially saturated without concomitant off-resonance RF irradiation effects on spin bath A, i.e., effectively $\Delta_{\mathrm{A}} \to \infty$ (in practice, achievement of this condition is usually attempted) the relevant equations (58a–c) reduce to (Caines *et al.*, 1991):

$$R^{\mathrm{A}} = \frac{M_{\mathrm{z}}^{\mathrm{A}}}{M_{\mathrm{z0}}^{\mathrm{A}}} = \frac{\left\{ \mathrm{R_A} \left[\left(\dfrac{R_{\mathrm{T}}}{f} + R_{\mathrm{B}} \right) + \dfrac{\omega_{1\mathrm{B}}^2 T_{2\mathrm{B}}}{1 + 4\pi^2 \Delta_{\mathrm{B}}^2 T_{2\mathrm{B}}^2} \right] + R_{\mathrm{B}} R_{\mathrm{T}} \right\}}{D'} \tag{59a}$$

$$R^{\mathrm{B}} = \frac{M_{\mathrm{z}}^{\mathrm{B}}}{M_{\mathrm{z0}}^{\mathrm{B}}} = \frac{\left\{ R_{\mathrm{B}}[(R_{\mathrm{T}} + R_{\mathrm{A}})] + \dfrac{R_{\mathrm{A}} R_{\mathrm{T}}}{f} \right\}}{D'} \tag{59b}$$

$$D' = (R_{\mathrm{T}} + R_{\mathrm{A}}) \left[\frac{R_{\mathrm{T}}}{f} + R_{\mathrm{B}} + \frac{\omega_{1\mathrm{B}}^2 T_{2\mathrm{B}}}{1 + 4\pi^2 \Delta_{\mathrm{B}}^2 T_{2\mathrm{B}}^2} \right] - \frac{R_{\mathrm{T}}^2}{f} \tag{59c}$$

A set of equations equivalent to Eqs. (59a–c) was derived by Grad and Bryant (1990) by solving for the steady-state solution of a subset of four of the six differential equations written above [Eqs. (57a, b, d, f)]. The elimination of Eqs. (57c, e) is equivalent to denying the occurrence of transverse magnetization components ($M_{\mathrm{x,y,A}}$), which is exactly the situation in the absence of off-resonance effects on spin bath A, i.e., $\Delta_{\mathrm{A}} \to \infty$.

For the case of complete saturation of the B proton spin bath and no off-resonance effects on spin bath A, i.e., setting $M_{\mathrm{z}}^{\mathrm{B}}$ and $M_{\mathrm{y}}^{\mathrm{A}}$ to zero in Eq. (57a), the steady-state solution for the intensity ratio of the Z spin bath is:

$$\frac{M_{\mathrm{z}}^{\mathrm{A}}}{M_{\mathrm{z0}}^{\mathrm{A}}} = \frac{R_{\mathrm{A}}}{R_{\mathrm{T}} + R_{\mathrm{A}}} \tag{60a}$$

when $R_T \gg R_A$, Eq. (60a) reduces to:

$$\frac{M_z^A}{M_{z0}^A} = \frac{R_A}{R_T} \tag{60b}$$

By contrast, for the case of *partial* saturation of the B proton spin bath, assuming negligible off-resonance saturation effects on the A proton spin bath by the RF irradiation field, and assuming that $R_T \gg R_A$, Eqs. (59a–c) yield the following:

$$\frac{M_z^A}{M_{z0}^A} = \frac{M_z^B}{M_{z0}^B} = \left(\frac{R_A}{f} + R_B\right)\left(R_B + \frac{\omega_{1B}^2 T_{2B}}{1 + 4\pi^2\Delta_B^2 T_{2B}^2}\right)^{-1} \tag{61}$$

Since $f = M_{z0}^B / M_{z0}^A$,

$$M_z^A = \frac{1}{f}(M_z)^B \tag{62}$$

and, hence, M_z^A is directly proportional to M_z^B, scaled by the factor $1/f$. Thus, when the condition of $R_T \gg R_A$ is fulfilled, the change of the A proton spin bath steady-state longitudinal magnetization, in response to the variation of the frequency offset of the RF irradiation field *partially* saturating the B proton spin bath, reflects the B proton spin bath steady-state longitudinal magnetization and, hence, its spectrum. Using analogous expressions, Grad and Bryant (1990) reached an identical conclusion, and termed the resulting line shape a "Z-spectrum." As noted by Grad and Bryant (1990), this approach offers considerable potential for the study of heterogeneous liquid–solid systems such as tissue, where dynamic range considerations often preclude the direct acquisition of the B spectrum. If off-resonance effects on the A proton spin bath by the RF irradiation field are not negligible, then the simple relationships between M_z^A and M_z^B described above are no longer valid.

Based on the theoretical analysis of Andree (1978), Fung and McGaughy (1980) have argued that cross-relaxation and other relaxation contributions enter into R_A and R_B because of chemical exchange between the bound and free states of the A spin bath. Thus, R_A and R_B, in terms of their analysis, which includes the consequences of chemical exchange, do not simply represent the spin–lattice relaxation rate constants for the uncoupled A and B proton spin baths. Inclusion of chemical exchange effects between free and bound A results in R_A assuming contributions from both the intrinsic relaxation rate of "uncoupled" free A spins and cross-relaxation with the B proton spin bath. However, when the mole fraction of the bond A spins *engaged in cross-relaxation* is small, R_A is well approximated by the intrinsic relaxation characteristics of the "uncoupled" A proton spin bath. Analogous arguments apply to R_B.

Simulations of the behavior of a proton two-phase system magnetically coupled by cross-relaxation using Eqs. (58) and (59) are shown in Figs. 36 and 37. Representative spin–lattice relaxation times (see figure legends) for rabbit kidney tissue (Wolff and Balaban, 1989), corrected for cross-relaxation effects (Grad and Bryant, 1990), are employed in the simulations. Figure 36 depicts the dispersion of M_z^A/M_{z0}^A with the variation in RF irradiation offset frequency (Δ_B) at different f values (constant $R_T/R_A = 100$) assuming the presence [Eq. (58)] and absence [Eq. (59)] of off-resonance effects of spin bath A. With increasing values of f the dispersion curves assume increasing proton spin bath B character. Off-resonance RF irradiation effects are apparent at relatively small offset frequencies and are dependent on f. At low f values the off-resonance effects become significantly more pronounced. At f values of 0.05 or larger [chicken thigh muscle has an f value of 0.19 (Edzes and Samulski, 1978)], off-resonance effects are negligible at RF offset frequencies of 5 kKz or greater at a B_2 field strength of 0.1 G, i.e., Eq. (59) which assumes that the absence of off-resonance effects is adequate to represent the expected experimental behavior. Using

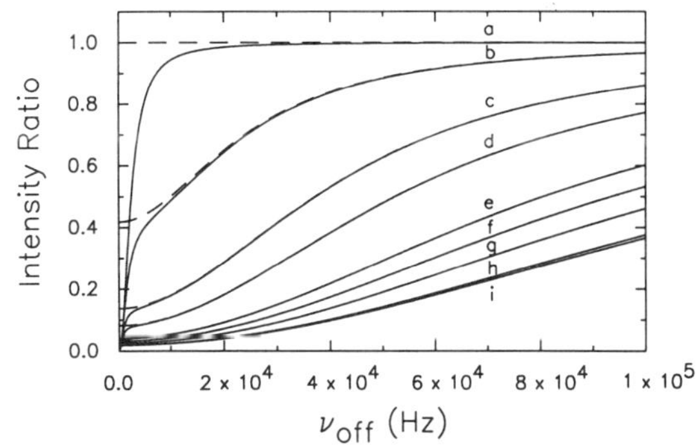

Figure 36. Computer-simulated ^{1}H intensity ratio curves of water (M_z^A/M_{z0}^A) versus RF irradiation offset frequency ($\Delta_A = \Delta_B$) as a function of f ($= M_{z0}^B/M_{z0}^A$) for a two proton spin bath (A and B) with partial saturation of proton spin bath B in the presence (solid lines) and absence (dashed lines) of off-resonance partial saturation effects of proton spin bath A using Eqs. (58) and (59), respectively. f was varied as follows: (a) 10^{-10}; (b) 0.01; (c) 0.05; (d) 0.1; (e) 0.3; (f) 0.5; (g) 1.0; (h) 10.0; (i) 100.0. The A proton spin bath represents the bulk water phase, while the B proton spin bath represents macromolecular protons. The simulations assume an off-resonance RF field strength of 0.1 G (Wolff and Balaban, 1989; Eng *et al.* 1991). The following magnetic relaxation parameters representative of rabbit kidney were assumed (Wolff and Balaban, 1989); $T_{1A} = 2.0$ sec, $T_{2A} = 60$ msec, $T_{1B} = 1.0$ sec, $T_{2B} = 10\ \mu$sec (Grad and Bryant, 1990); R_T was fixed at 100 sec^{-1}. The proton chemical shift values for the A and B spins were assumed to be identical. Taken from Caines *et al.* (1991).

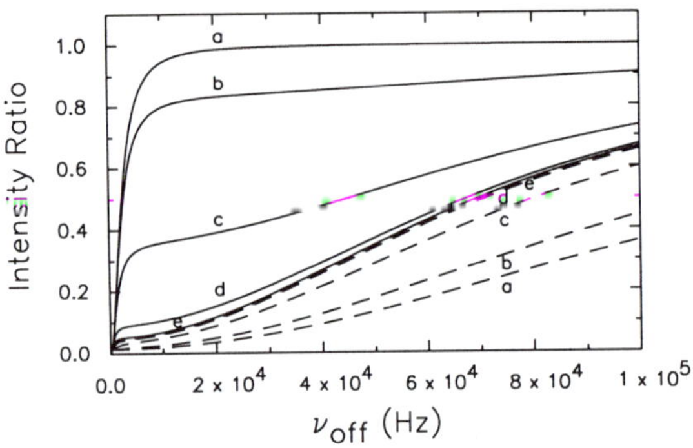

Figure 37. Computer-simulated ^{1}H intensity ratio curves of water (M_z^A/M_{z0}^A) and macromolecular protons (M_z^B/M_{z0}^B) versus RF irradiation offset frequency ($\Delta_A = \Delta_B$) as a function of cross-relaxation rate (R_T) for a two proton spin bath (A and B) calculated using Eq. (58). The A proton spin bath (solid lines) represents the bulk water phase, while the B proton spin bath (dashed lines) represents macromolecular protons. R_T was varied as follows: (a) 0.01 $\sec^{-1}$; (b) 0.1 $\sec^{-1}$; (c) 1.0 $\sec^{-1}$; (d) 10.0 $\sec^{-1}$; (e) 100 $\sec^{-1}$. The simulations assume an off-resonance RF field strength of 0.1 G (Wolff and Balaban, 1989; Eng *et al.*, 1991). The following magnetic relaxation parameters, representative of rabbit kidney, were assumed (Wolff and Balaban, 1989): $T_{1A} = 2.0$ sec, $T_{2A} = 60$ msec, $T_{1B} = 1.0$ sec, $T_{2B} = 10$ μsec (Grad and Bryant, 1990) and f was fixed at 0.2. The proton chemical shift values for the A and B spins were assumed to be identical. Line (e) represents the convergence of the M_z^A/M_{z0}^A and M_z^B/M_{z0}^B dispersion curves. This line is sometimes referred to as the Z-spectrum (Grad and Bryant, 1990). Taken from Caines *et al.* (1991).

Eq. (58) for the simulations shown in Fig. 37, the calculated dispersion curves representing M_z^A/M_{z0}^A and M_z^B/M_{z0}^B converge with increasing R_T/R_A, a result also predicted by Eq. (61). As shown in this figure, convergence of the dispersion curves requires R_T/R_A ratios in excess of 20 for the representative set of employed tissue relaxation parameters. Thus, an increase of this ratio fron near unity shifts the proton spin bath dispersion curves toward each other, from pure A and B dispersion curves, respectively. It is also clear from these simulations that the success of the off-resonance partial saturation experiment in eliciting a Z-spectrum requires that R_T be at least the same order of magnitude or larger than R_A. If it is not, then the Z-spectrum of the A spins contains virtually no information about the B proton spin system. With increasing R_T/R_A, the Z-spectrum contains an increasing proportion of B spin character and ultimately the A spin spectral response faithfully represents the B spin system.

To further explore the factors affecting the spectral characteristics of the macromolecular Z-spectrum, the intensity ratio dispersion curve behavior of the hypothetical aqueous protein solution was simulated at fast and slow rotational correlation times (Fig. 38). The smaller correlation time was chosen to represent an isotropically reorienting monomeric protein,

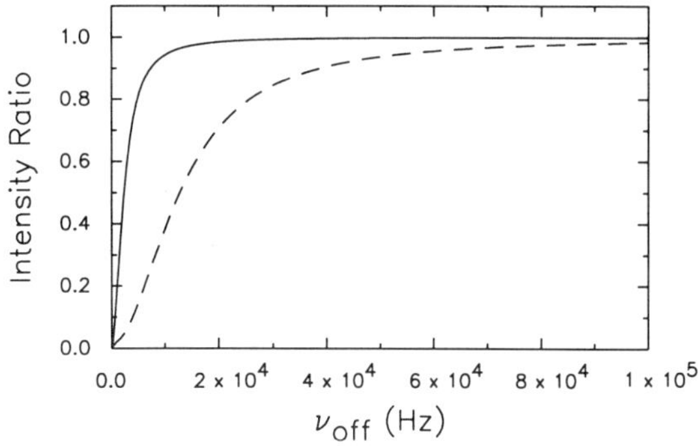

Figure 38. Computer-simulated water ^1H intensity ratio dispersion curves for water proton–macromolecular proton cross-relaxation at fast [34 nsec (——)] and slow [1000 nsec (– – –)] macromolecular rotational correlation time. The simulations assume a 68-kDa internally rigid spherical protein, with $\bar{v} = 0.72$ g/g and a hydration value of $\delta = 0.5$ g H_2O/g protein at a concentration of 300 g/kg ($f = 0.15$). Assumed magnetic resonance parameters include a B_2 field strength of 0.1 G at a B_0 field of 1.33 T, $T_{1A} = 0.9$ sec, $T_{2A} = 60$ msec, and a bound water exchange time (τ_{ex}) of 10 μsec. The average number of protein protons interacting with each water proton was assumed to be 5 (values ranging from 3 to 10 had little effect on the calculated dispersion curves). The macromolecular proton relaxation times T_{1B}, T_{2B}, R_T (effective intermolecular cross-relaxation rate constant, representing transfer from water protons to protein proteins), k_T (intermolecular cross-relaxation rate constant), and k_S (intramolecular spin diffusion rate constant) were calculated following previously published procedures (Kalk and Berendsen, 1976; Edzes and Samulski, 1978). Calculated values of T_{1B}, T_{2B}, R_T, k_T, and k_S, respectively, used in the simulations at the two correlation times were as follows: 0.4 sec, 4.3 msec, 4.0 sec^{-1}, 5.0 sec^{-1}, 60.7 sec^{-1} (34.3 nsec); 11.7 sec, 0.15 msec, 108.7 sec^{-1}, 133.3 sec^{-1}, 1785.7 sec^{-1} (1000 nsec). The rate-limiting step in both cases is intermolecular cross-relaxation.

whereas the larger time was selected to mimic the behavior of a highly cross-linked (aggregated) state of the same protein. The following macromolecular model was employed: spherical, 68-kDa protein, at a concentration of 300 g/kg water. Standard values for partial specific volume and average hydration were assumed (Squire and Himmel, 1979) (see legend, Fig. 38). The assumed monomeric model approximates the molecular weight and concentration of hemoglobin in the red blood cell. The calculated f value of 0.15 (assuming a mean amino acid residue weight of 110, an average value of 6 nonexchangeable protons per amino acid residue, and a uniform representation of the 20 naturally occurring amino acids) is similar to that found in vertebrate muscle tissue. The shorter rotational correlation time of 34 nsec, representing the monomeric protein state, was calculated using the Stokes–Einstein equation [Eq. (21)]. To represent the reorientation behavior of the highly aggregated state, a rotational correlation time of 1000 nsec was assumed. The basic considerations of the simulation model assume that a protein proton experiences intramolecular dipole–dipole relaxation by another protein proton at an internuclear distance of 1.78 Å (Kalk and Berendsen, 1976; Edzes and Samulski, 1978). The protein matrix is assumed to be rigid, i.e., no internal motion is present, and isotropic reorientation is assumed to occur (defined by τ_{rot}). Cross-relaxation between protein protons and water protons at the macromolecule–solvent interface was calculated assuming an intermolecular proton–proton distance of 2.7 Å (Kalk and Berendsen, 1976; Edzes and Samulski, 1978). The correlation time representing this process was defined by: $\tau_0^{-1} = \tau_{rot}^{-1} + \tau_{ex}^{-1}$ (Marshall, 1970), where τ_{ex} is the exchange lifetime of a hydration layer-bound water proton. For the purposes of this simulation, τ_{ex} was fixed at 10 μsec, a value representing intermediate to fast exchange (Edzes and Samulski, 1978). The details of the required calculations follow the procedures described by Kalk and Berendsen (1976) and Edzes and Samulski (1978). To eliminate the difficult task of accurately simulating the relaxation behavior of the bulk water proton on purely theoretical grounds, the simulations were carried out assuming the experimental value of T_{1A} in aqueous BSA solutions at the f value assumed in the simulation (Grad *et al.*, 1990), whereas for T_{2A} a value of 60 msec was assumed. The latter value is characteristic of water tissue protons (Wolff and Balaban, 1989). The simulated A spin bath dispersion curves (Z-spectra) presented in Fig. 38 dramatically illustrate the effect of increasing the rotational correlation time and, hence, ultimately reveal the important role of dipolar averaging in defining Z-spectral behavior. As shown in this figure, an off-resonance RF offset frequency of 5 to 10 kHz ($B_2 = 0.1$ G) produces a large difference in the intensity ratio behavior of water protons arising from fast ($\tau_{rot} =$ 34 nsec) and very slow ($\tau_{rot} = 1000$ nsec) reorientational motion. The calculations reveal that the rate-limiting step is the intermolecular transfer of Zeeman energy (see legend, Fig. 38). Other simulations (not shown) demonstrate convergence of the water (A) and protein (B) spin bath intensity ratio dispersion curves, thus indicating that a faithful Z-spectrum is attained. Furthermore, simulations performed at 1.33 and 7.05 T indicate virtually no field strength dependence within this field strength range, an expected result. The occurrence of internal motion within the protein matrix results in narrowing of the Z-spectrum, a point made previously by Edzes and Samulski (1978). The implications of the simulations are discussed in Section 7.3.2.1.

A more accurate simulation would involve the use of spectral density functions for translational instead of rotational motion in the calculation of cross-relaxation between the A and B proton spin baths (Fung and McGaughy, 1980). Such an approach has in fact been taken by Kimmich *et al.* (1990) for the calculation of water proton NMRD curves of aqueous protein solutions and tissue samples. This approach, which implicitly includes exchange contributions from bound water, could be utilized for the simulation of water proton dispersion behavior in the off-resonance RF irradiation experiment involving aqueous protein solutions and tissue samples.

Grad and Bryant (1990) derived the following expression for the width squared of the Z-spectrum at half height:

$$\Delta_{1/2}^2 = \frac{\omega_{1B}^2 f}{\pi^2}\left(\frac{(R_T/R_A)+1}{R_B T_{2B}\{(R_T/R_B)+f[(R_T/R_A)+1]\}}\right)+\frac{1}{\pi^2 T_{2B}^2} \tag{63}$$

Thus, they suggested and experimentally verified that a plot of $\Delta_{1/2}^2$ versus ω_{1B}^2 is linear and yields T_{2B} from the intercept, where the reciprocal of the slope in the limit of $R_T \gg R_A$ is:

$$\text{slope}^{-1} = \pi^2 T_{2B}\left(R_B + \frac{R_A}{f}\right) \tag{64}$$

The terms R_A (free of cross-relaxation contributions) and R_T may be measured by an inversion recovery experiment where the B proton spin bath is saturated (Grad *et al.*, 1990). This enables the determination of R_B if f is known from an independent measurement. Simulations (not shown) demonstrate that Eq. (63) provides an accurate representation of the expected dispersion behavior, i.e., width of the Z-spectrum at half height, provided that off-resonance effects on spin bath A can be neglected. However, the occurrence of significant off-resonance effects on spin bath A, such as would be encountered at low f values, or small offset frequencies coupled with large B_2 field strengths, renders this equation less accurate because a significant difference exists betwen the true and apparent amplitudes of the Z-spectrum at zero offset frequency (cf Fig. 36).

Incorporation of Cross-Relaxation into Off-Resonance Rotating Frame Spin–Lattice Relaxation Formalism

The off-resonance rotating frame spin–lattice relaxation formalism can be extended to include the effects of magnetization transfer between a proton two-phase system (Caines *et al.*, 1991). The time-dependent behavior of the longitudinal magnetization in each phase coupled by cross-relaxation is given by the following set of differential equations, valid at small Θ and when $\omega_e \ll \omega_0$:

$$\frac{d}{dt}M_z^A = -(R_T + R_A^{\text{off}})M_z^A + R_A^{\text{off}}M_{ze}^A + \frac{R_T}{f}M_z^B \tag{65a}$$

$$\frac{d}{dt}M_z^B = -\left(\frac{R_T}{f} + R_B^{\text{off}}\right)M_z^B + R_B^{\text{off}}M_{ze}^B + R_T M_z^A \tag{65b}$$

where $R_A^{off} = 1/T_{1\rho,A}^{off} = R_A + 1/T_{e,A}$, $R_B^{off} = 1/T_{1\rho,B}^{off} = R_B + 1/T_{e,B}$, and M_{ze}^{A} and M_{ze}^{B} are the z-components of the magnetization along the effective magnetic field (B_e) for spins A and B [Eqs. (51) and (52), respectively]. $1/T_{e,A,B}$ is defined below. The other symbols have their usual meaning. The steady-state solution for the spectral intensity ratio of spins A (R^A) and B (R^B) under continuous-wave RF irradiation is given by:

$$R^A = \frac{\left[R_A\left(\frac{R_T}{f} + R_B^{off}\right)\cos^2\Theta_A + R_B R_T \cos^2\Theta_B\right]}{D^{off}} \tag{66a}$$

$$R^B = \frac{\left[R_B(R_T + R_A^{off})\cos^2\Theta_B + R_A \frac{R_T}{f}\cos^2\Theta_A\right]}{D^{off}} \tag{66b}$$

where

$$D^{off} = R_A^{off} R_B^{off} + R_A^{off}\frac{R_T}{f} + R_B^{off} R_T \tag{66c}$$

The theoretical spectral intensity ratios given by Eqs. (66a, b) and the corresponding Eqs. (58a, b) derived from the modified Bloch formalism for cross-relaxation can be shown to be equivalent when $\cos^2\Theta \approx 1$. Equating the ratio M_z/M_0 from the Bloch and off-resonance rotating frame spin-lattice relaxation formalisms, we obtain:

$$\frac{1 + T_2^2\Delta\omega^2}{1 + T_2^2\Delta\omega^2 + \gamma^2 B_2^2 T_1 T_2} = \cos^2\Theta\left(\frac{T_{1\rho}^{off}}{T_1}\right) \tag{67}$$

The relationship between $T_{1\rho}^{off}$ and T_1, at small Θ, is given by:

$$\frac{1}{T_{1\rho}^{off}} = \frac{1}{T_1} + \frac{1}{T_e} \tag{68}$$

where $1/T_e$ is the contribution to $1/T_{1\rho}^{off}$ which results from the off-resonance effective field (James *et al.*, 1977).

Given that the B_2 RF field amplitude is much smaller than the static B_0 field, and the off-resonance experiment is generally performed with offset frequencies such that $\cos^2\Theta \approx 1$, the preceding Eqs. (67) and (68) reduce to:

$$\frac{1}{T_e} = \frac{T_2\omega_1^2}{1 + 4\pi^2\Delta^2 T_2^2} \tag{69}$$

where $\omega_1^2 = (\gamma B_2)^2$ and $2\pi\Delta = \Delta\omega$. Inserting Eq. (68) into Eqs. (66a, b) gives:

$$R^A = \frac{\left[R_A\left(\frac{R_T}{f} + R_B + \frac{1}{T_{e,B}}\right)\cos^2\Theta_A + R_B R_T \cos^2\Theta_B\right]}{D^{\text{off}}} \tag{70a}$$

$$R^B = \frac{\left[R_B\left(R_T + R_A + \frac{1}{T_{e,A}}\right)\cos^2\Theta_B + R_A\frac{R_T}{f}\cos^2\Theta_A\right]}{D^{\text{off}}} \tag{70b}$$

Using the expression for $1/T_e$ given by Eq. (69), Eqs. (70a, b) are equivalent to the corresponding Eqs. (58a, b) when the value of $\cos^2\Theta \approx 1$. Thus, the off-resonance rotating frame spin–lattice relaxation formalism and the Bloch formalism modified to include cross-relaxation are identical under the usual experimental conditions employed.

7.3. Applications

7.3.1. Background

Proton NMR relaxation experiments performed on heterogeneous biological systems, as noted above, have shown that magnetization transfer of Zeeman spin energy by way of dipole–dipole-mediated cross-relaxation occurs between the bulk water protons and macromolecular protons across the macromolecular surface-interface. The characteristic process of spin energy transfer is consistent with the following model which divides the overall magnetization transfer process into three steps (Kalk and Berendsen, 1976; Edzes and Samulski, 1978); (1) chemical exchange between the bulk water and protein-bound water transfers spin energy from the bulk water phase to the bound (hydrated) water at the macromolecular surface; (2) intermolecular cross-relaxation transfers spin energy at the protein–water interface by way of mutual spin flips between bound water protons and macromolecular protons situated at the macromolecular surface; (3) spin diffusion and intramolecular cross-relaxation transfers spin energy from the macromolecular surface throughout the macromolecular proton system.

Since chemical exchange and translational diffusion in step 1 is usually rapid, a single average spin–lattice relaxation rate for water protons is observed. In addition, intramolecular cross-relaxation in step 3 strongly influences the proton spin–lattice relaxation time of proteins. When $\omega_0^2\tau_0^2 \gg 1$, intramolecular cross-relaxation dominates and protein protons tend to relax with the same spin–lattice relaxation rate, i.e., a uniform spin temperature is established (Kalk and Berendsen, 1976). Studies of Sykes *et al.* (1978) support the latter findings. Thus, the transfer of spin energy by way of the three interactions described above will be limited by intermolecular cross-relaxation in step 2, and the magnetization transfer between water protons and macromolecular protons will behave as a two-phase system. In addition, the cross-relaxation transfer rate increases with increasing correlation time, i.e., with molecular weight (Kalk and Berendsen, 1976; Edzes and Samulski, 1978; Zhong *et al.*, 1990). In terms of the formalism presented in the preceding section, bulk water protons and macromolecular protons will be designated the A and B proton spin baths, respectively.

7.3.2. Magnetic Cross-Relaxation-Based Spectroscopy

7.3.2.1. Aqueous Protein Solutions. Bryant and co-workers (Grad and Bryant, 1990; Grad *et al.*, 1990, 1991) investigated the water proton spectral intensity ratio ($R^A = M_z^A/M_0^A$) offset frequency dispersion behavior (termed by them, the Z-spectrum) of aqueous monomeric and cross-linked BSA solutions. The simulations presented in Fig. 38 utilized parameters relevant to these experimental studies. The agreement between experiment and theory is reasonable considering the nature of the approximations. Using a B_2 field strength of 0.12 G, Grad and Bryant observed an intensity ratio (M_z^A/M_0^A) of 0.87 (offset frequency of 5 kHz) and 0.55 (offset frequency of 20 kHz) for monomeric and cross-linked BSA (4 mM monomer units), respectively, whereas the simulations for essentially the same conditions indicate intensity ratios of 0.82 and 0.70, respectively. The experimental studies suggest, and the simulations underscore the point, that the crucial condition for obtaining a Z-spectrum is a reduction of rapid macromolecular reorientational motion which leads to increased intermolecular cross-relation and macromolecular spin diffusion. Thus, when motional narrowing is prohibited by cross-linking (Grad and Bryant, 1990; Grad *et al.*, 1990, 1991) or noncovalently by intermolecular effects arising in very concentrated protein conditions such as encountered in eye lens homogenates (Morgan *et al.*, 1989; Rydzewski *et al.*, 1992), *spectral* line shapes become significantly broader. The important role of solvent water relaxation was experimentally established by diluting the ^1H spins with ^2H and noting an increase in the amplitude of the Z-spectrum at zero offset frequency (Grad and Bryant, 1990), thus demonstrating the enhanced sensitivity of the remaining water protons to intermolecular dipole–dipole relaxation by the macromolecular protons. In other experiments, Grad and Bryant (1990) substituted dimethylsulfoxide for water and obtained a Z-spectrum, thus revealing that proton exchange between water protons and macromolecular protons was not essential to the generation of cross-relaxation effects.

7.3.2.2. Mammalian Tissue Preparations. In companion studies, Grad *et al.* (1991) determined the Z-spectral behavior of different rat tissues. An objective of these experiments was the characterization of the offset dispersion response of a variety of organ tissue types excised from healthy rats, and hence, the gaining of information about the usually unobservable immobilized protons in the tissue. The Z-spectrum technique may be utilized in either a spectroscopic or imaging mode, since the water proton magnetization intensity (M_z^A/M_{z0}^A) is monitored in the magnetic resonance experiment. The latter is discussed below (see Section 7.3.3).

Grad *et al.* (1991) argued that the immobilized or solidlike components present in tissue were responsible for the generation of a Z-spectrum in the off-resonance RF irradiation experiment. The NMR spectral linewidth of the solidlike (immobilized) components present in the tissue are expected to be dominated by proton–proton magnetic dipole–dipole interactions (Kalk and Berendsen, 1976; Stoez *et al.*, 1978). Intermolecular dipole–dipole cross-relaxation interactions are anticipated to be similar, since the immobilized macromolecular components are likely to be fully hydrated. Different solidlike macromolecular species are thus likely to be characterized by the same spectral width and, hence, have similar spin diffusion rates. However, if a significant amount of intramolecular local or internal motion is present, averaging of dipole–dipole coupling will occur, thereby leading to a decrease in the spectral width. Thus, the Z-spectrum of the solidlike macromolecular components may prove to be sensitive to differences in internal macromolecular motional dynamics, a conclusion supported by spectral simulation. However, the success of achieving spatially localized cross-relaxation information is expected to depend on the extent of water proton spin bath magnetic localization and, hence, isolation from other tissue regions.

The particular offset irradiation frequency which generates the largest variation in the spectral intensity ratio among different tissue types can, therefore, be used in magnetic resonance imaging experiments as a means of generating a new form of magnetic resonance image contrast, which is ultimately based on macromolecular dynamics. Thus, the view that water tissue proton relaxation times simply reflect the motional freedom of water molecules is erroneous. For example, much of the eye lens water proton relaxation data in the literature have been interpreted in terms of free and bound water fractions, without consideration of cross-relaxation effects (Lerman *et al.*, 1982; Lerman and Moran, 1989; Pope *et al.*, 1982; Rácz *et al.*, 1979, 1983), thereby ignoring an appreciable relaxation contribution from the constituent lens proteins. The occurrence of cross-relaxation in lens tissue is demonstrated by the ability to obtain a lenticular water proton dispersion curve by off-resonance RF irradiation. An example of such a curve, obtained from human lens tissue, is shown in Fig. 39. Compared to the Z-spectrum obtained from an aqueous solution of cross-linked BSA by Grad and Bryant (1990), the off-resonance dispersion behavior of the human lens indicates the presence of solidlike material that is very effective in cross-relaxation with lenticular water protons. This interpretation is consistent with the NMR studies of Morgan *et al.* (1989) and Rydzewski *et al.* (1992) which demonstrate directly the occurrence of solidlike protein domains in different species of lens tissue. It is also interesting to note that at offset frequencies of less than ~4.8 kHz the lenticular water proton dispersion curve is dominated by off-resonance effects, whereas at larger frequencies magnetization transfer effects are preponderant. It is well known that changes occur in lens protein structure with age and cataract formation (for a review see Harding and Crabbe, 1984), which are likely to profoundly affect both internal motional as well as reorientational dynamics and, hence, lenticular water relaxation.

Simulations, like those presented in Figs. 36 to 38, or *ex vivo* Z-spectral determinations, such as those described below, may be used as a guide for establishing optimal image contrast parameters for use in magnetic resonance imaging experiments. Experimentally, measurements performed at 10 kHz offset ($B_2 = 0.12$ G) generally resulted in producing the largest variation in the spectral intensity ratio and, hence, would be expected to generate the maximum image contrast (Grad *et al.*, 1991). The off-resonance

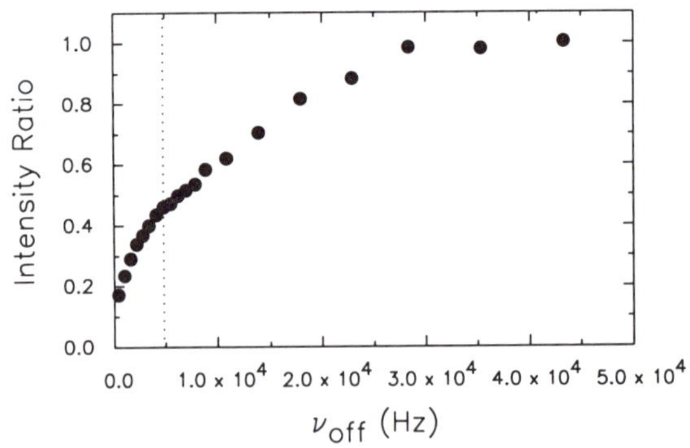

Figure 39. Proton intensity ratio dispersion plot of bulk water signal obtained from a clear 70-year-old human eye lens. The proton spectra were obtained at 7.05 T employing an off-resonance RF field strength of 0.98 G applied for 5.0 sec prior to acquisition of the FID. The temperature was maintained at 22°C. At off-resonance irradiation frequencies of less than 4.8 kHz, indicated by the dotted line, off-resonance irradiation effects on the water resonance dominate, whereas at greater irradiation frequencies magnetization transfer effects are preponderant.

studies which compared kidney, liver, and adrenal gland showed negligible differences and each had an intensity ratio of ~0.7 at 10 kHz offset ($B_2 = 0.12$ G). However, for the other tissues examined the approximate intensity ratio values at 10 kHz offset had a greater variation: bone, 0.5; cerebral cortex, 0.7; eye, 0.8; muscle, 0.45; cord, 0.65; and cerebellum, 0.75. Each of the tissue intensity ratio curves was characterized by a different Z-spectrum amplitude. It is not unreasonable to anticipate that the Z-spectrum of healthy tissues will differ from diseased tissue, or tissue experiencing pathophysiological insult. An excellent tissue candidate for the verification of this expectation is the cataractous eye lens. In cataract, the constituent lens proteins, the crystallins, are usually characterized by an enhanced degree of aggregation, in many instances the result of intermolecular disulfide cross-linking (for a review see Harding and Crabbe, 1984). Thus, it is anticipated that the spatially localized Z-spectral map elicited in an imaging experiment (*vide infra*) of a cataractous lens will be different in the clear versus opaque regions of different lens fiber cells. If this expectation can be confirmed [such experiments are currently in progress (Schleich *et al.*, 1990)], then the proton off-resonance irradiation technique which leads to a Z-spectrum of the immobilized components could foreshadow a viable means for the detection of lenticular precataractous regions, i.e., localized macromolecular protein domains which have altered motional dynamics, but have yet to begin their presumed irreversible course toward opacification. The ability to detect such regions would provide an opportunity for therapeutic drug intervention, and subsequent monitoring of treatment efficacy.

7.3.3. Magnetic Cross-Relaxation-Based Imaging

The recent application of saturation transfer methodology to proton magnetic resonance imaging experiments for the study of magnetization transfer in biological tissue has resulted in a unique form of contrast-enhanced images relative to conventional T_1 or T_2 weighted images (Wolff and Balaban, 1989; Eng *et al.*, 1991). Tissue contrast in proton magnetic resonance-generated images is a direct result of the relaxation characteristics and spin density of water protons in specific tissue environments. The relaxation properties of water in biological tissues under favorable circumstances can be dominated by cross-relaxation-mediated magnetization transfer effects with macromolecules (*vide supra*).

In saturation transfer-based spectroscopy and imaging experiments, the free bulk water proton signal amplitude is directly affected by the irradiation power, duration, and offset frequency of the RF irradiation field employed to saturate the restricted proton environment magnetization. Irradiation power and offset frequency of the saturating RF irradiation are carefully chosen to minimize off-resonance effects on the free bulk water magnetization. Typical RF irradiation parameters are 0.07 to 0.1 G applied at an offset frequency of 5 to 10 kHz.

Assuming a two-phase proton system, the rate of magnetization transfer between protons in "free" bulk water (^1H_f) of relatively unrestricted motion and protons of macromolecules with restricted motion (^1H_r) is given by (Wolff and Balaban, 1989; Eng *et al.*, 1991):

$$k_{for} = \frac{1}{T_{1fsat}}\left(1 - \frac{M_{sf}}{M_{0f}}\right) \tag{71}$$

where k_{for} is the forward pseudo-first-order rate constant for magnetization transfer from ^1H_f to ^1H_r. The spin–lattice relaxation time T_{1fsat}, and the magnetization M_{sf} of the ^1H_f are determined in the presence of ^1H_r saturation, and M_{0f} is the ^1H_f magnetization in the absence of ^1H_r saturation. Several other relationships also apply to this experiment (Wolff and Balaban, 1989; Eng *et al.*, 1991). These include:

$$\frac{M_{sf}}{M_{0f}} = \frac{T_{1fsat}}{T_{1f}} = \frac{\tau_{1f}}{T_{1f}} + \frac{\tau_{1f} k_{rev} M_{sr}}{M_{0f}} \tag{72}$$

and

$$\frac{1}{T_{1fobs}} = \frac{1}{T_{1f}} + \frac{X_r}{T_{1r} + (1/k_{rev})} \tag{73}$$

where M_{sr} is the magnetization of ^1H_r in the presence of RF saturation and

$$\frac{1}{\tau_{1f}} = k_{for} + \frac{1}{T_{1f}} \tag{74}$$

T_{1fobs} is the observed spin–lattice relaxation time for the free water protons in the presence of magnetization transfer, T_{1f} is the spin–lattice relaxation time constant for the free water protons in the absence of cross-relaxation contributions ($= R_A^{-1}$), X_r is the mole fraction of the restricted proton pool, and k_{rev} ($= k_{for}/f$) is the pseudo-first-order rate constant for magnetization transfer from ^1H_r to ^1H_f. Equation (73) is valid when $X_r < 0.3$ and when the proton chemical shift values for ^1H_r and ^1H_f are equal.

Typical values for k_{for} (^1H_f to ^1H_r) are approximately 1 sec^{-1} for rabbit kidney and 3 sec^{-1} for skeletal muscle *in vivo*.

By calculating a simple pixel-by-pixel ratio (M_{sf}/M_{0f}) of an image produced with presaturation irradiation relative to a control image produced without this irradiation, Wolff and Balaban (1989) were above to obtain a qualitative view of the degree of magnetization transfer in different tissue environments, thereby establishing a technique with a new form of image contrast. However, proton magnetization arising from urine, blood vessels, and fat was not significantly affected by the saturating preirradiation RF field. The initial experiments of Balaban and co-workers were confined to rabbit kidney and skeletal muscle.

In a further study, Balaban and co-workers (Eng *et al.*, 1991) developed the methodology for obtaining quantitative magnetization transfer rate constant images from off-resonance irradiation experiments. The magnetization transfer rate constant image was obtained by calculating k_{for} for each pixel in the image from T_{1fsat}, M_{sf}, and M_{0f} on a pixel-by-pixel basis using Eq. (71). A k_{for} image of the kidney indicated that the free water in the inner medulla had a significantly slower magnetization transfer rate with macromolecular protons than that in the cortex. Although the k_{for} image was consistent with the simple ratio image (M_{sf}/M_{0f}) of this tissue, the magnitude of the contrast between the cortex and medulla was significantly greater for the k_{for} image than for the M_{sf}/M_{0f} image.

In their formalism representing the magnetization transfer contrast imaging experiment, Balaban and co-workers (Wolff and Balaban, 1989; Eng *et al.*, 1991) explored the dependence of M_{sf}/M_{0f} upon the duration and strength of the off-resonance B_2 field. Thus, by starting with the differential equation expressing the time dependence of M_{sf}:

$$\frac{dM_{sf}}{dt} = -\frac{M_{sf}}{\tau_{1f}} + \frac{M_{0f}}{T_{1f}} + k_{rev}M_{sr} \tag{75}$$

and assuming that M_{sr} is constant, the following expression was derived by them by invoking the steady-state condition at infinite time:

$$\frac{M_{sf}}{M_{0f}} = \frac{\tau_{1f}}{T_{1f}} + \left(1 - \frac{\tau_{1f}}{T_{1f}}\right) m_{zratio}(r) \tag{76}$$

where $M_{zratio}(r) = M_{sr}/M_{0r}$. In the presence of *total* ^1H_r saturation $M_{sr} = 0$ and the time dependence of M_{sf}/M_{0f} is the familiar equation for saturation-transfer:

$$\frac{M_{sf}}{M_{0f}} = \frac{\tau_{1f}}{T_{1f}} + k_{for}\tau_{1f} \exp\left(-\frac{t}{\tau_{1f}}\right) \tag{77}$$

This equation shows that upon instantaneous saturation of ^1H_r the free water proton magnetization will approach a new steady-state value with a time constant of τ_{1f}. This equation assumes the complete absence of off-resonance effects. Similar expressions were derived in Section 6 for chemical exchange.

To account for a small amount of off-resonance effects (termed by them "nonspecific RF bleedover") estimated to be ~15% at a B_2 field strength of 0.16 G applied 5 kHz off-resonance from ^1H_r, they introduced a correction factor $m_{zratio}(f)$, equal to M_{sf}/M_{0f}, into the second term of Eq. (75). Substituting $m_{zratio}(f)M_{0f}/T_{1f}$, which effectively replaces M_{0f} with M_{sf}, modifies Eq. (75) to:

$$\frac{M_{sf}}{M_{0f}} = \left(\frac{\tau_{1f}}{T_{1f}}\right) m_{zratio}(f) + \left(1 - \frac{\tau_{1f}}{T_{1f}}\right) m_{zratio}(r) \tag{78}$$

In the notation adopted in this chapter, the modified Eq. (75) can be written in the form:

$$\frac{d}{dt} M_z^A = -(R_T + R_A) M_z^A + R_A M_{ze}^A + \left(\frac{R_T}{f}\right) M_z^B \tag{79}$$

where M_{ze}^A is the z-component of effective magnetization aligned along the effective field. The following definitions relate the modified form of Eq. (75) and Eq. (79): $1/\tau_{1f} = R_A + R_T$, $k_{rev} = R_T/f$, $1/T_{1f} = R_A$, $M_{sf} = M_z^A$, $M_{sr} = M_z^B$, and $m_{zratio}(f)M_{0f} = M_{ze}^A$. However, this formalism neglects off-resonance irradiation relaxation behavior by equating R_A to $1/T_{1f}$. To properly include such effects, R_A should be equated to $1/T_{1\rho}^{off}$, as described in Section 6.2. The relaxation time constant $T_{1\rho}^{off}$ is a function of both B_2 and ν_{off}, and thus it is not surprising to note the marginal fit in Fig. 3 (a plot of M_{sf}/M_{0f} versus B_2^2) of Eng *et al.* (1991).

As noted by Grad and Bryant (1990), the evaluation of T_{2r} by Wolff and Balaban (1989) failed to include cross-relaxation between the bulk water and macromolecular proton spin systems. Thus, Grad and Bryant (1990) suggest that the derived value of T_{2r} of 0.2 msec should be reduced to 10 μsec.

Other experiments performed by Eng *et al.* (1991) involving random 90° phase shifts, which would destroy any spin-locked transverse magnetization, arising from off-resonance irradiation effects, indicated that the predominant contribution to the reduction of the bulk free water magnetization originates from magnetization transfer effects. Experiments involving the use of deuterated and tritiated water revealed that the macromolecular "interaction site" is not exchangeable. Model lipid systems also display magnetization transfer effects (Wolff *et al.*, 1990), and suggest the importance of both the macromolecular correlation time and the presence of a macromolecular functional group to which water can hydrogen bond.

To obtain enhanced image contrast in standard gradient recalled echo proton magnetic resonance images, Wolff *et al.* (1991) added to the conventional sequence a low-power off-resonance RF irradiation field. Compared to conventional gradient recalled images, the magnetization transfer contrast

(MTC)-enhanced images were found to have significantly increased contrast between different tissue types. An example of a magnetization transfer-enhanced cat head image relative to a conventional image is shown in Fig. 40. Clearly evident is the enhanced contrast afforded by magnetization transfer ensuing from the saturation of the immobile proton pool.

The off-resonance rotating frame spin–lattice relaxation experiment may also be incorporated into a conventional imaging sequence. Preliminary experiments by Rommel and Kimmich (1990) demonstrate the feasibility of such an approach. Basically, the preparation sequence is an inversion–recovery sequence in which low-power off-resonance RF irradiation is

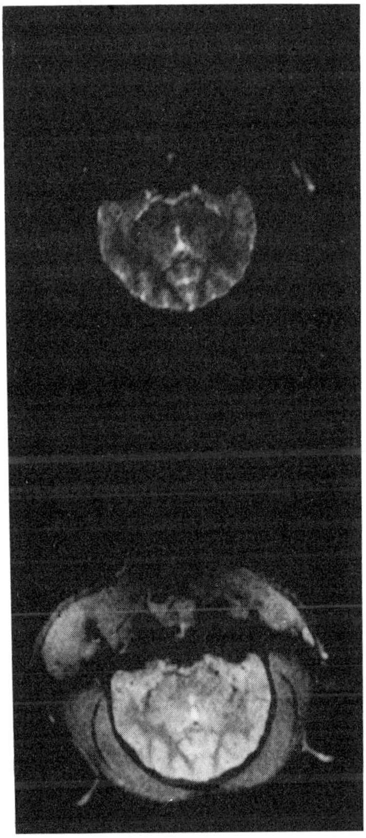

Figure 40. Gradient recalled echo proton magnetic resonance images of a cat brain obtained at 4.7 T in the absence (top) and presence (bottom) of low-power off-resonance RF irradiation (0.1 G, 10 kHz off-resonance) applied for 400 msec prior to the imaging sequence. The effect of magnetization transfer is clearly evident. The muscle and white matter are the areas most affected by the irradiation, with CSF, blood vessels, and fat showing little or no effect. Adapted from Wolff *et al.* (1991); images courtesy of Dr. R. S. Balaban.

applied in the time interval between the 180° inversion and 90° "read" pulse (see Cornell and Pope, 1974). For imaging experiments, the 90° "read" pulse is replaced by a conventional gradient-based imaging sequence. Preliminary studies with phantoms and *in vivo* mouse tumor studies are encouraging. Image contrast in such an experiment is expected to be dependent on both off-resonance and cross-relaxation effects, governed by the theoretical considerations elaborated in the preceding sections.

ACKNOWLEDGMENTS. The authors thank Drs. Robert Balaban and Robert Bryant for providing preprints in advance of publication. We also thank Dr. Arthur Stevens for incisive commentary and critical review, and James Loo for outstanding technical assistance. T.S. recalls an hour-long spirited conversation, which marked the genesis of Sections 6 and 7, with Bob Balaban at his 1989 Biophysical Society Meeting poster. The preparation of this chapter was supported by National Institutes of Health grant EY 04033.

REFERENCES

Abragam, A., 1961, *The Principles of Nuclear Magnetism*, Oxford University Press, London, p. 59.

Alger, J. R., and Shulman, R. G., 1984, *Q. Rev. Biophys.* **17**:83.

Andree, P. J., 1978, *J. Magn. Reson.* **29**:419.

Bain, A. D., Ho, W. P. Y., and Martin, J. S., 1981, *J. Magn. Reson.* **43**:328.

Bauer, D. R., Opella, S. J., Nelson, D. J., and Pecora, R., 1975, *J. Am. Chem. Soc.* **97**:2580.

Bax, A., 1983, *J. Magn. Reson.* **52**:76.

Beaulieu, C. F., Clark, J. I., Brown, R. D., Spiller, M., and Koenig, S. H., 1988, *Magn. Reson. Med.* **8**:45.

Beaulieu, C. F., Brown, R. D., III, Clark, J. I., Spiller, M., and Koenig, S. H., 1989, *Magn. Reson. Med.* **10**:362.

Bendel, P., and James, T. L., 1982, *J. Magn. Reson.* **48**:76.

Bettelheim, F. A., and Siew, E. L., 1983, *Biophys. J.* **41**:29.

Bittl, J. A., DeLayre, J., and Ingwall, J. S., 1987, *Biochemistry* **26**: 6083.

Blicharski, J. S., 1972, *Acta Phys. Pol.* **A41**:223.

Brauer, M., and Sykes, B. D., 1981, *Biochemistry* **20**:2060.

Brauer, M., and Sykes, B. D., 1987, *in Phosphorus NMR in Biology* (Burt, C. T., ed.), CRC Press, Boca Raton, Fla., p. 153.

Brindle, K. M., 1988, *Prog. Nucl. Magn. Reson. Spectrosc.* **20**:257.

Brooks, D. G., 1989, B.A. thesis, Concentration effects on protein rotational diffusion, University of California, Santa Cruz.

Brown, J. C., Pusey, P. N., Goodwin, J. W., and Ottewill, R. H., 1975, *J. Phys. A Math. Gen.* **8**:664.

Brown, T. R., 1980, *Philos. Trans. R. Soc. London Ser. B* **289**:441.

Bryant, R. G., and Shirley, W. M., 1980, *Biophys. J.* **32**:3.

Caines, G. H., and Schleich, T., 1991, *J. Magn. Reson.* **95**:457.

Caines, G. H., Schleich, T., Morgan, C. F., and Farnsworth, P. N., 1990, *Biochemistry* **29**:7547.

Caines, G. H., Schleich, T., and Rydzewski, J. M., 1991, *J. Magn. Reson.* **95**:558.

Christensen, M., and Jacobsen, J. P., 1988, *Magn. Reson. Med.* **7**:197.

Cornell, B. A., and Pope, J. M., 1974, *J. Magn. Reson.* **16**:172.

Daniel, C., and Woods, F. S., 1980, *Fitting Equations to Data*, Wiley, New York.

Dubin, S. B., Clark, N. A., and Benedek, G. B., 1971, *J. Chem. Phys.* **54**:5188.

Edzes, H. T., and Samulski, E. T., 1977, *Nature* **265**:521.

Edzes, H. T., and Samulski, E. T., 1978, *J. Magn. Reson.* **31**:207.

Eldar, H., and Degani, H., 1989, *Magn. Reson. Med.* **11**:121.

Eng, J., Ceckler, T. L., and Balaban, R. S., 1991, *Magn. Reson. Med.* **17**:304.

Forsen, S., and Hoffman, R. A., 1963a, *J. Chem. Phys.* **39**:2892.

Forsen, S., and Hoffman, R. A., 1963b, *Acta Chem. Scand.* **17**:1787.

Fung, B. M., and McGaughy, T. W., 1980, *J. Magn. Reson.* **39**:413.

Gore, J. C., Brown, M. S., Zhong, J., and Armitage, I. M., 1989, *J. Magn. Reson.* **83**:246.

Grad, J., and Bryant, R. G., 1990, *J. Magn. Reson.* **90**:1.

Grad, J., Mendelson, D., Hyder, F., and Bryant, R. G., 1990, *J. Magn. Reson.* **86**:416.

Grad, J., Mendelson, D., Hyder, F., and Bryant, R. G. 1991, *Magn. Reson. Med.* **17**:452.

Hallenga, K., and Koenig, S. H., 1976, *Biochemistry* **15**:4255.

Harding, J. J., and Crabbe, M. J. C., 1984, in *The Eye* (Davson, H., ed.), 3rd ed., Vol. 1b, Academic Press, New York, p. 244.

Hartzel, C. J., Whitefield, M., Oas, T. G., and Drobny, G. P. 1987, *J. Am. Chem. Soc.* **109**:5966.

Heatly, F., 1986, *Annu. Rep. NMR Spectrosc.* **17**:179.

Hull, W. E., and Sykes, B. D., 1975a, *J. Mol. Biol.* **98**:121.

Hull, W. E., and Sykes, B. D., 1975b, *J. Chem. Phys.* **63**:867.

Jacquinot, J. F., and Goldman, M., 1973, *Phys. Rev.* **B8**:1944.

James, T. L., 1980, *J. Magn. Reson.* **39**:141.

James, T. L., 1984, in *Phosphorus-31 NMR* (Gorenstein, D. G., ed.), Academic Press, New York, p. 349.

James, T. L., Matson, G. B., Kuntz, I. D., and Fisher, R. W., 1977, *J. Magn. Reson.* **28**:417.

James, T. L., Matson, G. B., and Kuntz, I. D., 1978, *J. Am. Chem. Soc.* **100**:3590.

James, T. L., Matthews, R., and Matson, G. B., 1979, *Biopolymers* **18**:1763.

Jones, G. P., 1966, *Phys. Rev.* **148**:332.

Kalk, A., and Berendsen, H. J. C., 1976, *J. Magn. Reson.* **24**:343.

Kimmich, R., Nusser, W., and Gneitting, T., 1990, *Colloids Surfaces* **45**:283.

Koenig, S. H., 1975, *Biopolymers* **14**:2421.

Koenig, S. H., 1980, in *Water in Polymers* (Rowland, S. P., ed.), American Chemical Society, Washington, D.C., p. 157.

Koenig, S. H., 1988, *Biophys. J.* **53**:91.

Koenig, S. H., and Brown, R. D., III, 1987, in *NMR Spectroscopy of Cells and Organisms* (Gupta, R. K., ed.), Vol. II, CRC Press, Boca Raton, Fla., p. 75.

Koenig, S. H., Bryant, R. G., Hallenga, K., and Jacob, G. S., 1978, *Biochemistry* **17**:4348.

Koenig, S. H., Brown, R. D., III, Adams, D., Emerson, D., and Harrison, G. G., 1984, *Invest. Radiol.* **2**:76.

Koenig, S. H., Beaulieu, C. F., Brown, R. D., III, and Spiller, M., 1990, *Biophys. J.* **57**:461.

Kumosinski, T. F., and Pessen, H., 1989, in *NMR in Agriculture* (Pfeffer, P., and Gerasimowicz, W. V., eds.), CRC Press, Boca Raton, Fla., p. 219.

Lerman, S., and Moran, M., 1989, *Exp. Eye Res.* **48**:454.

Lerman, S., Megaw, J., Gardner, K., Long, R., Ashley, D., and Goldstein, J. H., 1982, *Invest. Ophthalmol. Vis. Sci.* **23**:218.

Lester, C. C., and Bryant, R. G., 1991, *Magn. Reson. Med.* **22**:143.

Levy, R. M., Karplus, M., and McCammon, J. A., 1981, *J. Am. Chem. Soc.* **103**:994.

Lindstrom, T. R., and Koenig, S. H., 1974, *J. Magn. Reson.* **15**:344.

Lindstrom, T. R., Koenig, S. H., Boussios, T., and Bertles, J. F., 1976, *Biophys. J.* **16**:679.

McCain, D. C., 1987, in *Phosphorus NMR in Biology* (Burt, C. T., ed.), CRC Press, Boca Raton, Fla., p. 25.
McCain, D. C., and Markley, J. L., 1986, *J. Am. Chem. Soc.* **108**:4259.
Marshall, A. G., 1970, *J. Chem. Phys.* **52**:2527.
Morgan, C. F., Schleich, T., Caines, G. H., and Farnsworth, P. N., 1989, *Biochemistry* **28**:5065.
Morgan, C. F., Schleich, T., Caines, G. H., and Michael, D., 1990a, *Biopolymers* **29**:469.
Morgan, C. F., Schleich, T., and Caines, G. H., 1990b, *Biopolymers* **29**:501.
Oas, T. G., Hartzell, C. J., Dahlquist, F. W., and Drobny, G. P., 1987, *J. Am. Chem. Soc.* **109**:5962.
Pessen, H., and Kumosinski, T. F., 1985, *Methods Enzymol.* **117**:219.
Pope, J. M., Chandra, S., and Balfe, J. D., 1982, *Exp. Eye Res.* **34**:57.
Rácz, P., Tompa, K., and Pócsik, I., 1979, *Exp. Eye Res.* **28**:129.
Rácz, P., Tompa, K., Pócsik, I., and Bánki, P., 1983, *Exp. Eye Res.* **36**:663.
Raeymaekers, H. H., Eisendrath, H., Verbeken, A., van Haverbeke, Y., and Muller, R. N., 1989, *J. Magn. Reson.* **85**:421.
Rommel, E., and Kimmich, R., 1990, *Magn. Reson. Med.* **12**:390.
Rydzewski, J. M., and Schleich, T., 1992, to be published.
Rydzewski, J. M., Wang, X.-W., Stevens, A., Serdahl, C., and Schleich, T., 1992, *Exp. Eye Res.*, in press.
Schauer, G., Kimmich, R., and Nusser, W., 1988, *Biophys. J.* **53**:397.
Schleich, T., Morgan, C. F., and Caines, G. H., 1989, *Methods Enzymol.* **176**:386.
Schleich, T., Caines, G. H., and Rydzewski, J. M., 1990, pending patent application.
Seelig, J., 1978, *Biochim. Biophys. Acta* **515**:105.
Shirley, R. G., and Bryant, R. G., 1982, *J. Am. Chem. Soc.* **104**:2910.
Simpson, J. H., and Carr, H. Y., 1958, *Phys. Rev.* **111**:1201.
Sophianopoulos, A. J., and van Holde, K. E., 1964, *J. Biol. Chem.* **239**:2516.
Squire, P. G., and Himmel, M. E., 1979, *Arch. Biochem. Biophys.* **196**:165.
Stark, R. E., Jelinski, L. W., Ruben, D. J., Torchia, D. A., and Griffin, R. G., 1983, *J. Magn. Reson.* **55**:266.
Stevens, A., Caines, G. H., and Schleich, T., 1992, to be published.
Stoez, J. D., Redfield, A. G., and Malinowski, D., 1978, *FEBS Lett.* **91**:320.
Surewicz, W. K., and Mantsch, H. H., 1988, *Biochim. Biophys. Acta* **952**:115.
Sykes, B. D., Hull, W. E., and Snyder, G. H., 1978, *Biophys. J.* **21**:137.
Tanaka, T., Ishimoto, C., and Chylack, L. T., Jr., 1977, *Science* **197**:1010.
Wilbur, D. J., Norton, R. S., Clouse, A. O., Addleman, R., and Allerhand, A., 1976, *J. Am. Chem. Soc.* **98**:8250.
Wills, P. R., Nichol, L. W., and Siezen, R. J., 1980, *Biophys. Chem.* **11**:71.
Woessner, D. E., 1962, *J. Chem. Phys.* **37**:647.
Wolff, S. D., and Balaban, R. S., 1989, *Magn. Reson. Med.* **10**:135.
Wolff, S. D., Fralix, T. A., Simon, S. A., and Balaban, R. S., 1990, *9th Annu. Meet. Soc. Magn. Reson. Med. Abstr.* p. 350.
Wolff, S. D., Eng, J., and Balaban, R. S., 1991, *Radiology* **179**:133.
Zhong, J., Gore, J. C., and Armitage, I. M., 1989, *Magn. Reson. Med.* **11**:295.
Zhong, J., Gore, J. C., and Armitage, I. M., 1990, *Magn. Reson. Med.* **13**:192.

3

NMR Methods in Studies of Brain Ischemia

Lee-Hong Chang and Thomas L. James

1. INTRODUCTION

In vivo NMR spectroscopy has gained world-wide attention since the publication of the use of surface coils to noninvasively observe metabolic changes in whole animals (Ackerman *et al.*, 1980). It has provided an attractive method for continuously monitoring metabolic changes in short time intervals in the physiological and pathophysiological states of an organ or a tissue of an individual animal (Shulman, 1979; Gadian, 1982; James and Margulis, 1984; Radda, 1986; Cohen, 1987). Due to relatively low sensitivity, NMR methods detect only those metabolites whose concentrations are $\geqslant 1.0$ mM and that are not tightly bound to macromolecules or immobile cellular structures. A large number of NMR-observable nuclei have been used as naturally occurring or artificial probes for *in vivo* studies.

Ischemia is defined as the reduction of blood flow going into a tissue. It is often accomplished experimentally by occluding blood vessels, therefore depriving that tissue of oxygen and nutrients. Inducing brain ischemia in animals produces experimental models that simulate stroke, thus rendering feasible studies of the mechanism of that pathological state. Oxygen deprivation in brain tissue leads to a cascade of metabolic events that characterize ischemic brain injury (Siesjö and Wieloch, 1985). Several major

---

Lee-Hong Chang • Department of Pharmaceutical Chemistry, University of California, San Francisco, California 94143. **Thomas L. James** • Department of Pharmaceutical Chemistry and Department of Radiology, University of California, San Francisco, California 94143.

Biological Magnetic Resonance, Volume 11: In Vivo Spectroscopy, edited by Lawrence J. Berliner and Jacques Reuben. Plenum Press, New York, 1992.

metabolic changes have been identified, including: (1) phosphorus energy metabolite levels, (2) lactate content, (3) intracellular pH, (4) water homeostasis, (5) electrolyte homeostasis, and (6) free fatty acid content. Neurotransmitters or neuromodulators, such as glutamate, have recently been implicated to mediate neurotoxicity in ischemic brain injury (Choi and Rothman, 1990). Development of NMR methodologies has facilitated the monitoring of these effects noninvasively and sequentially. For example, ^{31}P NMR spectroscopy monitors changes in phosphorus energy metabolites and intracellular pH (Moon and Richards, 1973). ^{1}H NMR spectroscopy detects changes in lactate, glutamate, and amino acid levels (Behar *et al.*, 1984; Rothman *et al.*, 1984b; Gadian *et al.*, 1986; Change *et al.*, 1987b). ^{1}H NMR imaging follows the development of brain edema (i.e., increase in water content) (Brant-Zawadzki *et al.*, 1984; Moseley *et al.*, 1990). Metal cation NMR (e.g., ^{23}Na and ^{39}K) distinguishes between intracellular and extracellular compartmentation of cations such as Na^+ and K^+ (Springer, 1987). Indirect measurement of Mg^{2+} (Gupta and Moore, 1980; Gupta and Yushok, 1980) and Ca^{2+} (Smith *et al.*, 1983) via chelating agents has also been reported. ^{13}C NMR has been used to observe free fatty acid content (Moonen *et al.*, 1985). A variety of spectroscopic and imaging techniques have been shown to be capable of measuring cerebral blood flow (Ackerman *et al.*, 1987; Müller and Seelig, 1987; Eleff *et al.*, 1988; Le Bihan *et al.*, 1988; Villringer *et al.*, 1988; Ewing *et al.*, 1989).

In this chapter, we will focus our discussion on our approach and experience with *in vivo* ^{1}H and ^{31}P NMR spectroscopy and spatially localized spectroscopy in brain ischemia studies. Our discussion will emphasize the technical aspects of *in vivo* NMR spectroscopy rather than mechanistic insights into tissue metabolism obtained by *in vivo* NMR spectroscopy. Several recent reviews have appeared in the literature on similar subjects (Avison *et al.*, 1986; Prichard and Shulman, 1986; Williams and Gadian, 1986; Matson and Weiner, 1988; Pettegrew, 1990).

2. ANIMAL PREPARATION AND PHYSIOLOGICAL MONITORING

Many experimental models have been developed in numerous species of animals for studies of ischemic brain injury. They can be either global or focal (regional), complete or incomplete, permanent or transient (Garcia, 1984). The choice of a suitable model for ischemic study has not always been easy, because many of these models give variable results, probably reflecting the physiological variability between individual animals. Choice of a model also depends on the objective of the experiment. The variability is multifactorial and complicated (Graham, 1988); therefore it is very important to perform the experiment in a rigorously controlled physiological state, so the results can be more reproducible and better interpreted.

In our lab, a horizontal bore magnet has often been used, since it allows the animals to be placed in a comfortable, natural position (Gonzalez-Mendez *et al.*, 1984). For various reasons, vertical bore magnets have also been used and shown not to have adverse effects on the physiological condition of the animals. In most cases here, animals are anesthetized and ventilated with 1–1.5% isoflurane in 70% N_2O/30% O_2 mixture, and paralyzed with pancuronium (1 mg/kg body wt) to restrain the animals from jumping out of the magnet. As there is evidence that different anesthetic agents exert varying degrees of protection against brain injury (Warner, 1989), it is important to report the anesthesia protocol used. Barbiturates, reported to have the most protective effects (Karlsson and Steen, 1989), should be used with care in brain injury studies. Arterial pressure, blood pH, blood gases, and blood glucose are monitored through a femoral arterial catheter and controlled via ventilation. Although surface coils give some spatial localization, it is often useful to remove scalp, to retract temporalis muscles in order to avoid ^{31}P and ^{1}H signal contamination from these tissues, and to reflect periosteum (a fatty membrane on top of the skull) to reduce lipid contamination in ^{1}H NMR spectroscopy (Chang *et al.*, 1987b). These surgical procedures always produce a pool of body fluid covering the exposed skull. Care must be taken to stop surrounding tissues from exuding fluids near the surface coil. We have found that electrocoagulation and application of a small amount of SuperGlue help. A well-groomed, dry preparation aids shimming of the B_0 magnetic field homogeneity and gives well-resolved resonance signals.

Body temperature can be maintained with a plastic water jacket. Rectal temperature is monitored by a digital thermometer through a thin, nonmagnetic thermocouple probe (e.g., Omega Engineering, Inc., Type E). If desired, an electroencephalogram (EEG) can be monitored through very thin copper leads glued to the temporal region of the exposed skull a good distance (roughly 10 mm) away from the surface coil, to avoid interference with NMR signal detection. Whenever these temperature probes or EEG leads are used, they must be properly grounded, braided, and shielded so they do not interfere with NMR detection.

3. SURFACE COIL PROBE DESIGN

3.1. The Design of a Surface Coil

A surface coil in its simplest form is a single-turn flat circular coil (Ackerman *et al.*, 1980), which can be regarded as a single-turn solenoid coil. For studies of living systems, the surface coils are placed directly in front of the region of interest (ROI) thus rendering the observation of the NMR signal to a localized volume near the surface of an organ or a tissue. The surface coil functions both as the transmitter that delivers the radiofrequency (RF) energy (B_1) to the sample and as the receiver that detects a minute NMR signal from the sample. It also provides the necessary orthogonal configuration of B_0 and B_1, good filling factor, and excellent sensitivity (Gadian, 1982). One of the unique features of the surface coil is its B_1 excitation profile (Fig. 1), which is extremely inhomogeneous, being approximately crescent-shaped in a volume roughly defined by the coil radius (the sensitive volume). The B_1 field gradient produced by the surface coil can be exploited in applications to rotating frame zeugmatography (Hoult, 1979), localized spectroscopy (Garwood *et al.*, 1984), and other techniques that require a B_1 field gradient (Karczmar *et al.*, 1989). On the other hand, since the signal generated by different parts of the sample is not uniform, sensitivity varies according to spatial distribution, making quantitation of the metabolites within the sample difficult. However, there are two ways to tackle the B_1 field inhomogeneity problem. First, one can use RF pulses that are relatively insensitive to B_1 field inhomogeneity such as elaborated composite pulses (Tycko *et al.*, 1985; Hetherington *et al.*, 1986) and adiabatic fast-passage pulses (Silver *et al.*, 1984; Bendall and Pegg, 1986; Hardy *et al.*, 1986; Ugurbil *et al.*, 1988). Second, one can use a large coil for excitation and a smaller coil for detection together with suitable coil decoupling schemes (Bendall *et al.*, 1986; Edelstein *et al.*, 1986).

In practice, an NMR probe consists of the surface coil itself plus a tuning and matching circuit. The surface coil is usually made of AWG #16 to #20 solid copper wire encased in heat-shrinkable tubing. The coil can be molded to various shapes, e.g., circular, elliptical, square, or concentric,

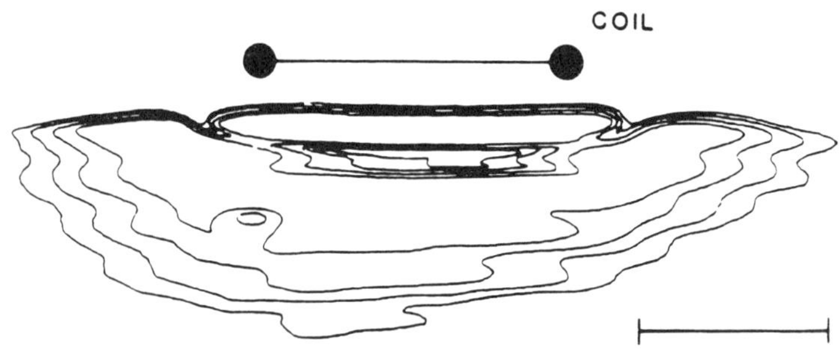

Figure 1. Two-dimensional, cross-sectional image of the B_1 excitation profile measured in a homogeneous aqueous solution beneath a surface coil. The contours define regions of magnetic field strength. A thin high-magnetic-flux region can be seen between the coil and the crescent-shaped sensitive volume. (Adapted from Gonzalez-Mendez *et al.*, 1985.)

to fit the shape of the organ under study. Higher-order configurations have also been devised (Hyde *et al.*, 1986; Roemer *et al.*, 1990). Thin copper tape (Balaban *et al.*, 1986; Edelstein *et al.*, 1986) or copper tubing (Bottomley *et al.*, 1989) have also been used. The basic probe tuning circuit (Hoult and Richards, 1976; Fukushima and Roeder, 1981), shown in Fig. 2A, usually consists of a tuning capacitor (C_1) in parallel with the surface coil and matching capacitor (C_2) in series with the surface coil. These high-Q (quality factor), nonmagnetic, tunable air capacitors (Johanson, Boonton, N.J. 07005) are suitable for routine, low-power spectroscopic applications. The probe should be tuned to the resonance frequency of the observed nucleus and matched to the impedance (50 Ω) of the spectrometer to ensure maximum power transmission. This can be accomplished by repeatedly adjusting the two capacitors using the reflected power detected from a simple setup consisting of a Wavetek sweep generator and a Tektronix oscilloscope. The Q, quality factor, of the coil can then be estimated by (Gadian, 1982; Murphy-Boesch, 1984; Chang *et al.*, 1987a):

$$Q = 2 \times (\nu / \Delta\nu_{0.71}) \tag{1}$$

where ν is the resonance frequency and $\Delta\nu_{0.71}$ is the full width at 0.71 peak height (half-power level).

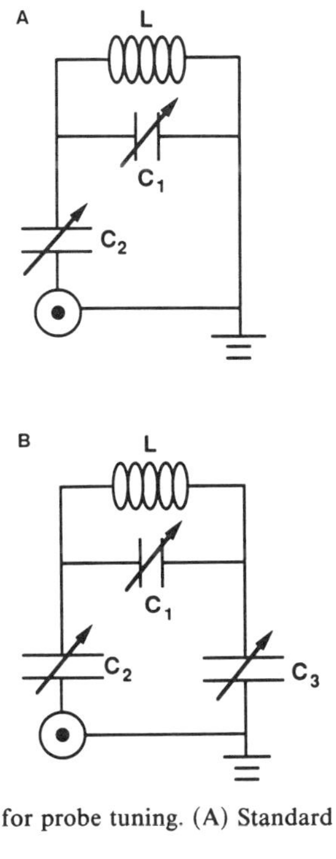

Figure 2. Circuit diagram for probe tuning. (A) Standard and (B) balanced-matched.

3.2. The Balanced–Matched Probe Circuit

Biological tissues have an intrinsically high concentration of ions, and consequently high conductivity. They can produce a significant amount of noise and large losses in the NMR probe, thereby reducing signal-to-noise ratios (S/N) (Hoult and Lauterbur, 1979; Gadian and Robinson, 1979). These losses of sensitivity are primarily caused by the currents induced in the tissue by the magnetic and electric fields generated by the coil. There is no advantage in reducing the magnetic losses by reducing the magnetic field, since S/N is linearly proportional to magnetic field for *in vivo* NMR studies (Hoult and Lauterbur, 1979). However, it is advantageous to reduce dielectric losses produced by the electric field. One procedure that has been successfully used is to place a properly grounded Faraday shield (e.g., a thin copper sheet) between the sample and coil (Hoult and Lauterbur, 1979; Gadian and Robinson, 1979). However, this is often cumbersome to implement and does not gain much in S/N (Gadian, 1982). The other measure is to use a two-turn surface coil in combination with a so-called balanced-matched tuning circuit (Murphy-Boesch and Koretsky, 1983; Murphy-Boesch, 1984). In the balanced-matched tuning scheme (Fig. 2B), an extra matching capacitor (C_3) is added to the standard circuit (Fig. 2A) to balance the coil with respect to ground, thus reducing the losses incurred by loading the coil with a conductive sample (Murphy-Boesch and Koretsky, 1983). In a tuned circuit, the two matching capacitors (C_2 and C_3) are approximately of equal value and twice as large as the single matching capacitor of the standard circuit.

3.3. The Balanced–Matched Double-Tuned Probe Circuit

In a pulsed Fourier transform (FT) NMR experiment, much time is lost while waiting for the magnetized nuclear spins to relax to equilibrium before they can be excited again. A more efficient way of performing the experiment is to observe two or more types of nuclear spins (e.g., ^{1}H and ^{31}P) in a time-shared fashion (Schnall *et al.*, 1985; Chang *et al.*, 1987a). This would enable one to monitor both nuclei simultaneously so that changes in intracellular pH, high-energy phosphates, and lactate levels could be better correlated. Similarly, other nuclei (see Introduction) could also be concurrently monitored to provide a more complete metabolic picture of the tissue than could be accomplished using a single-tuned coil. Thus, the efficiency of the scan process is greatly increased by increasing the amount of information obtained per unit time from the sample. Another potential use for a double-tuned coil would be for ^{13}C spectroscopy with simultaneous ^{1}H decoupling (or ^{1}H observation with ^{13}C decoupling), so that metabolic pathways of ^{13}C-labeled substrates could be followed (Stronski *et al.*, 1986).

A variety of double-tuned probe circuits used in *in vivo* spectroscopy have been published (Schnall *et al.*, 1985; Chang *et al.*, 1987a; Gonnella and Silverman, 1989). Many of these circuits incorporate a large number of lumped elements, making them complicated and cumbersome to build and use. The efficiency and sensitivity of the probes are also degraded. We have routinely used a balanced-matched, double-tuned probe that is simple to construct and has high sensitivity (Chang *et al.*, 1987a). The circuit diagram of the probe is depicted in Fig. 3A. The sample coil, which can be a surface coil, has an inductance L_1. C_1, C_2, C_3 and C_4 are tunable capacitors, and L_2 is an inductor. Disregarding the trap circuit elements L_2 and C_2 for the moment, the circuit is a balanced-matched single-tuned circuit (compare to Fig. 2B). The introduction of the trap circuit with elements L_2 and C_2 makes a coil resonate at two different frequencies (Schnall *et al.*, 1985). The inductor with inductance L_2 is optimized for the desired frequencies (e.g., for ^{1}H-^{31}P, ^{1}H-^{23}Na). C_1 tunes and C_3 matches the lower resonance; C_2 tunes and C_4 matches the higher resonance. C_3 and C_4 are the two balanced-matched capacitors that allow us to tune and match both frequencies better and also reduce the dielectric losses when the coil is loaded with a biological sample. In our experience, an extra ceramic chip capacitor C_5 (Dielectric Laboratories Inc., Cazenovia, N.Y. 13035), which is about 10 pF (depending on the size of the coil and the frequencies), placed near and in parallel with the sample coil helps tuning and matching (Fig. 3B). This probe is a single-port design that gives good sensitivity but suffers from poor isolation between the two channels. It is also possible to break the two frequencies into two different channels (i.e., make it a double-port design) with better isolation between the two channels for studies that require simultaneous irradiation at one frequency and detection at another frequency (e.g., ^{13}C detection with ^{1}H decoupling). This design is illustrated in Fig. 3C. Although

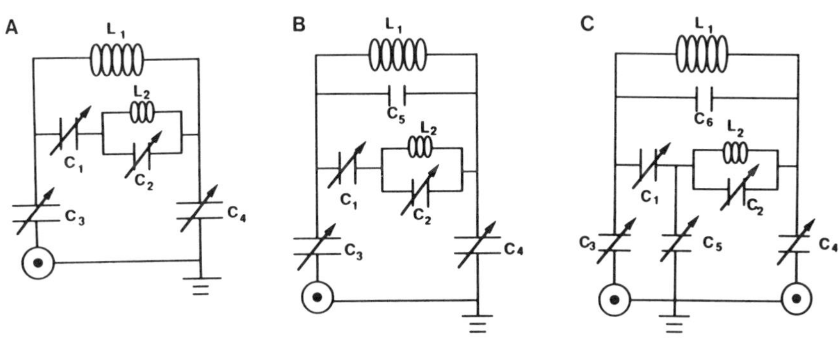

Figure 3. Circuit diagram for double-tuned probe tuning. (A) Single-port, balanced-matched; (B) a modification of A; and (C) double-port, balanced-matched.

double-tuned probes allow one to observe signals from two frequencies at the same time, the sensitivity of each frequency is not as good as that from probes single-tuned to the two respective frequencies. Multiple-tuned probe circuits have also been described (Schnall *et al.*, 1988; Eleff *et al.*, 1988).

4. ^{31}P NMR SPECTROSCOPY

4.1. What Can Be Observed?

^{31}P has to date been the most popular nucleus for *in vivo* NMR studies of tissue metabolism. It is 100% naturally abundant and has a sensitivity roughly 1/15th that of ^{1}H. The chemical shift range of tissue metabolites is approximately 30 ppm. Surprisingly, it is possible to obtain spectra containing relatively few but very important intermediates of energy metabolism with very good S/N in a few minutes. Although sensitivity is linearly proportional to magnetic field, there is probably no advantage in going to ultrahigh field for ^{31}P due to linewidth increases from chemical shift heterogeneity and from chemical shift anisotropy relaxation effects. The spin–lattice and spin–spin relaxation times of phosphorus metabolites from several different tissues have been examined *in vivo* at several different magnetic field strengths (Evelhoch *et al.*, 1985).

A partially saturated ^{31}P spectrum obtained from a rat brain at 5.6 T using a one-pulse experiment is shown in Fig. 4A. Several resonance peaks are discernible. Three resonances arise from the α, β, and γ phosphates of ATP (and any other nucleoside triphosphates). The α- and γ-phosphate resonances also contain contributions from the α- and β-phosphate resonances, respectively, of any nucleoside diphosphates. The α-phosphate peak of ATP (second peak from the right) is further overlapped by NAD/NADP signals. The β-phosphate resonance of ATP, relatively free from contaminating resonances, is used to quantitate changes in ATP concentration. A distinctly sharp phosphocreatine (PCr) peak is observed with a PCr/ATP ratio of 1.5–1.7 in an uncontaminated (without signals from facial muscles), fully relaxed, brain ^{31}P spectrum. The inorganic phosphate (P_i) peak has a highly pH-dependent chemical shift, enabling intracellular pH measurements (Moon and Richards, 1973). The phosphomonoester (PME) peak is poorly resolved and may contain signals from AMP, sugar phosphates (glucose-6-phosphate, fructose-6-phosphate), phosphocholine, and phosphoethanolamine. The phosphodiester (PDE) peak is also poorly resolved, possibly containing glycerophosphorylcholine and glycerophosphorylethanolamine. Further, it is quite magnetic field-dependent (Murphy *et al.*, 1989) and has been identified as primarily relatively immobile phospholipid bilayer with a small contribution from a motionally averaged

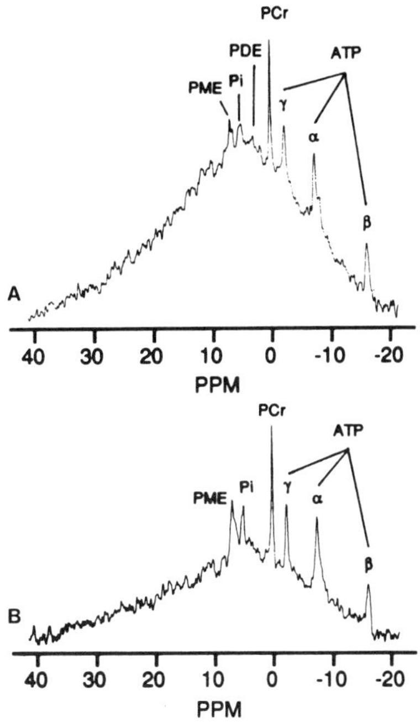

Figure 4. ^{31}P NMR spectra from a rat brain *in vivo* at 5.6 T. (A) Spectrum obtained by one-pulse experiment showing large broad component, and (B) spectrum obtained by selective off-resonance irradiation exhibiting reduced broad components. Signal assignments are as follows: PME, phosphomonoesters; P_i, inorganic phosphate; PDE, phosphodiesters; PCr, phosphocreatine; α-, β-, γ-ATP, respectively, α-, β-, γ-phosphates of adenosine triphosphate.

macromolecule(s). At low magnetic field, brain spectra show large PDE peak, but at high field (i.e., >4 T), this PDE signal is not apparent due to significant chemical shift anisotropy effects. Finally, a large broad component lies beneath the whole spectrum and has been attributed to bone (Ackerman *et al.*, 1980) and immobile membrane phospholipids (Gonzalez-Mendez *et al.*, 1984).

4.2. Quantitation of ^{31}P Metabolites

In an NMR spectrum, resonance peak areas reflect the metabolite concentrations. Since a spectrum with a flat baseline helps the quantitation of peak areas, several methods have been attempted to remove the broad component in the brain ^{31}P spectrum. One of these methods is the so-called convolution difference (Ackerman *et al.*, 1980; Campbell *et al.*, 1973) in which the free induction decay (FID) is multiplied by $[\exp(-t/T_A) - K\exp(-t/T_B)]$ to enhance resolution and eliminate broad components, where T_A is a time constant that would give a line broadening of 10–30 Hz, T_B is another time constant that would give a line broadening of 400–600 Hz which simulates the broad component, and K is a constant ranging from 0.8 to 1.0. Another technique is to convert the intrinsic NMR Lorentzian lineshape to Gaussian lineshape (Ernst, 1966) so that resolution is enhanced and the broad component is eliminated (Moseley *et al.*, 1990). This is accomplished by multiplying the FID by $\exp(-at - bt^2)$ where a is negative and b positive. The FID has originally the Lorentzian form of $\exp(-t/T_2^*)$ so that if $a = -t/T_2^*$, a Gaussian lineshape with $\exp(-bt^2)$ will result. Since a Gaussian has much less extensive wings as compared to a Lorentzian lineshape, resolution is enhanced; proper choice of b can lead to a further reduction in linewidth at half-height. Third, the broad component can also be removed by selective off-resonance irradiation (Gonzalez-Mendez *et al.*, 1984), e.g., at a frequency 20 ppm downfield from the PCr peak, which does not affect the intensities of the mobile phosphorus metabolite resonances (Fig. 4B) while largely eliminating resonances from relatively immobile phosphorus nuclei (James *et al.*, 1978; James and Matson, 1979). We routinely use the off-resonance irradiation method since the digital manipulation methods necessarily generate some distortion of the spectral lines, and great care must be taken when interpreting the resulting signal intensities (Gadian, 1982; Gonzalez-Mendez *et al.*, 1984). The peak areas are usually obtained by fitting the peaks with a curve-fitting routine to separate overlapping peaks (e.g., GE/Nicolet curve analysis program or NMR1 of New Methods Research, Inc.). Since *in vivo* resonance peaks have a Gaussian contribution to the lineshape due to inhomogeneous line broadening, peaks are usually curve-fitted with a mixed Lorentzian/Gaussian function.

In principle, NMR signal intensity, i.e., the area under the resonance peak, is directly proportional to the concentration of a metabolite. The issue of determination of absolute concentrations has often been avoided despite its importance in comparisons of normal and pathophysiological states of an experimental animal model or in diagnosis of human disease; rather, relative changes in peak intensities are often utilized. Values of molar concentration provide more information than ratios of peak areas or heights in many applications, e.g., for determination of rate constants or free energy of hydrolysis. Accurate measurements of the absolute concentrations of tissue metabolites require detailed knowledge of the volume of sample from which the signals are detected and the relative coil sensitivity over that volume (sensitive volume). Quantitation of metabolites by *in vivo* NMR spectroscopy is further complicated by B_1 inhomogeneity over the sensitive volume of the RF coil because of coil geometry, especially for surface coils in which the sensitive volume of the coil becomes ill-defined. Despite its difficulties, several intriguing protocols have been proposed. For example, water signal intensity has been used as an internal standard for obtaining the concentrations of phosphorus metabolites (Thulborn and Ackerman, 1983). In other cases, external phosphorus references have been used in combination with relatively homogeneous, large volume coils together with an appropriate B_1 map of the sensitive volume to quantitate localized phosphorus metabolites (Bottomley *et al.*, 1988; Roth *et al.*, 1989).

4.3. Measurement of Intracellular pH

In 1973, it was found that ^{31}P NMR spectroscopy could be used to measure intracellular pH (pH_i) (Moon and Richards, 1973). It has been widely used since then. Most studies make use of the pH dependence of the P_i chemical shift, together with the assumption that the P_i signal is intracellular in origin. The pH_i can be estimated by measuring the chemical shift difference between the P_i and PCr resonances and converting it to a pH value via:

$$\mathrm{pH} = \mathrm{p}K_a + \log[(\delta - \delta_A)/(\delta_B - \delta)] \tag{2}$$

where $\mathrm{p}K_a = -\log K_a$, K_a is the proton dissociation constant for the monoanionic species, δ is the chemical shift difference between P_i and PCr peaks, δ_A and δ_B are the chemical shifts of the diprotonated and monoprotonated phosphate species, respectively. A suitable titration curve of P_i in a solution of proper ionic strength and counterion concentrations can be used to obtain $\mathrm{p}K_a$, δ_A, and δ_B. We have used 6.77, 3.27, and 5.78 for $\mathrm{p}K_a$, δ_A, and δ_B, respectively (Seo *et al.*, 1983). Slightly different values have been used by other groups (e.g., Bailey *et al.*, 1981; Petroff *et al.*, 1985a; Taylor *et al.*, 1983). The equation is most accurate in the pH range of 6.1 to 7.5. Aside from P_i being essentially 100% monoanionic below 6.0, rendering the calculation inaccurate, the chemical shift of PCr is also greatly influenced by acidic pH. The P_i calibration curve is also affected by ionic strength, temperature, chemical shift reference, and cations, especially magnesium. Several other pH-dependent phosphorus metabolites have also been used as alternatives including 2,3-diphosphoglycerate (Moon and Richards, 1973), fructose-1-phosphate (Cohen *et al.*, 1978), glucose-6-phosphate (Navon *et al.*, 1979), 2-deoxyglucose-6-phosphate (Navon *et al.*, 1977), ATP (Pettegrew *et al.*, 1988), and phosphoethanolamine (Corbett *et al.*, 1988). Several ^{19}F-labeled compounds have also been introduced as possible pH indicators (Deutch and Taylor, 1987). ^{1}H NMR spectroscopy has also been applied to determine pH_i by observing the imidazole protons of histidine (Rabenstein and Isab, 1982), or the histidine residue of carnosine (Yoshizaki *et al.*, 1981; Pan *et al.*, 1988) or anserine (Williams *et al.*, 1985).

5. ^{1}H NMR SPECTROSCOPY

5.1. ^{1}H NMR Resonances *in Vivo*

^{1}H is the most sensitive of the naturally occurring, abundant nuclei in biological systems. However, the combination of a small chemical shift range of roughly 15 ppm in which only about 5 ppm contains signals from tissue metabolites, the presence of a large number of proton-containing compounds in the cell, and the scalar coupling of the ^1H resonances gives rise to complicated spectra of overlapping broad peaks. Furthermore, there is the dynamic range problem of detecting millimolar signals in the presence of the intense H_2O signal (approximately 80 M proton concentration in most tissues) and significant, broad lipid signals. Water suppression techniques to overcome the dynamic range problem are necessary to observe ^1H metabolites of less than 10 mM concentration. However, the intense H_2O signal is advantageous for magnetic field shimming, enabling adjustment of room temperature shim currents in the presence of the biological sample. *In vivo* NMR spectroscopy has successfully been used to study lactate, amino acids, and other compounds of intermediary metabolism.

Suppression of the H_2O signal in FT NMR spectroscopy is complicated (Hore, 1989; Meier and Marshall, 1990). A Hahn spin-echo pulse sequence (90°–τ–180°–τ–ACQ, where τ is the echo delay time) was used in the early 1980s to observe small mobile metabolites in the cells (Agris and Campbell, 1982). The H_2O signal was sufficiently suppressed in this study because of the relatively long T_1 and short T_2 relaxation times of the water protons at very high magnetic field (11 T). A combination of presaturation of the H_2O resonance with a Hahn spin-echo sequence (presaturation 90°–τ–180°–τ–ACQ) has been used to detect metabolites in the brain using a surface coil (Behar *et al.*, 1984). Since a surface coil has been used, the "90°" pulse is inhomogeneous, having a distribution of pulse angles centered about 90°; consequently, the 90° pulse is sometimes designated as a θ pulse and the 180° pulse as a 2θ pulse. The significant lipid signals arising from the scalp can be eliminated by surgically removing the scalp from the sensitive volume of the surface coil. The use of the spin-echo pulse sequence also minimizes broad resonances in the spectrum from protons with short T_2, such as lipids. A more complete elimination of lipid signal can be achieved by further removing the periosteum (a fatty membrane surrounding the cranium) (Chang *et al.*, 1990b).

5.2. *In Vivo* ^1H NMR Pulse Sequences

In addition, lipid signals arising from intact scalp can be eliminated by editing techniques, such as homonuclear ^1H double resonance difference spectroscopy (Rothman *et al.*, 1984a, b). With the spin-echo sequence, multiplet resonances which arise from spin–spin (or J) coupling are phase modulated relative to a singlet (noncoupled) resonance (Campbell *et al.*, 1974). A delay τ is chosen ($1/2J$ for doublets and quartets, $1/4J$ for triplets) so that the selected multiplet is 180° out of phase (inverted) relative to a singlet. The phase modulation can be removed by selective decoupling of the resonance to which the multiplet resonance is coupled (Brown *et al.*, 1977). Therefore, in the presence of decoupling, the selected multiplet is in phase with the singlet. If scans from the pulse sequence without decoupling are subtracted from those obtained with decoupling, one obtains a spectrum containing only the selected multiplet with elimination of all noncoupled resonances (singlets). The method is most useful in editing the lactate β-CH_3 doublet which is due to coupling to the α-CH. Choice of the optimum echo delay in the spin–echo sequence needs special consideration. The coupling constant between the methyl and methine protons is 7.0 Hz (Chang and Cohen, unpublished data; Bolinger *et al.*, 1987). Accordingly, the echo delay ($1/2J$) should be 71.4 msec. However, it has been shown that the optimum delay will be less than $1/2J$ when the linewidths are large, due to short T_2, relative to the coupling constants, which is often encountered *in vivo* (Lohman and Bulthuis, 1985). Several different values for τ have appeared in the literature: mostly 68 msec or its multiples (Rothman *et al.*, 1984b; Hetherington *et al.*, 1985; Williams *et al.*, 1986; Cohen *et al.*, 1990a), 64 msec (Moseley *et al.*, 1990), and 60 msec (Campbell *et al.*, 1974; Behar *et al.*, 1984; Chang *et al.*, 1987b). It is also possible to eliminate lipid signals without removing scalp by extending the echo delay to 204 (3×68) msec at the expense of lower S/N (Cohen *et al.*, 1990a). The ^{1}H resonance peaks that can be observed from a rat brain at 4.7 T by using a presaturation spin–echo sequence with a τ of 68 msec are shown in Fig. 5. The pulse width that gives the largest tissue water signal intensity is used for the 90° pulse. Chemical shift scale is referenced to the methyl resonance of the N-acetyl moiety of N-acetylaspartate (NAA) at 2.02 ppm. The most prominent peaks are N-(C=O)CH_3 of N-acetylaspartate, γ-CH_2 from glutamate and glutamine (2.34 ppm), N-CH_3 of phosphocreatine and creatine (3.03 ppm), N-$(CH_3)_3$ of choline (3.21 ppm) and choline-containing species, and β-CH_3 of lactate (1.33 ppm), inverted due to phase modulation. The

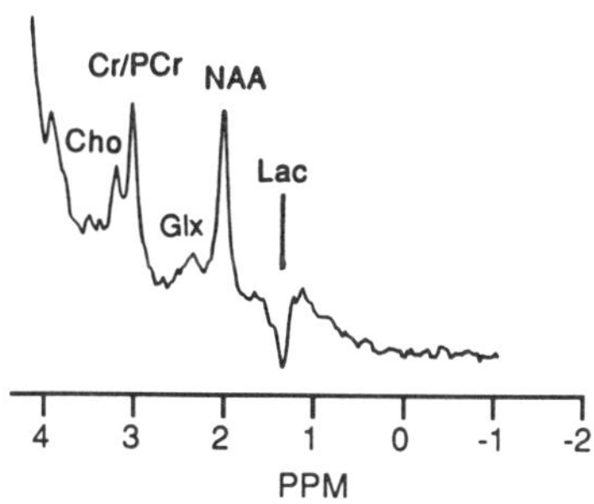

Figure 5. ^{1}H NMR spectrum of a rat brain *in vivo* with regional ischemia at 4.7 T. Signal assignments are as follows: Cho, choline; Cr/PCr, creatine/phosphocreatine; Glx, glutamate and glutamine; NAA, N-acetylaspartate; Lac, lactate.

levels of all of these metabolites may be altered in a variety of diseased states, such as stroke or tumor.

Binomial water suppression pulses (Hore, 1983) have also been implemented in the spin-echo sequence for spectral editing of *in vivo* metabolite signals, e.g., $1\bar{3}3\bar{1}$-τ-$2\bar{6}6\bar{2}$-τ-ACQ (Hetherington *et al.*, 1985). It has been noted that this sequence does not produce an inverted lactate -CH_3 signal because its coupling partner, the lactate -CH resonance, does not experience a 180° excitation; this has the same effect as selective decoupling (Brindle *et al.*, 1984). In another example, two alternate pulses are used: $1\bar{3}3\bar{1}$-τ-180°-τ-ACQ and $1\bar{3}3\bar{1}$-τ-$2\bar{6}6\bar{2}$-τ-ACQ (Williams *et al.*, 1986). The difference between alternate scans gives an edited spectrum with only the selected signal. It should be noted that the frequency response of the binomial pulse sequences in the presence of the inhomogeneous B_1 field produced by a surface coil is strongly dependent on the spatial localization of the sample (Blondet *et al.*, 1987). In addition, phase anomalies have been observed in a spin-echo experiment when the solvent nuclear magnetization is perturbed by the unbalanced 180° refocusing pulses (Edzes *et al.*, 1990). An ^{1}H-observe/^{13}C-decouple spectral editing technique has also been reported (Rothman *et al.*, 1985). A pulse sequence which uses a selective homonuclear polarization transfer technique for spectral editing has been presented (Hardy and Dumoulin, 1987). Multiple quantum spectroscopy has also been applied in spectral editing, e.g., the use of zero quantum coherence transfer (Sotak and Freeman, 1988) or double quantum coherence transfer (Sotak *et al.*, 1988; Doddrell *et al.*, 1989; Knüttel and Kimich, 1989; Nosel *et al.*, 1989). Two-dimensional correlation spectroscopy (COSY) (Berkowitz *et al.*, 1988; Cohen *et al.*, 1989) and one-dimensional soft COSY (Stryjewski *et al.*, 1990) have also been used *in vivo.*

5.3. Quantitation of ^{1}H Metabolites

The most distinctive peak in the ^{1}H spectrum of a rat brain is the methyl resonance of the N-acetyl moiety of *N*-acetylaspartate. NAA is found only within the nervous system, and its biological function is largely speculative (Birken and Oldenforf, 1989). However, it has been convenient to use NAA as an internal standard in an ^{1}H spectrum for observing changes of other metabolites, such as lactate via lactate/NAA ratios (Petroff *et al.*, 1985b; Crockard *et al.*, 1987).

With assumptions to be discussed, the absolute lactate concentration (μmole/g wet wt brain tissue) can also be estimated by (Chang *et al.*, 1987b, 1990b):

$$[\text{Lactate}] = F \times \frac{\text{peak area of lactate}}{\text{peak area of NAA}} \tag{3}$$

where F is a correction factor derived from correlating the peak areas determined by *in vivo* NMR spectroscopy with the results of *in vitro* chemical assays of brain extracts. The correction factor is needed to compensate for the use of a surface coil, B_1 inhomogeneity, pulse sequence, differential T_1 and T_2 effects on NAA and lactate, as well as other unknown factors. For the coil and pulse sequence utilized in our lab, $F = 19.6$. The equation is valid provided that NAA distributes homogeneously in the sensitive volume and remains constant throughout the distressed physiological states. A recent report comparing *in vivo* NMR data with tissue extract data demonstrated that the concentration of NAA in rabbit brain is stable even 1 hr postmortem (Petroff *et al.*, 1988). It has also been found by localized spectroscopy that certain human brain infarcts do not contain NAA (Bruhn *et al.*, 1989). This observation is apparently due to the fact that the human lesions studied were several days old. In our experience, during the first 4 hr following ischemia, the NAA concentration remains constant. It has also been shown that the NAA concentrations are different in the gray and white matter of human brain (Frahm *et al.*, 1989). Most studies from this lab utilize a coil which observes signal emanating from nearly the entire rat cerebrum; so the critical issue is whether or not the total NAA concentration is unchanged and whether or not the purported NAA signal indeed is due solely to NAA. It has been indicated that the peak at 2.02 ppm does not arise from NAA only; instead, other *N*-acetyl-containing species, such as sialac acid-containing glycolipids and glycoproteins in cell membrane, may contribute to *in vivo* NAA signal (Gyulai *et al.*, 1987; Ligeti *et al.*, 1987). It should also be noted that the NAA observed *in vivo* is "free," i.e., not tightly bound to give short T_2. However, the distribution of NAA in the rat cerebrum is fairly homogeneous (Birken and Oldenforf, 1989), since the distinction between the gray and white matter is not clearly defined in the rat cerebrum. Alternatively, as long as the total pool of phosphocreatine and creatine remains constant, the peak at 3.03 ppm, which somewhat overlaps the choline methyl resonance, has been used as internal standard (Pan *et al.*, 1988; Williams *et al.*, 1988).

The correction factor F should be determined for each surface coil and each pulse sequence used. Because of the inhomogeneous B_1 field produced by a surface coil, the detection efficiency of the lactate signal by *in vivo* NMR spectroscopy is less than optimal, especially with the spin–echo pulse sequence employing an inhomogeneous 180° refocusing pulse. More homogeneous 90° and 180° pulses, such as composite pulses employed in a spin–echo sequence (Hetherington *et al.*, 1986), would increase lactate detection efficiency. It has also been shown that when a binomial water suppression pulse is used, nearly 90% of lactate can be detected *in vivo* (Williams *et al.*, 1988). However, in this study, the crucial calibration lactate concentration (7.5 mM/g tissue) determined *in vitro* on the brain *postmortem* is only about half of what others have found (15 mM/g tissue; Hetherington *et al.*, 1985; Chang *et al.*, 1987b) which might give rise to the high detection efficiency.

6. TIME-SHARED ^{1}H AND ^{31}P SPECTROSCOPY

In order to increase sample and information throughput, multinuclear NMR spectroscopy can be performed in a time-shared manner (see Section 3.3). In an efficiently interleaved experiment, an ^{1}H pulse sequence and a ^{31}P pulse sequence are applied alternatively. Thus, while excited ^{1}H nuclei relax toward equilibrium, ^{31}P signals can be acquired; and vice versa, while waiting for the excited ^{31}P nuclei to relax, ^{1}H signals can be observed. This process is continued repetitively until good S/N are attained. When ^{1}H and ^{31}P spectroscopy are carried out simultaneously in the same time block, ^{1}H metabolites such as lactate as well as ^{31}P metabolites and intracellular pH can be better correlated (Chang *et al.*, 1990b; Ligeti *et al.*, 1987). In practice, the NMR spectrometer needs to be configured to switch the spectrometer frequency between the ^{1}H and ^{31}P frequencies as illustrated in Fig. 6. Since the double-tuned probe that we used does not have very good isolation between the two channels, the ^{1}H and ^{31}P preamplifiers are protected by low- and high-pass filters, respectively. As shown in Fig. 6, when the ^{31}P protocol is being carried out, the spectrometer is set to the

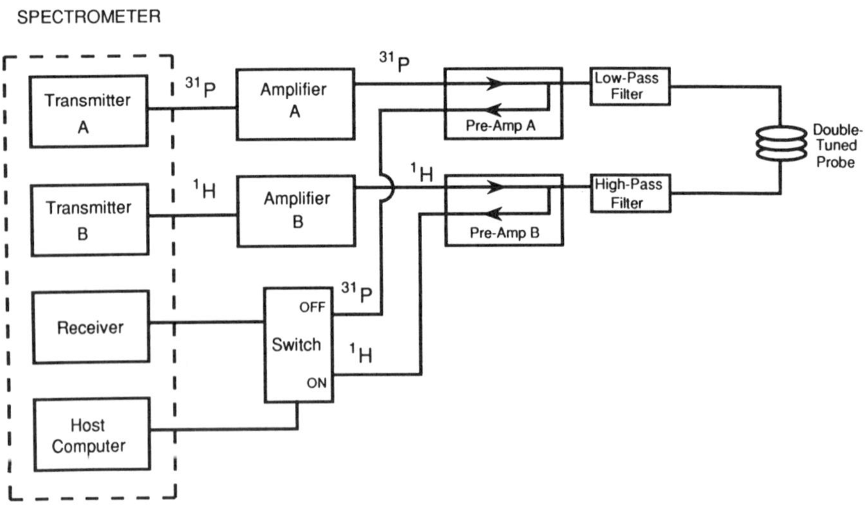

Figure 6. Block diagram of the spectrometer configuration for carrying out time-shared ^{1}H and ^{31}P NMR spectroscopy.

regular configuration. But, when the ^{1}H protocol is performed, the spectrometer is switched by the host computer to transmit and receive ^{1}H frequencies. These modifications may not be necessary for other spectrometers.

7. SPECTROSCOPIC IMAGING FOR STUDIES OF REGIONAL ISCHEMIA

Studying tissue metabolism in regional ischemia, in which a small region of the brain is rendered ischemic, requires spatially localized spectroscopy to avoid a misleading metabolic picture (Cohen *et al.*, 1990b). Among the many spatial localization techniques (Aue, 1986; Bottomley, 1989), we have chosen one-dimensional (1D) surface-coil spectroscopic imaging (SI) in which spatial information is phase-encoded by a pulsed magnetic field gradient (Brown *et al.*, 1982; Kumar *et al.*, 1975). Spectroscopic imaging is a relatively simple method with minimal chemical shift artifacts. Minimization of chemical shift artifacts is of great concern at high magnetic field (4.7 T and higher), and especially difficult for nuclei that have low gyromagnetic ratios (e.g., ^{31}P), as well as for metabolites that span a wide chemical shift range (e.g., ^{19}F). The use of a surface coil assists in localization, improves S/N, and consequently shortens data collection time (De Graaf *et al.*, 1988; Freeman *et al.*, 1989). This allows one to use localized spectroscopy to study dynamic metabolic changes in relatively short time intervals giving excellent temporal and spatial resolution (Cohen *et al.*, 1990b). Good S/N can be obtained from voxel sizes smaller than 0.15 cm^3 in less than 10 min. The method is reminiscent of rotating frame spectroscopy (Hoult, 1979; Garwood *et al.*, 1984), but gives more flexibility and better spatial localization. When 1D surface-coil spectroscopic imaging is used, well-defined flat slices can be easily selected from top to bottom, left to right, or front to back across an organ according to the direction of the field gradient used. The method is also applicable to many different nuclei such as ^{1}H or ^{31}P. Therefore, it is possible to conveniently interleave ^{1}H or ^{31}P or even more nuclei in a time-shared manner (Chang *et al.*, 1990a). Furthermore, in a relatively short time interval, one will be able to simultaneously observe metabolic changes from both normal and injured regions in one experiment with good S/N. Motion artifacts and diffusion effects due to pulsed field gradients are very minimal. In addition, when ^{1}H NMR imaging is also carried out using an imaging coil to give anatomical structure, spectroscopic information can be correlated with anatomy.

The 1D spatially localized, spectroscopic imaging pulse sequences we used are shown in Fig. 7. The two pulse sequences can be used either separately or interleaved in a time-shared manner. For ^{1}H spectroscopic

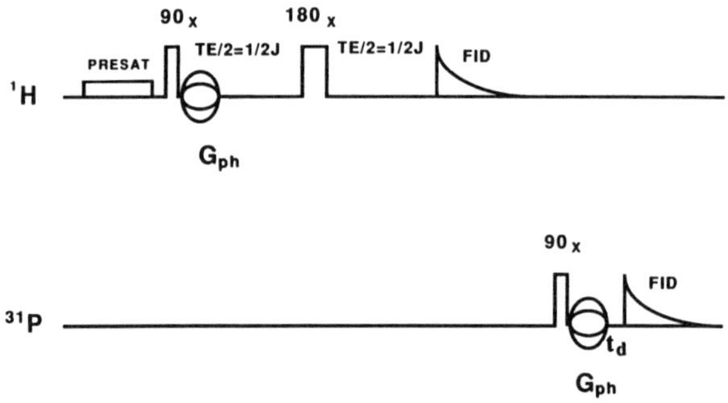

Figure 7. Time-shared, one-dimensional, ^1H and ^{31}P spectroscopic imaging pulse sequence. (From Chang *et al.*, 1990a.)

imaging, a presaturation-90°-τ-180°-τ-ACQ spin-echo pulse sequence is used. Suppression of water signal is achieved by a long (0.1 sec), low-power presaturation pulse. Spatial encoding is achieved by a phase-encoding method with the application of a sine-bell-shaped magnetic field gradient (G_{ph}) immediately after the 90° pulse. The number of phase-encoding steps determines the number of slices (NS) obtained. The magnetic field gradient increment determines the slice thickness (ST) and field of view (FOV):

$$\mathrm{ST\,(cm)} = \frac{1}{(\gamma/2\pi) \times G_{\mathrm{step}} \times t \times (2/\pi) \times \mathrm{NS}} \tag{4}$$

where γ is the gyromagnetic ratio, $\gamma/2\pi$ is given in Hz/G, G_{step} is the phase-encoding gradient step size in G/cm, t is the phase-encoding time in seconds, and $2/\pi$ corrects for the use of a sine-bell-shaped gradient instead of a rectangular gradient. The field gradient direction determines the slice orientation. For example, if the magnetic field gradient in the x direction were used, slices perpendicular to the plane of the surface coil and parallel to the long axis of the magnet (z direction) are produced. Echo delay TE/2 is set to 68 msec in order to selectively invert the lactate methyl resonance (Cohen *et al.*, 1990b). This relatively long delay also allows the eddy currents produced by the pulsed gradients to settle down.

For ^{31}P spectroscopic imaging, a 90° pulse is followed immediately by the phase-encoding field gradient and a short delay (t_d) to allow eddy currents to disappear. Typically, we use 1 msec phase-encoding time and 400 μsec delay for t_d. The effects of eddy currents on the linewidth and resonance frequency of the signals are not apparent in our experiments with such short delays. If shielded gradient coils (Mansfield and Chapman, 1986) were used, one might not even need to have this short delay. Moreover, there is usually an instrument delay DE, equal to one dwell time, before signal acquisition; but this instrument delay can be omitted, so the delay before signal acquisition is shortened thus reducing baseline artifacts. Spectral width is also set as small as possible, so the dwell time can be long in order to avoid losing too many points at the beginning of the FID. The ^{1}H and ^{31}P spectroscopic imaging pulse sequences are then interleaved in a time-shared manner as previously described (Section 6).

Figure 8 shows spectra obtained by 1D interleaved ^{1}H and ^{31}P spectroscopic imaging (Chang *et al.*, 1990a) from a rat brain with an occlusion of the right middle cerebral artery (a regional ischemia model) (Longa *et al.*, 1989). Spatially localized spectra in Fig. 8A–E are for ^{1}H from the right side to the left side of the brain and those in Fig. 8A′–E′ are for ^{31}P. There is no broad component underlying the usual brain phosphorus resonance peaks due to the phase-encoding delay in the pulse sequence that eliminates broad resonance signals. As shown in Fig. 8A, B, the inverted lactate signal can be observed in the ischemic region of the brain with Lac/NAA of 0.99 and 0.89, respectively; and there is comparatively little lactate in the control side (Fig. 8C–E) with Lac/NAA of 0.35, 0.25, and 0.20, respectively. Spatially localized ^{1}H spectroscopy also helps in eliminating background lipids. Similarly, intracellular pH is lower in the ischemic region with pH_i less than 5.8 (Fig. 8A′, B′) than in the control side with pH_i of 6.98, 6.93, and 6.85 (Fig. 8C′–E′). These results indicate that only the right side of the brain is ischemic, and the left side is still normal. This kind of spatial resolution is not evident without using localized spectroscopy.

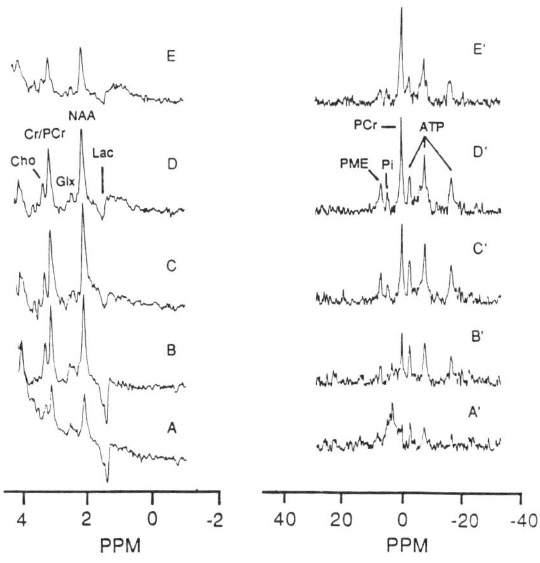

Figure 8. Spatially localized ^{1}H and ^{31}P spectra obtained by time-shared ^{1}H and ^{31}P spectroscopic imaging at 4.7 T from a rat brain with regional ischemia. Peak assignments are the same as those of Figs. 4 and 5. (From Chang *et al.*, 1990a.)

However, there are some problems associated with spectroscopic imaging. (1) It does not allow one to shim on a specified small region of interest. But if a surface coil is used, a reasonably small volume can be nicely shimmed. (2) Because of the use of pulsed magnetic field gradients, eddy currents will occur. This can be alleviated by implementing a spin-echo in the pulse sequence with long echo delay (Fig. 7) and/or using self-shielded gradient coils that produce very little eddy currents. (3) The pulse sequence used for ^{31}P spectroscopic imaging (Fig. 7) misses several initial data points of the FID. The number of points lost can be reduced by the use of strong field gradients with short phase-encoding time and minimal eddy currents. Linear prediction methods can be applied to extrapolate the initial points of the FID (Derby *et al.*, 1989; Twieg *et al.*, 1989). Spin-echo spectroscopic imaging (Haselgrove *et al.*, 1983) is another option if one can overcome the signal loss due to short T_2 of the phosphorus metabolites and the complications of J modulation due to phosphorus couplings. (4) The B_1 inhomogeneity produced by a surface coil and the variable voxel sizes caused by the excitation profile of a surface coil complicate the quantitation of metabolites from different slices. The solutions involve the use of more homogeneous B_1 pulses such as adiabatic pulse (Brown *et al.*, 1989) and/or the use of a large coil for excitation and a smaller coil for detection (Edelstein *et al.*, 1986).

8. CONCLUSION

For the past decade, we have witnessed a dramatic increase in the applications of *in vivo* NMR spectroscopy to studies of tissue metabolism, in both humans and animals. *In vivo* NMR spectroscopy provides an unparalleled modality for noninvasive study of tissue metabolism in very complicated living systems. Although it promises to be a valuable clinical tool in diagnostic medicine, NMR study of tissue metabolism in an animal model provides valuable insights *per se* and is the first step in delineating the technique for clinical applications. All of the spectroscopic methods need to be evaluated, and the accumulated metabolic information interpreted in terms of the pathophysiological state. Clinical protocols need to be established for diagnosis and therapy. It is hoped that in the coming decade, *in vivo* NMR spectroscopy will evolve as both a research probe and a clinical tool.

REFERENCES

Ackerman, J. J. H., Grove, T. H., Wong, G. G., Gadian, D. G., and Radda, G. K., 1980, *Nature* **283**:167.

Ackerman, J. J. H., Ewy, C. S., Becker, N. N., and Shalowitz, R. A., 1987, *Proc. Natl. Acad. Sci. USA* **84**:4099.

Agris, P. F., and Campbell, I. D., 1982, *Science* **216**:1325.
Aue, W. P., 1986, *Rev. Magn. Reson. Med.* **1**:21.
Avison, M. J., Hetherington, H. P., and Shulman, R. G., 1986, *Annu. Rev. Biophys. Biophys. Chem.* **15**:377.
Bailey, I. A., Williams, S. R., Radda, G. K., and Gadian, D. G., 1981, *Biochem. J.* **196**:171.
Balaban, R. S., Koretsky, A. P., and Katz, L. A., 1986, *J. Magn. Reson.* **68**:556.
Behar, K. L., Rothman, D. L., Shulman, R. G., Petroff, O. A. C., and Prichard, J. W., 1984, *Proc. Natl. Acad. Sci. USA* **81**:2517.
Bendall, M. R., and Pegg, D. T., 1986, *J. Magn. Reson.* **67**:376.
Bendall, M. R., Connelly, A., and McKendry, J. M., 1986, *Magn. Reson. Med.* **3**:157.
Berkowitz, B. A., Wolff, S. D., and Balaban, R. S., 1988, *J. Magn. Reson.* **79**:547.
Birken, D. L., and Oldenforf, W. H., 1989, *Neurosci. Biobehav. Rev.* **13**:23.
Blondet, P., Decorps, M., Confort, S., and Albrand, J. P., 1987, *J. Magn. Reson.* **75**:434.
Bolinger, L., Seeholzer, S., Kofron, J., and Leigh, J. S., 1987, private communication.
Bottomley, P. A., 1989, *Radiology* **170**:1.
Bottomley, P. A., Charles, H. C., Roemer, P. B., Flamig, D., Engeseth, H., Edelstein, W. A., and Mueller, O. M., 1988, *Magn. Reson. Med.* **7**:319.
Bottomley, P. A., Hardy, C. J., Roemer, P. B., and Mueller, O. M., 1989, *Magn. Reson. Med.* **12**:348.
Brant-Zawadzki, M., Bartkowski, H. M., Ortendahl, D. A., Pitts, L. H., Hylton, N. M., Nishimura, M. C., and Crooks, L. E., 1984, *Am. J. Neuroradiology* **5**:125.
Brindle, K. M., Porteous, R., and Campbell, J., 1984, *J. Magn. Reson.* **56**:543.
Brown, F. F., Campbell, I. D., Kchel, P. W., and Rabenstein, D. L., 1977, *FEBS Lett.* **82**:12.
Brown, T. R., Kincaid, B. M., and Ugurbil, K., 1982, *Proc. Natl. Acad. Sci. USA* **79**:3523.
Brown, T. R., Buchtal, S. D., Murphy-Boesch, J., Nelson, S. J., and Taylor, J. S., 1989, *J. Magn. Reson.* **82**:629.
Bruhn, H., Frahm, J., Gyngell, M. L., Merboldt, K. D., Hanicke, W., and Sauter, R., 1989, *Magn. Reson. Med.* **5**:126.
Campbell, I. D., Dobson, C. M., Williams, R. J. P., and Xavier, A. V., 1973, *J. Magn. Reson.* **11**:172.
Campbell, I. D., Dobson, C. M., Williams, R. J. P., and Wright, P. F., 1974, *FEBS Lett.* **57**:96.
Chang, L.-H., Chew, W. M., Weinstein, P. R., and James, T. L., 1987a, *J. Magn. Reson.* **72**:168.
Chang, L.-H., Pereira, B. M., Weinstein, P. R., Keniry, M. A., Murphy-Boesch, J., Litt, L., and James, T. L., 1987b, *Magn. Reson. Med.* **4**:575.
Chang, L.-H., Cohen, Y., Weinstein, P. R., Chileuitt, L., and James, T. L., 1990a, *Magn. Reson. Imag.*, **9**:223.
Chang, L.-H., Shirane, R., Weinstein, P. R., and James, T. L., 1990b, *Magn. Reson. Med.* **13**:6.
Choi, D. W., and Rothman, S. M., 1990, *Annu. Rev. Neurosci.* **13**:171.
Cohen, S. M., (ed.), 1987, *Physiological NMR Spectroscopy: From Isolated Cells to Man, Ann. N.Y. Acad. Sci.* **508**.
Cohen, S. M., Ogawa, S., Rottenberg, H., Glynn, P., Yamane, T., and Shulman, R. G., 1978, *Nature* **273**:554.
Cohen, Y., Chang, L.-H., Litt, L., and James, T. L., 1989, *J. Magn. Reson.* **85**:203.
Cohen, Y., Chang, L.-H., Litt, L., Kim, F., Severinghaus, J. W., Weinstein, P. R., Davis, R. L., Germano, I., and James, T. L., 1990a, *J. Cereb. Blood Flow Metab.* **10**:277.
Cohen, Y., Sanada, T., Pitts, L. H., Chang, L.-H., Nishimura, M. C., Weinstein, P. R., Litt, L., and James, T. L., 1990b, *Magn. Reson. Med.* **17**:225.
Corbett, R. J., Laptook, A. R., Hassan, A., and Nunnally, R.-L., 1988, *Magn. Reson. Med.* **6**:99.
Crockard, H. A., Gadian, D. G., Frackowiak, R. S., Proctor, E., Allen, K., Williams, S. R., and Ross-Russell, R. W., 1987, *J. Cereb. Blood Flow Metab.* **7**:394.
De Graaf, A. A., Bovee, W. M. M. J., Deutz, N. E. P., and Chamuleau, R. A. F. M., 1988, *Magn. Reson. Imag.* **6**:255.

Derby, K., Hawryszko, C., and Tropp, J., 1989, *Magn. Reson. Med.* **12**:235.

Deutch, C. J., and Taylor, J. S., 1987, *Ann. N.Y. Acad. Sci.* **508**:33.

Doddrell, D. M., Brereton, I. M., Moxon, L. N., and Garroway, G. J., 1989, *Magn. Reson. Med.* **9**:132.

Edelstein, W. A., Hardy, C. J., and Mueller, O. M., 1986, *J. Magn. Reson.* **67**:156.

Edzes, H. T., Zuurbier, C. J., Huijing, P. A., and Bulthuis, J., 1990, *J. Magn. Reson.* **87**:395.

Eleff, S. M., Schnall, M. D., Ligetti, L., Osbakken, M., Subramanian, V. H., Chance, B., and Leigh, J. S., 1988, *Magn. Reson. Med.* **7**:412.

Ernst, R., 1966, *Adv. Magn. Reson.* **2**:1.

Evelhoch, J. L., Ewy, C. S., Siegfried, B. A., and Ackerman, J. J. H., 1985, *Magn. Reson. Med.* **2**:410.

Ewing, J. R., Branch, C. A., Helpern, J. A., Smith, M. B., Butt, S. M., and Welch, K. M. A., 1989, *Stroke* **20**:259.

Frahm, J., Buhn, H., Gyngell, M. L., Merboldt, K. D., Hanicke, W., and Sauter, R., 1989, *Magn. Reson. Med.* **11**:47.

Freeman, D., Barker, P. B., Parivar, F., Benninghoven, S., Jones, L. W., Moress, E. A., and Ross, B., 1989, *Transplantation* **48**:202.

Fukushima, E., and Roeder, S. B. W., 1981, *Experimental Pulse NMR: A Nuts and Bolts Approach*, Addison-Wesley, Reading, Mass.

Gadian, D. G., 1982, *NMR and Its Applications to Living Systems*, Oxford University Press, London.

Gadian, D. G., and Robinson, F. N. H., 1979, *J. Magn. Reson.* **34**:449.

Gadian, D. G., Proctor, E., Williams, S. R., Cady, E. B., and Gardiner, R. M., 1986, *Magn. Reson. Med.* **3**:150.

Garcia, J. C., 1984, *Stroke* **15**:5.

Garwood, M., Schleich, T., Matson, G. B., and Acosta, G., 1984, *J. Magn. Reson.* **60**:268.

Gonnella, N. C., and Silverman, R. F., 1989, *J. Magn. Reson.* **85**:24.

Gonzalez-Mendez, R., Litt, L., Koretsky, A. P., Von Colditz, J., Weiner, M. W., and James, T. L., 1984, *J. Magn. Reson.* **57**:526.

Gonzalez-Mendez, R., Moseley, M. E., Murphy-Boesch, J., Chew, W. M., Litt, L., and James, T. L., 1985, *J. Magn. Reson.* **65**:516.

Graham, D. I., 1988, *J. Cereb. Blood Flow Metab.* **8**:769.

Gupta, R. K., and Moore, R. D., 1980, *J. Biol. Chem.* **255**:3987.

Gupta, R. K., and Yushok, W. D., 1980, *Proc. Natl. Acad. Sci. USA* **77**:2487.

Gyulai, L., Schnall, M., McLaughlin, A. C., Leigh, J. S., and Chance, B., 1987, *J. Cereb. Blood Flow Metab.* **7**:543.

Hardy, C. J., and Dumoulin, C. L., 1987, *Magn. Reson. Med.* **5**:58.

Hardy, C. J., Edelstein, W. A., and Vatis, D., 1986, *J. Magn. Reson.* **66**:470.

Haselgrove, J. C., Subramanian, V. H., Leigh, J. S., Gyulai, L., and Chance, B., 1983, *Science* **220**:1170.

Hetherington, H. P., Avison, M. J., and Shulman, R. G., 1985, *Proc. Natl. Acad. Sci. USA* **82**:3115.

Hetherington, H. P., Wishart, D., Fitzpatrick, S. M., Cole, P., and Shulman, R. G., 1986, *J. Magn. Reson.* **66**:313.

Hore, P. J., 1983, *J. Magn. Reson.* **55**:283.

Hore, P. J., 1989, *Methods Enzymol.* **176**:64.

Hoult, D. I., 1979, *J. Magn. Reson.* **33**:183.

Hoult, D. I., and Lauterbur, P. C., 1979, *J. Magn. Reson.* **34**:425.

Hoult, D. I., and Richards, R. E., 1976, *J. Magn. Reson.* **24**:71.

Hyde, J. S., Jesmanowicz, A., Froncisz, W., Kneeland, J. B., and Grist, T. M., 1986, *J. Magn. Reson.* **70**:512.

James, T. L., and Margulis, A. R., (eds.), 1984, *Biomedical Magnetic Resonance*, Radiology Research and Education Foundation, San Francisco.

James, T. L., and Matson, G. B., 1979, *J. Magn. Reson.* **33**:345.

James, T. L., Matson, G. B., and Kuntz, I. D., 1978, *J. Am. Chem. Soc.* **100**:3590.

Karczmar, G. S., Tavares, N. J., and Moseley, M. E., 1989, *Radiology* **172**:363.

Karlsson, B. R., and Steen, P. A., 1989, *J. Neurosurg. Anesthesiol.* **1**:278.

Knüttel, A., and Kimich, R., 1989, *Magn. Reson. Med.* **10**:404.

Kumar, A., Welti, D., and Ernst, R., 1975, *J. Magn. Reson.* **18**:69.

Le Bihan, D., Breton, E., Lallemand, D., Aubin, M. L., and Vignaud, J., 1988, *Radiology* **168**:497.

Ligeti, L., Osbakken, M. D., Subramanian, H. V., Kovach, A. G. B., Leigh, J. S., and Chance, B., 1987, *Magn. Reson. Med.* **4**:441.

Lohman, J. A. B., and Bulthuis, J., 1985, *J. Magn. Reson.* **63**:593.

Longa, E. Z., Weinstein, P. R., Carlson, S., and Cummins, R., 1989, *Stroke* **20**:84.

Mansfield, P., and Chapman, B., 1986, *J. Magn. Reson.* **66**:573.

Matson, G. B., and Weiner, M. W., 1988, in *Magnetic Resonance Imaging* (Stark, D. D., and Bradley, W. G., eds.), Mosby, St. Louis, p. 201.

Meier, J. E., and Marshall, A. G., 1990, in *Biological Magnetic Resonance* (Berliner, L. J., and Reuben, J., eds.), Plenum Press, New York, p. 199.

Moon, R. B., and Richards, J. H., 1973, *J. Biol. Chem.* **248**:7276.

Moonen, C., Oberhaensli, R., Galloway, G., Rajagopalan, B., and Radda, G., 1985, *Clin. Sci.* **68**(Suppl. 2):29.

Moseley, M. E., Cohen, Y., Mintorovitch, J., Chileuitt, L., Shimizu, H., Kucharczyk, J., Wendland, M. F., and Weinstein, P. R., 1990, *Magn. Reson. Med.* **14**:330.

Müller, S., and Seelig, J., 1987, *J. Magn. Reson.* **72**:456.

Murphy, E. J., Rajagopalan, B., Brindle, K., and Radda, G. K., 1989, *Magn. Reson. Med.* **12**:282.

Murphy-Boesch, J., 1984, in *Biomedical Magnetic Resonance* (James, T. L., and Margulis, A. R., eds.), Radiology Research and Education Foundation, San Francisco, p. 47.

Murphy-Boesch, J., and Koretsky, A. P., 1983, *J. Magn. Reson.* **54**:526.

Navon, G., Ogawa, S., Shulman, R. G., and Yamane, T., 1977, *Proc. Natl. Acad. Sci. USA* **74**:87.

Navon, G., Shulman, R. G., Yamane, T., Eccleshall, T. R., and Lam, K. B., 1979, *Biochemistry* **18**:4487.

Nosel, W., Trimble, L. A., Shen, J. F., and Allen, P. S., 1989, *Magn. Reson. Med.* **11**:398.

Pan, J. W., Hamm, J. R., Rothman, D. L., and Shulman, R. G., 1988, *Proc. Natl. Acad. Sci. USA* **85**:7836.

Petroff, O. A. C., Prichard, J. W., Behar, K. L., Alger, J. R., den Hollander, J. A., and Shulman, R. G., 1985a, *Neurology* **35**:781.

Petroff, O. A. C., Prichard, J. W., Behar, K. L., Rothman, D. L., Alger, J. R., and Shulman, R. G., 1985b, *Neurology* **35**:1681.

Petroff, O. A. C., Ogino, T., and Alger, J. R., 1988, *J. Neurochem.* **51**:163.

Pettegrew, J. W., 1990, *NMR: Principles and Applications to Biomedical Research*, Springer-Verlag, Berlin.

Pettegrew, J. W., Withers, G., Panchalingam, K., and Post, J. F. M., 1988, *Magn. Reson. Imag.* **6**:135.

Prichard, J. W., and Shulman, R. G., 1986, *Annu. Rev. Neurosci.* **9**:61.

Rabenstein, D. L., and Isab, A. A., 1982, *Anal. Biochem.* **121**:423.

Radda, G. K., 1986, *Science* **233**:640.

Roemer, P. B., Edelstein, W. A., Hayes, C. E., Souza, S. P., and Mueller, O. M., 1990, *Magn. Reson. Med.*, **16**:192.

Roth, K., Hubesch, B., Meyerhoff, D. J., Naruse, S., Gober, J. R., Lawry, T. J., Boska, M. D., Matson, G. B., and Weiner, M. W., 1989, *J. Magn. Reson.* **81**:299.

Rothman, D. L., Arias-Mendoza, F., Shulman, G. I., and Shulman, R. G., 1984a, *J. Magn. Reson.* **60**:430.

Rothman, D. L., Behar, K. L., Hetherington, H. P., and Shulman, R. G., 1984b, *Proc. Natl. Acad. Sci. USA* **81**:6330.

Rothman, D. L., Behar, K. L., Hetherington, H. P., den Hollander, J. A., Bendall, M. R., Petroff, O. A. C., and Shulman, R. G., *Proc. Natl. Acad. Sci. USA* **82**:1633.

Schnall, M. D., Subramanian, V. H., Leigh, J. S., and Chance, B., 1985, *J. Magn. Reson.* **65**:122.

Schnall, M. D., Yoshizake, K., Chance, B., and Leigh, J. S., 1988, *Magn. Reson. Med.* **6**:15.

Seo, Y., Murakami, H., Watari, Y., Imai, Y., Yoshizaki, K., Nishikawa, H., and Morimoto, T., 1983, *J. Biochem.* **94**:729.

Shulman, R. G., 1979, *Biological Applications of Magnetic Resonance*, Academic Press, New York.

Siesjö, B. K., and Wieloch, T., 1985, *Br. J. Anaesth.* **57**:47.

Silver, M. S., Joseph, R. I., and Hoult, D. I., 1984, *J. Magn. Reson.* **59**:347.

Smith, G. A., Hesketh, R. T., Metcalfe, J. C., Feeney, J., and Morris, P. K., 1983, *Proc. Natl. Acad. Sci. USA* **80**:7178.

Sotak, C. H., and Freeman, D. M., 1988, *J. Magn. Reson.* **77**:382.

Sotak, C. H., Freeman, D. M., and Hurd, R. E., 1988, *J. Magn. Reson.* **78**:355.

Springer, C. S., 1987, *Annu. Rev. Biophys. Biophys. Chem.* **16**:375.

Stromski, M. E., Arias-Mendoza, F., Alger, J. R., and Shulman, R. G., 1986, *Magn. Reson. Med.* **3**:24.

Stryjewski, D., Oschkinat, H., and Leibfritz, D., 1990, *Magn. Reson. Med.* **13**:158.

Taylor, D. J., Bore, P. J., Styles, P., Gadian, D. G., and Radda, G. K., 1983, *Mol. Biol. Med.* **1**:77.

Thulborn, K. R., and Ackerman, J. J. H., 1983, *J. Magn. Reson.* **55**:357.

Twieg, D. B., Meyerhoff, D. J., Hubesch, B., Roth, K., Sappey-Marinier, D., Boska, M. D., Gober, J. R., Schaefer, S., and Weiner, M. W., 1989, *Magn. Reson. Med.* **12**:291.

Tycko, R., Cho, H. M., Schneider, E., and Pines, A., 1985, *J. Magn. Reson.* **61**:90.

Ugurbil, K., Garwood, M., and Rath, A., 1988, *J. Magn. Reson.* **80**:448.

Villringer, A., Rosen, B. R., Belliveau, J. W., Ackerman, J. L., Lauffer, R. B., Buxton, R. B., Chao, Y. S., Wedeen, V. J., and Brady, T. J., 1988, *Magn. Reson. Med.* **6**:164.

Warner, D. S., 1989, *J. Neurosurg. Anesthesiol.* **1**:290.

Williams, S. R., and Gadian, D. G., 1986, *Q. J. Exp. Physiol.* **71**:335.

Williams, S. R., Gadian, D. G., Proctor, E., Sprague, D. B., Talbot, D. G., Young, I. R., and Brown, F. F., 1985, *J. Magn. Reson.* **63**:406.

Williams, S. R., Gadian, D. G., and Proctor, E., 1986, *J. Magn. Reson.* **66**:562.

Williams, S. R., Proctor, E., Allen, K., Gadian, D. G., and Crockard, H. A., 1988, *Magn. Reson. Med.* **7**:425.

Yoshizaki, K., Seo, Y., and Nishikawa, H., 1981, *Biochim. Biophys. Acta* **678**:283.

4

Shift-Reagent-Aided ^{23}Na NMR Spectroscopy in Cellular, Tissue, and Whole-Organ Systems

Sandra K. Miller and Gabriel A. Elgavish

1. INTRODUCTION

Nondestructive observation of intracellular sodium (Na_i) levels is of utmost clinical and biochemical importance. In viable cells, a 10- to 20-fold concentration gradient between Na_i and extracellular sodium (Na_0) is maintained. Deviation from this gradient is an important indicator of detrimental processes. Therefore, it may be a valuable parameter for the assessment of cell viability and reversibility of normal function following damage, e.g., an ischemic insult.

Via the various sodium transport mechanisms (Na^+-H^+ antiport, Na^+-Ca^{2+} exchange, Na^+-K^+ ATPase, and others), sodium levels and fluxes have an important role in the regulation of intracellular pH, of intracellular calcium levels, and, therefore, of cellular function. In muscle cells, e.g., in the myocardium, changes in Na_i levels are coupled with contractile function. Real-time, nondestructive monitoring of Na_i, therefore, would be a powerful research tool enhancing our biochemical understanding of cellular processes

---

Sandra K. Miller • Division of Cardiovascular Disease, Department of Medicine, University of Alabama at Birmingham, Birmingham, Alabama 35294. *Present address of Dr. Miller:* Frederik Philips NMR Research Center, School of Medicine, Department of Radiology, Emory University, Atlanta, Georgia 30322. **Gabriel A. Elgavish** • Division of Cardiovascular Disease, Department of Medicine and Department of Biochemistry, University of Alabama at Birmingham, Birmingham, Alabama 35294.

Biological Magnetic Resonance, Volume 11: In Vivo Spectroscopy, edited by Lawrence J. Berliner and Jacques Reuben. Plenum Press, New York, 1992.

in health and disease. Once development of the methodology reached the stage of safe human application, its use in the clinical context could offer a sensitive *in vivo* marker of localized variations in tissue viability.

^{23}Na is a relatively sensitive (about 50% more sensitive than ^{31}P) NMR-observable nuclide, and therefore the intrinsic noninvasive character of NMR could be utilized to achieve the goal stated above. Unfortunately, however, the frequency positions ("chemical shift") of intra- and extracellular sodium are identical. Together with the fact that the total tissue quantity of Na_0 is much larger than that of Na_i, this precluded NMR observation of Na_i by NMR, until Degani and Elgavish (1978) demonstrated that anionic paramagnetic complexes of gadolinium used as relaxation reagents could be utilized to differentiate the sodium signals from the two compartments. This was followed in 1982 by the introduction of paramagnetic dysprosium complexes as shift reagents for metal cations (Gupta and Gupta, 1982; Pike and Springer, 1982). Combined with the intrinsic features of NMR spectroscopy, the ability of an aqueous shift reagent to differentiate signals from intracellular and extracellular spaces provided a nondestructive, continuous method to monitor intracellular cation concentration. Subsequently, ^{23}Na NMR studies of various biological systems quickly followed.

Several review articles on ^{23}Na NMR in biological systems have been published (Berendsen and Edzes, 1973; Civan and Shporer, 1978; Nissen *et al.*, 1989; Springer, 1987; Boulanger and Vinay, 1989; Gupta, 1987; Elgavish, 1989). The Berendsen and Edzes (1973) and Civan and Shporer (1978) articles are excellent reviews containing details of ^{23}Na NMR theory as well as the foundations for application of ^{23}Na NMR spectroscopy to biological systems. Both were written before the development of methods for separating the intra- and extracellular Na signals. In the Berendsen and Edzes article, total tissue Na determinations were described along with an excellent theoretical discussion of the nature of Na in tissue. Civan and Shporer dealt extensively with the issue of the NMR spectral visibility of Na. The Springer review focuses on the measurement of metal ion compartmentation in tissues using NMR. This review also contains an extensive list of review articles dating back to 1973 on compartmentation and related subjects.

The Boulanger and Vinay review summarizes the NMR methods used to measure sodium in biological tissues and provides a summary of the results. The Gupta article is a review exclusively of shift-reagent-aided ^{23}Na NMR of intact cells and tissues. The Elgavish review concerns itself with shift-reagent-aided ^{23}Na NMR spectroscopy of isolated perfused rat hearts and is limited to some methodology and a few selected examples of applications from the author's laboratory. In the present review we present the theory and methodology, and an attempt at a comprehensive view of the literature, of shift-reagent-aided ^{23}Na NMR spectroscopy on single and multiple cell systems subdivided into the various organ and cell systems. Also, we include the latest information on several new advances in the measurement of Na and Na_i specifically.

2. THEORY

2.1. General

^{23}Na is a quadrupolar nuclide with a spin quantum number (I) of 3/2, a natural isotopic abundance of 100%, an NMR receptivity of 9.27 relative to 100 for ^{1}H, and an average concentration of 44 mM in the human body. Springer (1987) defines an index called BIONMR receptivity, which is the product of the tissue concentration and the NMR receptivity of a particular nuclide, with the index for ^{1}H normalized to 100. The values obtained for ^{23}Na and ^{31}P are 4.08×10^{-3} and 25.74×10^{-3}, respectively. One should relate, however, to the BIONMR receptivity index with great caution. Since this index lumps together the contributions from all chemical species of a given nuclide, and it does not take into account linewidth and T_1, using it for any practical comparison of different nuclides is impossible in most cases. A case in point is the comparison between ^{23}Na and ^{31}P. According to the above BIONMR receptivities, the ease with which a ^{31}P NMR spectrum should be obtained in a typical biological sample would be about sixfold higher than that for ^{23}Na. In fact, in real life experiments, the converse is true.

For nuclides with $I \geqslant 1$, more than one allowed NMR transition exists. ^{23}Na, possessing an $I = 3/2$ spin, has three transitions: a center line reflecting the transition between energy levels $m = \pm 1/2$, and two satellite lines for transitions $m = 3/2$ to $m = 1/2$ and $m = -1/2$ to $m = -3/2$. In the absence of any additional effective interaction, the energy gaps that give rise to these three transitions are equal, and therefore give rise to a spectrum of a single resonance with an intensity reflecting the superposition of all three transitions.

In the solid state, the quadrupolar effect lifts the degeneracy among the three transitions and thus splits the spectrum into three resonances.

The T_2 relaxation is given by:

$$\left(\frac{1}{T_2}\right)_{\mathrm{I}} = [(e^2qQ)^2/20](1+\eta^2/3)\tau_c\left[\frac{1}{1+\omega_0^2\tau_c^2}+\frac{1}{1+4\omega_0^2\tau_c^2}\right] \quad (1)$$

$$\left(\frac{1}{T_2}\right)_{\mathrm{II}} = [(e^2qQ)^2/20](1+\eta^2/3)\tau_c\left[1+\frac{1}{1+\omega_0^2\tau_c^2}\right] \quad (2)$$

where e^2qQ is the magnitude of the nuclear quadrupolar interaction (in rad/sec), η is the dimensionless asymmetry factor defined as $(V_{xx} - V_{yy})/V_{zz}$, the V components are second derivatives of the electrical potential and constitute the principal components of the electric field gradient, τ_c is the correlation time, and ω_0 is the Larmor frequency. The $(1/T_2)_{\text{I}}$ decay represents 40% of the signal and can be attributed to the transition between the +1/2 and −1/2 Zeeman energy levels. The remaining 60% decays at a rate of $(1/T_2)_{\text{II}}$ and can be attributed to the 3/2 to 1/2 and −1/2 to −3/2 transition. Thus, in the solid state, and other ordered systems, the quadrupolar interaction splits the ^{23}Na NMR spectrum into three transitions with a ratio of 3:4:3 for the integrated areas under the signals (Fig. 1). The outer resonances typically possess considerably shorter T_2's than the inner transition, and since the natural linewidth of NMR signals is proportional to $1/T_2$, they are broadened to such an extent that they are usually

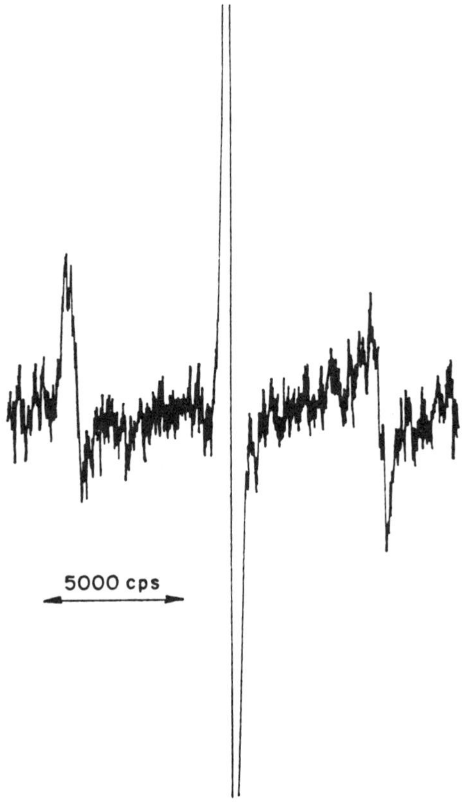

Figure 1. Derivative of the ^{23}Na signal obtained from sodium linoleate in water. In addition to the central signal, satellite lines are easily distinguished. (From Shporer and Civan, 1972.)

not detected under conventional conditions of spectral acquisition. This results in a "visibility" of only 40% of the total signal that would have been observed under homogeneous conditions. A more detailed discussion of this phenomenon may be found below (Section 2.3) and in Civan and Shporer (1978) and Berendsen and Edzes (1973).

The two components of the longitudinal relaxation rate are given by

$$\left(\frac{1}{T_1}\right)_{\text{I}} = [(e^2qQ)^2/10](1 + \eta^2/3)\tau_c\left[\frac{1}{1 + \omega_0^2\tau_c^2}\right] \tag{3}$$

$$\left(\frac{1}{T_1}\right)_{\text{II}} = [(e^2qQ)^2/10](1 + \eta^2/3)\tau_c\left[\frac{1}{1 + 4\omega_0^2\tau_c^2}\right] \tag{4}$$

Only 20% of the signal intensity decays with $(1/T_1)_{\text{I}}$ and is not readily detectable while the remaining 80% decays at a rate of $(1/T_1)_{\text{II}}$. In nonviscous, homogeneous solutions, conditions of "extreme narrowing" apply and the above equations converge to

$$\frac{1}{T_1} = \frac{1}{T_2} = C \cdot \tau_c \tag{5}$$

where the constant C is a function of the relevant magnetic interactions.

2.2. Bulk Magnetic Susceptibility

The bulk magnetic susceptibility shift (BMS) is a phenomenon that affects both NMR spectroscopy and NMR imaging, and can lead to misinterpretation of data (Chu *et al.*, 1990b). The chemical shift of a nucleus (δ) is composed of the following contributions

$$\delta = \delta_\sigma + \delta_s + \delta_\chi \tag{6}$$

where δ_σ is the host molecule's electronic shielding effect, δ_s is the solvent's electronic shielding effect, and δ_χ is the BMS effect. Variations in BMS across a sample may occur for different reasons, including a transition from an aqueous solution to air or from tissue to a cavity, but one of the most prominent effects may be induced by the presence of paramagnetic metal ions. The change (Δ_p) in the chemical shift of a nucleus induced by the presence of a moiety that contains a paramagnetic metal ion is defined as

$$\Delta_p = \delta_{\text{obs}} - \delta_d \tag{7}$$

where δ_{obs} and δ_d are the observed shift in the presence of the paramagnetic moiety and the diamagnetic shift observed in its absence, respectively. Δ_p consists of four contributions:

$$\Delta_p = \delta_{cf} + \delta_{hc} + \delta_{hd} + \delta_\chi \tag{8}$$

where δ_{cf} is the shift due to the formation of the complex with the paramagnetic moiety, δ_{hc} is the hyperfine contact shift, δ_{hd} is the hyperfine dipolar term, and δ_X is the paramagnetic contribution to the BMS shift. It is important to note that while δ_{cf} and δ_{hc} are transmitted through chemical bonds only, both δ_{hd} and δ_X are through-space effects. The dipolar term, however, because of its inverse cubic decay with the distance from the center of the paramagnetic metal ion, has a practical effect only when the paramagnetic metal ion and the observed nucleus reside in the same chemical complex. The paramagnetic BMS effect does not require direct complexation. The BMS term, which results from the partial alignment of the magnetic moments of the paramagnetic molecules by the external magnetic field, is defined as

$$\delta_\chi = \frac{\nu_i - \nu_0}{\nu_0} \times 10^6 = \frac{B_i - B_0}{B_0} \times 10^6 \tag{9}$$

where ν_i and ν_0 are the Larmor frequencies and using the Larmor equation, $\nu = \gamma \cdot B/2\pi$, they can be replaced by the magnetic flux densities, B_i and B_0, where γ is the magnetogyric ratio of the observed nucleus. B_i and B_0 are, respectively, the flux densities acting on this nucleus in the presence and in the absence of the BMS effect but otherwise in the same environment. The flux density in the presence of BMS is given by

$$B_i = (1 + D_i)B_0 + I_i \tag{10}$$

where the first and second terms are the homogeneous and inhomogeneous BMS contributions, respectively. The parameter D_i, in the homogeneous term, depends on the magnetic susceptibilities inside and outside of the compartment, on the geometric shape of the compartment, and on its orientation relative to the direction of the field of the external magnet. The inhomogeneous term, I_i, depends on all of the above and, additionally, on the spatial position of the nucleus within the compartment. Chu *et al.* (1990b) compiled expressions of D_i and I_i derived for relatively simple geometric shapes of compartments, e.g., sphere, cylinder, and coaxial cylinder. The expressions for D_i were different between inside and outside, and for cylinders with their axis aligned parallel or perpendicular to the direction of the external magnetic field. I_i had nonzero expressions for the outside compartments.

An illustration of quite large effects of the type discussed above, even in a diamagnetic system, is shown in Fig. 2. The spectrum was produced by filling a phantom, consisting of both a capillary tubing placed parallel to the external magnetic field and another tubing perpendicular to it, with a 5 M solution of NaCl. A spectral separation of 5.1 ppm is evident, in good agreement with the value calculated on the basis of the theoretical expressions mentioned above.

The effects of BMS shifts for ^{23}Na NMR in a compartmentalized phantom, in the presence of the paramagnetic shift reagents $DyCl_2$, DyTTHA, $Dy(PPP)_2$, or $Dy(CA)_3$ (where CA is the chelidamate anion, i.e., 4-hydroxypyridine-2,6-dicarboxylate), are shown in Fig. 3. The phantom used consisted of two coaxial capillary tubes, the outer annulus filled with a NaCl solution and the inner tube filled with a NaCl solution plus one of the four dysprosium compounds listed above. This phantom was then placed either parallel (foreground spectra) or perpendicular (background spectra) to the external magnetic field. The annular peaks in the spectra from the perpendicular phantom are inhomogeneously broadened as predictable from the theory. Also, the annular resonances in all spectra from both

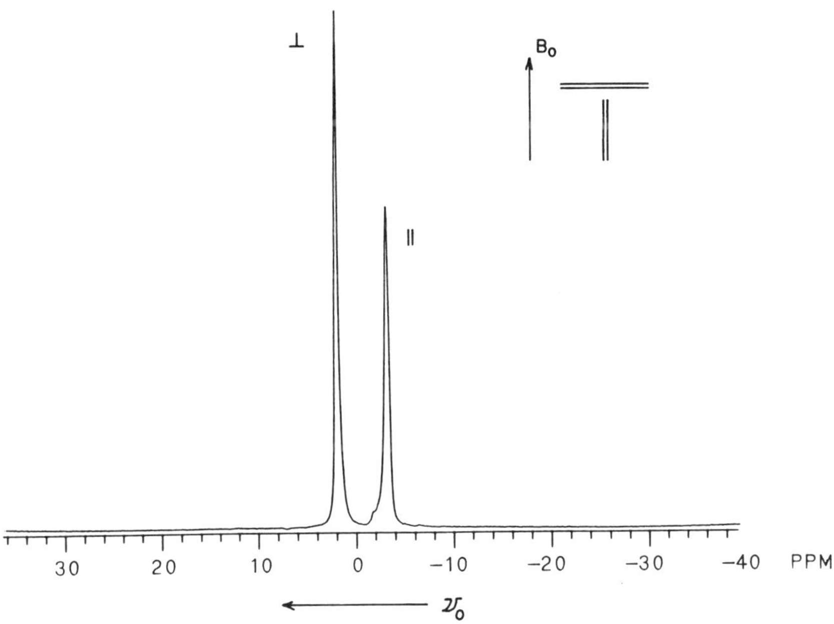

Figure 2. The ^{23}Na NMR (79.4 MHz) spectrum of a stationary capillary phantom sample with 5 M NaCl present in each capillary. (From Chu *et al.*, 1990b.)

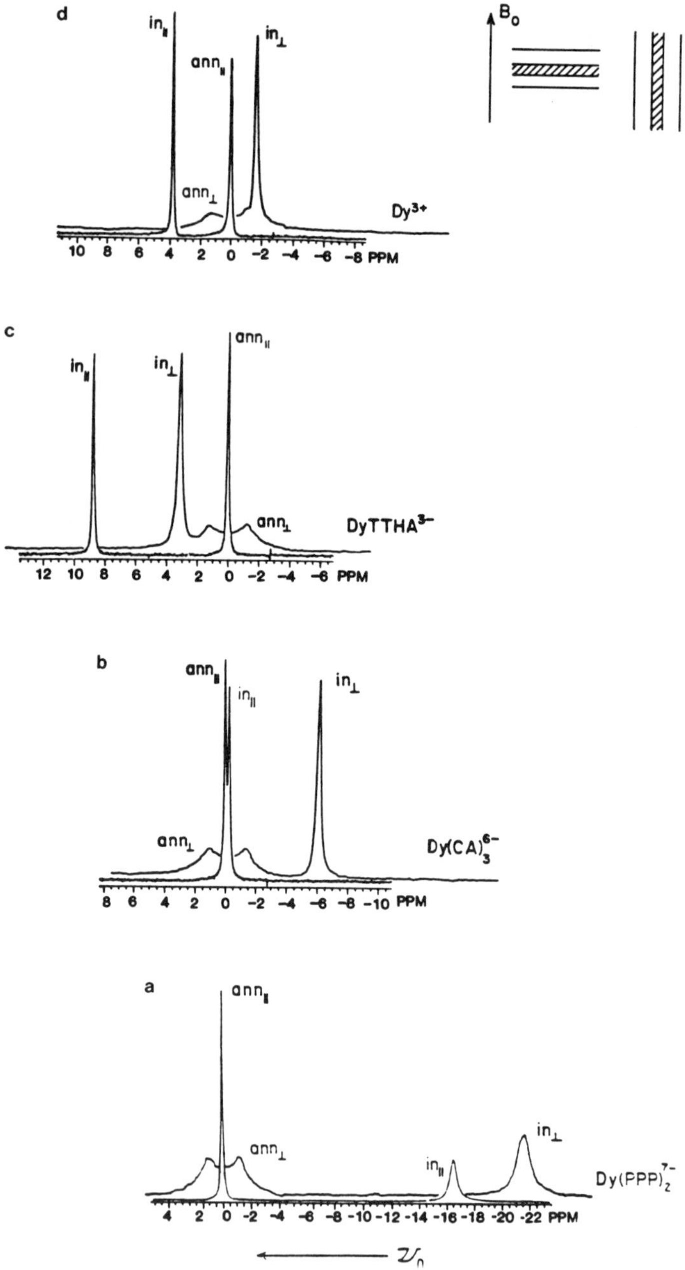

Figure 3. Pairwise spectra are plotted on identical scales and superimposed. (a) The spectra of the $Dy(PPP)_2$-containing sample. (b) The spectra of the $Dy(CA)_6^{3-}$-containing sample. (c) The spectra of the $Dy(TTHA)^{3-}$-containing sample. (d) The spectra of the Dy^{3+}-containing sample. (From Chu *et al.*, 1990b.)

configurations are centered at zero since this would be the case in a physiologically realistic situation. The inner signals are affected solely by BMS shifts in the uncomplexed $DyCl_3$ system, and undergo mixed BMS and hyperfine effects in the case of the three chelates. The large upfield shifts of the inner signals with $Dy(PPP)_2$ indicate a predominantly hyperfine effect in this system.

Chu *et al.* (1990b) also summarize various methods for the analysis of the susceptibility terms and the separation of the BMS and hyperfine contributions. Most of the methods deal with examining the sample at two different orientations relative to B_0. Mixed hyperfine and BMS effects may be encountered in organ or tissue spectroscopy where complex compartment shapes and configurations may lead to potentially misleading shift data. The assignment of shifted and unshifted signals to "pure" extra- and intracellular compartments based on the tacit assumption that the observed paramagnetic shifts are induced exclusively by the hyperfine mechanisms which require direct molecular interaction clearly ignores BMS effects. This may lead sometimes to potential overinterpretation of data, e.g., Naritomi *et al.* (1987) (see Section 4.8).

2.3. Spectral Visibility

The first measurements of tissue sodium by ^{23}Na NMR were carried out by Cope (1965, 1967). He observed that the integrated peak area of the tissue signal was only about 30–40% of the area of the signal from the same sample when the NMR measurement followed ashing and resolubilization. Cope, therefore, hypothesized that 60–70% of the intracellular sodium ion pool was immobilized by binding and thus rendered "NMR invisible," while the remaining free sodium ions gave rise to the observable pool. Later (Cope, 1970), this two-component hypothesis was further investigated by spin–echo NMR of Na^+ in rat muscle, brain, and kidney. The results were interpreted to indicate that 60–75% of Na was broadened to NMR invisibility by complexing with macromolecules, while the second, smaller fraction remained free in the aqueous solution in tissue and therefore yielded the usual sharp spectrum. Cope observed two relaxation components: a slow component with T_1 and T_2 of 10–15 msec (4–5 times shorter than in an aqueous NaCl solution), and a fast component which represented approximately 65% of the total tissue Na concentration with a T_2 less than 1 msec. The latter was similar to the value for Na complexed to a synthetic ion-exchange resin, where a fraction of Na dissolved in water showed a sharp NMR signal, while a fraction complexed to the ion-exchange resin had a markedly shorter relaxation time. This complexed fraction produced a steady-state signal which was so broad and of such low intensity that it became practically undetectable.

At that point it was quite clear that in tissue and other, tissuelike, systems, ^{23}Na NMR gave rise to two spectral components with very different relaxation rates and relative intensities. The explanation given by Cope, based on the existence of two chemically distinct fractions of sodium ions, however, was cast in doubt by Shporer and Civan (1972). These authors suggested that all experimental data could be better explained by the different Zeeman transitions split by the quadrupolar interaction, and that all of the intracellular Na nuclei present would be characterized by these transitions without any necessity for chemically distinct fractions. Subsequently, this alternative hypothesis was supported by detecting a two-component T_2 behavior, but only a single T_1, for intracellular sodium in muscle (Berendsen and Edzes, 1973; Shporer and Civan, 1974).

The model that can explain satisfactorily the above-described spectral behavior of tissue sodium is based on fast exchange between a pool of free sodium ions within the bulk isotropic cellular medium and a pool of sodium immobilized by binding to macromolecules. These two pools are not equivalent to those in the Cope hypothesis as their relative sizes *are not* the parameters that give rise to the relative contributions of the visible and "invisible" components of tissue sodium. These relative contributions are simply the result of the 3:2 ratio of the combined intensities of the two broad outer transitions to the intensity of the sharp central transition in the NMR transition manifold of the sodium nucleus split by quadrupole interactions.

In fact, since the fast exchange regime results in a single population of sodium ions with regard to NMR parameters, the second pool can be quite small in relative size (Berendsen and Edzes, 1973). Shporer and Civan (1974; see also Civan and Shporer, 1978) calculated that an immobilized fraction as small as 0.36–0.91% of sodium ions can give rise to the observed 40% visibility and two-component T_2 behavior. An alternative mechanism would consist of sodium ions freely diffusing among domains of ordered polyelectrolytes containing free Na^+ (Berendsen and Edzes, 1973). In either case, it should be pointed out that additional weak binding, without appreciable immobilization, of even relatively large fractions of Na^+ to various intracellular sites should not interfere with the above mechanisms that create the quadrupolar effect responsible for the observed phenomena of partial visibility and two-component relaxation.

Furthermore, variations in the relative sizes of the immobilized versus free sodium ion pools do not have any effect on the percentage of visibility. The 40% visibility results solely from the relative theoretical intensities of the sharp central transition versus the broad, usually undetected, two outer transitions. Once the below 1% immobilized fraction has rendered the outer transitions broad enough to escape detection, further increase in the population of this fraction will only make the outer transitions even broader, but will not change the percent visibility. Consequently, on the basis of the above theory, the visibility factor in different experimental systems should be either 0.4 or 1.0, but neither any intermediate value, nor a value below 0.4. Overall visibility factors intermediate between 0.4 and 1.0 may be observed, however, if the experimental system consists of several compartments with the sodium ions in slow exchange among them. If some of these compartments, with a total sodium ion population of $Na_{1.0}$, are characterized by a visibility factor of 1.0, while others, with a total sodium ion population of $Na_{0.4}$, are characterized by a visibility factor of 0.4, then the observed visibility V_{obs} is given by

$$V_{obs} = \frac{0.4 \cdot Na_{0.4} + Na_{1.0}}{Na_{0.4} + Na_{1.0}} \tag{11}$$

Clearly, depending on the relative values of $Na_{0.4}$ and $Na_{1.0}$, V_{obs} may be any value between 0.4 and 1.0. It may not always be possible to identify the different physically distinct compartments that give rise to the observed weighted-average visibility factor. Particularly, when an intermediate visibility of intracellular sodium is determined separately from total tissue visibility, the identity of the subcellular compartments that give rise to it may be difficult to ascertain. One should take into account, however, that if the Na_i signal is assigned on the basis of an unshifted resonance in the presence of a shift reagent, the contributing compartments may not all necessarily be intracellular; some may be extracellular compartments inaccessible to the particular shift reagent under the conditions of the experiment.

Another mechanism that in pulse NMR may be responsible, in the presence of quadrupole splitting, to reduced visibility is due to the fact that unless a very strong rotating H_1 field is used to produce the sampling pulse, each of the three sodium transitions will behave as a separate fictitious spin with an effective gyromagnetic ratio of $\alpha\gamma$, where γ is the normal gyromagnetic ratio of the quadrupolar nucleus, and with a modified transition dipole (Abragam, 1961). For a spin 3/2 nucleus, like ^{23}Na, $\alpha = \sqrt{3}$ and thus the effective gyromagnetic ratio is $\gamma_{Na} \cdot \sqrt{3}$. The intensity of the sharp, central 1/2 to $-1/2$ transition becomes only 20% of the total intensity in the absence of the quadrupolar interaction and the $\pi/2$ pulsewidth is exactly one-half of normal (Schmidt, 1972; Fukushima and Roeder, 1981).

Joseph and Summers (1987) used a method called the flip-angle effect (FAE), based on the above theory, to detect quadrupole splitting of ^{23}Na in tissue. The method involves a study of the rate of rotation of the flip angle as a function of varying radiofrequency amplitude. The spectral component that results from the central transition of a quadrupole-split manifold would reach maximum intensity at half the RF amplitude compared with that for an unsplit resonance. Their results on porcine skeletal muscle and cat brain indicate the absence of quadrupole splitting in these tissues, while, in contrast, they were able to demonstrate the presence of FAE in a model system of sodium linoleate.

Na visibilities in the range of 20 to 100% have been reported for various tissues (Nissen *et al.*, 1989; Boulanger *et al.*, 1985; Pike *et al.*, 1984; Shinar and Navon, 1984; Gupta and Gupta, 1982). The percent visibility may be tissue specific but even for the same tissue there are conflicting reports (Boulanger *et al.*, 1985; Ogino *et al.*, 1985; Ouwerkerk *et al.*, 1989; Nissen *et al.*, 1989).

Nissen *et al.* (1989) examined the visibility issue in RBC by a competition study with addition of small amounts of K^+ and Mg^{2+}. This led to large increases in the Na_i signal attributed to the release of previously invisible Na^+. Less release was obtained with Li^+ or choline, and this was explained by ion specificity in the competition for the association sites responsible for decreased visibility. Also, an increase in the T_1 of Na was noted after the addition of other ions.

Boulanger *et al.* (1985) compared the sodium concentration obtained by NMR with that determined by flame photometry and found that more than 20% is invisible in ghosts and cells. When cells are hemolyzed, the NMR-detectable increases depend on gel formation. The loss of signal in ghosts, cells, and undiluted hemolysates is attributed to partial binding of Na^+ to intracellular components. This binding is operative only when these components exist in a gel state.

Pike *et al.* (1984) examined the visibility in whole blood and washed erythrocytes in the presence of $Dy(PPP)_2$, TmTTHA, DyTTHA, or $Tm(PPP)_2$. Na_i and Na_0 concentrations, calculated on the basis of the absolute intensities of the corresponding NMR signals and using the measured hematocrit, were compared with the values obtained by centrifugation and flame emission photometry and were found to be the same within 5%. Thus, no NMR invisibility was detected for any ^{23}Na NMR signal in this system.

Shinar and Navon (1984) examined Na_i concentrations in human and canine erythrocytes determined by ^{23}Na NMR and volume measurements using ^{59}Co NMR of the extracellular probe $Co(CN)_6^{3-}$. The Na T_2 was found to be significantly shorter than the T_1, indicating binding to macromolecules. Also, T_1 was monoexponential while T_2 behaved biexponentially. Nevertheless, it was shown that neither binding to the inner side of the membrane nor binding to hemoglobin contributed to the observed transversal relaxation enhancement. Also, while the Na_i concentrations were calculated with the tacit assumption of 100% visibility, they were within the known range of literature values, thus implying that indeed no NMR invisibility occurred in this system.

It is evident that even in a relatively simple system, the erythrocyte, which lacks internal organelles such as nucleus and mitochondrion, different extent of visibility has been demonstrated by different investigators. Thus, neither in theory nor in practice has the issue of ^{23}Na NMR visibility been settled to date.

2.4. Lanthanide-Induced Shift and Relaxation

2.4.1. Induced Shift

Nuclei of ligands in paramagnetic complexes are magnetically coupled to the unpaired electronic spin of the central ion by the electron–nuclear hyperfine interaction. One of the consequences of this interaction is large chemical shifts (relative to the same system not coupled to a paramagnetic lanthanide ion) in the NMR spectra of the ligands. There are two major independent mechanisms of hyperfine interaction. The contact interaction results from a finite probability of finding an unpaired electronic spin in an atomic *s* orbital. The dipolar interaction takes place via space and causes shifts only if the magnetic susceptibility of the central ion is anisotropic. Otherwise it is effectively averaged out and no dipolar shifts are observed (this is the case, e.g., with S-state ions such as Gd^{3+}, where the only paramagnetic shifts are of contact and BMS origin). Thus, excluding BMS effects (see Section 2.2), the net paramagnetic shift, Δ_p, is the sum of the hyperfine contact (δ_{hc}) and the hyperfine dipolar (δ_{hd}) contributions:

$$\delta_p = \delta_{hc} + \delta_{hd} \tag{12}$$

2.4.1.1. Contact Shift. The first systematic study of chemical shifts in lanthanide complexes was carried out by Lewis *et al.* (1962) who measured and interpreted the shifts induced in the ^{17}O NMR signal position of water oxygens by lanthanide aquo complexes. The sign variation of the observed shifts along the lanthanide series suggested that there was no direct overlap between the $4f$ and the oxygen $2s$ wave functions, but rather a promotion of a bonding electron to an empty antibonding orbital leaving a net spin at the oxygen nucleus. This pathway of contact interaction is referred to as spin polarization in contrast to delocalization or direct overlap. The contact shift (in frequency units) can be expressed as

$$\delta_{hc} = A \cdot \langle S_z \rangle \tag{13}$$

where A is the hyperfine coupling constant (in frequency units) and $\langle S_z \rangle$ is the projection of the total electron spin magnetization of the lanthanide on the direction of the external magnetic field. To a first approximation (and excluding Eu^{3+} and Sm^{3+}) it is given (Lewis *et al.*, 1962) by

$$\langle S_z \rangle = g_L(g_L - 1) \cdot J(J+1) \cdot (\beta H_0/3kT) \tag{14}$$

where g_L is the Landé g-factor, J is the resultant electronic spin angular momentum (in units of $h/2\pi$) of the ground energy level, β is the Bohr magneton, and H_0 is the intensity of the external magnetic field. For Eu^{3+} and Sm^{3+}, higher energy levels have to be included. More recently, Golding and Halton (1972) introduced a second-order perturbation into the calculations of $\langle S_z \rangle$ of trivalent lanthanides. This resulted in corrections of up to 10% in the $\langle S_z \rangle$ values (at 300 K) with a more complex temperature dependence.

The ^{17}O shifts in 1 molal solutions of lanthanide perchlorates (Reuben and Fiat, 1969a) my be plotted against $\langle S_z \rangle$ (Reuben and Elgavish, 1979). The observed relatively good linear correlation means that the hyperfine coupling constant A for the water oxygen in the aquo complexes does not change appreciably along the lanthanide series. The temperature dependence of the ^{17}O shifts in aqueous solutions of $Dy(ClO_4)_3$ has been shown to conform to the Curie law contained in Eq. (14) (Reuben and Fiat, 1969a).

2.4.1.2. Dipolar Shift. Investigation of proton shifts in aqueous solutions of lanthanide perchlorates showed that the sign variation of the shifts along the series was irregular and definitely different from that observed with the ^{17}O shifts (Reuben and Fiat, 1967a). This observation led to the suggestion that the proton shifts were dominated by dipolar, rather than contact, contributions. It was also found that the proton shifts markedly departed from Curie's law (Reuben and Fiat, 1967b). A concise theory of dipolar shifts in lanthanide complexes in solution has been presented by Bleaney (1972). Bleaney's theory and its subsequent extension and generalizations (Golding and Pyykkö, 1973) provide positive predictions with regard to both the relative dipolar shifts along the lanthanide series and their temperature dependence. There are two cardinal points in this theory. Ignoring for the moment the $4f^5$ and $4f^6$ configurations of Sm^{3+} and Eu^{3+}, respectively, these points can be stated as follows.

(1) In a lanthanide complex the ground state (specified by its quantum number J) of the ion has $2J + 1$ levels split by the ligand field, which may be appreciably populated at room temperature. The magnetic properties of an isolated group of levels may be very anisotropic. However, the lifetime of the ion in any one of them is very short at room temperature. In fact, electron spin relaxation times of the order of 10^{-13} sec have been estimated from nuclear relaxation measurements. Therefore, in an NMR experiment only the magnetic moment averaged over a time long compared with the electron spin relaxation time has to be considered since the reorientation time of the complex itself is much longer. This average is composed of the contributions of the $2J + 1$ levels of the ground J manifold, each weighted by the corresponding relative population. As a first approximation it is assumed that at the temperature of the usual NMR experiment $kT \gg \Delta E_J$,

where ΔE_J is the energy span of the levels within the manifold. In this case, an equal distribution of population is assumed, and the equally weighted average of the magnetic moments over the entire ground manifold of a lanthanide ion results in an isotropic magnetic susceptibility. Thus, the dipolar shift in this first-order approximation vanishes.

(2) Higher-order terms of the populations of the energy levels take into account the deviation from equal populations at a finite temperature. This second-order approximation yields a T^1-dependent population distribution. These population values then serve as weighting factors in calculating the average magnetic moment. The latter, combined with Curie's law, results in an anisotropic, T^{-2}-dependent susceptibility. This second-order anisotropic susceptibility is the source of the dipolar nuclear resonance shifts in lanthanide complexes.

For the case of an axially symmetric complex the dipolar shift (in frequency units) can be expressed as a product of several factors:

$$\delta_{\text{hd}} = K \cdot G \cdot L \cdot Q \tag{15}$$

where

$$K = \frac{\beta^2}{60(kT)^2} \cdot \nu_0, \qquad G = \frac{3\cos^2\Theta - 1}{r^3}, \qquad L = 2A_2^0\langle r^2\rangle,$$

$$Q = g_{\text{L}}^2 \cdot J(J+1) \cdot (2J-1) \cdot (2J+3)\langle J\|\alpha\|J\rangle$$

ν_0 is the resonance frequency. The first factor, K, includes the dependence on the inverse square of the absolute temperature. The dependence on molecular structure is included in the second factor, G, where r is the distance between the central ion and a given ligand nucleus and Θ is the angle between the vector r and the principal axis of symmetry of the complex. The ligand field coefficient in the third factor, L, is defined according to Abragam and Bleaney (1970) and for a given complex is assumed to be a constant along the lanthanide series. The last factor, Q, has been calculated by Bleaney (1972). For a given ligand nucleus in an isostructural series of complexes, the relative shifts, induced by the different lanthanides along the lanthanide series, will be proportional to factor Q.

It is convenient to express the dipolar shift in axially symmetric systems by the following shorthand notation:

$$\delta_{\text{i,hd}} = D \cdot G_{\text{i}} \tag{16}$$

where $G_{\text{i}} = (3\cos^2\Theta_{\text{i}} - 1)/r_{\text{i}}^3$ is the geometrical function specific for a given ligand nucleus, i, and D is a temperature-dependent factor characteristic of the lanthanide ion.

2.4.2. Induced Relaxation

Nuclear relaxation is generated by the time-dependent terms of the nuclear spin Hamiltonian. The magnitude of the relaxation effect in paramagnetic lanthanide complexes is predominantly dipolar in origin and is governed by the intensity of the electron–nuclear interaction and by the rate at which this interaction is interrupted. Thus, the relaxation rates of ligand nuclei are determined by the molecular structure and by the molecular dynamics in solution and may in turn be used for studying these factors.

The electron spin relaxation time, T_{1e}, in the lanthanide cations is of the order of 10^{-13} sec (except Gd^{3+}). Thus, T_{1e} is always much shorter than any rotational reorientation time, τ_r. Consequently, the electron–nuclear dipolar interaction in a low-molecular-weight lanthanide complex (except Gd^{3+}) will contribute to the nuclear relaxation rate of a given ligand nucleus according to (Reuben and Fiat, 1969b):

$$1/T_{1M} = 1/T_{2M} = (20/15) \cdot \gamma_N^2 \cdot g_N^2 \cdot \beta^2 \cdot J(J+1) \cdot r^{-6} \cdot T_{1e} \tag{17}$$

where T_{1M} and T_{2M} are, respectively, the intrinsic longitudinal and transverse relaxation times of a given ligand nucleus in its lanthanide-bound state, γ_N is the nuclear magnetogyric ratio, r is the nucleus–Ln^{3+} distance, g_N is the nuclear g-factor, J is the resultant electronic spin angular momentum (in units of $h/2\pi$), and β is the Bohr magneton.

In relatively high-molecular-weight complexes, a so-called "susceptibility term" (Gueron, 1975; Vega and Fiat, 1976) becomes an additional, significant contribution to an extent that depends on molecular size, as manifest in the molecular rotational time, τ_r. The dependence on τ_r of the relative importance of this term versus the term shown in Eq. (17) is depicted in Mukidjam *et al.* (1987). The contribution of the contact interaction to the nuclear relaxation in lanthanide complexes can be shown to be usually negligible. Note [Eq. (17)] the inverse sixth-power dependence on the lanthanide–nucleus distance that, in favorable cases, allows accurate determination of relative distances within the complexed ligand.

In the case of Gd^{3+}, whose T_{1e} is relatively long, more than one dynamic mechanism contributes to the overall correlation time, τ_c. Thus, the approximations that lead to the relatively simple formula in Eq. (17) cannot be applied, and the full correlation function should rather be used, yielding

$$\begin{aligned} 1/T_{1M} &= 2 \cdot C\left(\frac{3\tau_c}{1+\omega_N^2\tau_c^2} + \frac{7\tau_c}{1+\omega_e^2\tau_c^2}\right) \\ 1/T_{2M} &= C\left(4\tau_c + \frac{3\tau_c}{1+\omega_N^2\tau_c^2} + \frac{13\tau_c}{1+\omega_e^2\tau_c^2}\right) \end{aligned} \tag{18}$$

where

$$C = \frac{1}{15}\gamma_N^2 \cdot g_N^2 \cdot \beta^2 \cdot J(J+1)r^{-6}, \qquad 1/\tau_c = 1/T_{1e} + 1/\tau_r + 1/\tau_M$$

and ω_N and ω_e are the resonance frequencies of the observed nucleus and of the unpaired electron, respectively, at the operating external magnetic field, and τ_M is the lifetime of the observed ligand nucleus in the gadolinium coordination sphere. Note, however, that within a given gadolinium complex the ratios of the relaxation rates of different ligand nuclei will be proportional to the inverse sixth power of the respective distance ratios without the need to calculate the above terms or to make any assumptions about them.

3. EXPERIMENTAL METHODS AND PROBLEMS

3.1. Shift-Reagent Method

3.1.1. Mechanism of Shift-Reagent Action

The method of induced differentiation in the NMR spectra of aqueous cationic species either by paramagnetic lanthanide shift reagents or by the corresponding gadolinium complex which acts as a relaxation reagent was introduced by Elgavish and Reuben (1977). The shift method was first applied by Gupta and Gupta (1982) to the problem of the separation of the ^{23}Na NMR signals of Na_i and Na_0 in cells and tissues. This is currently the method most frequently used for observation of the Na_i signal. The Na_i and Na_0 signals can be separated in the ^{23}Na NMR spectrum using a lanthanide shift reagent. These reagents do not cross the cell membrane and thus interacting only with the extracellular ions they cause a paramagnetic shift in the frequency position of the Na_0 signal away from the unshifted position of Na_i.

The discrimination between the ^{23}Na NMR resonances in the two compartments, extracellular and intracellular, is enabled by the fact that the appropriate shift reagents are membrane impermeable, so that only the extracellular resonance is affected by the paramagnetic chelate. The magnetically important moiety in a paramagnetic chelate, serving as a shift reagent, is the lanthanide metal ion. The lanthanides are the 15 elements in the periodic table between atomic numbers 57 and 71. Their trivalent cations display a calciumlike chemistry and form multidentate complexes—chelates—with a large number of polyvalent organic and inorganic anions that serve as ligands. Thirteen of the lanthanide ions possess unpaired electrons in their inner (4*f*) electron shell. This property gives rise to the unique magnetic behavior called paramagneticity. This paramagneticity consists of creating a local magnetic field around the lanthanide ion. The effective strength of this field is quite large, and sometimes is comparable to that of the external field generated by the magnet of the NMR instrument. Unlike the latter, however, the paramagnetic effect of the lanthanide ion decays within very short distances, typically becoming negligible beyond about 1 nm. Consequently, for an NMR-observed nucleus to be significantly affected, this nucleus has to be within such a short distance from the paramagnetic center. In the case of the positively charged sodium ion, this, most likely, necessitates the formation of ion pairs with negatively charged functional groups of a ligand complexed to the lanthanide ion (see Fig. 41, p. 234).

During the time an individual sodium ion resides in the proximity of the lanthanide metal ion, its NMR spectral parameters will be strongly affected (see Section 2.4). The intrinsic chemical shift position of the bound sodium is shifted by up to hundreds of ppm in the ^{23}Na NMR spectrum, its T_1 and T_2 relaxation times shortened, and its linewidth extensively broadened. Shift reagents, however, are used at a concentration typically 10- to 30-fold lower (4–12 mmole/liter) than the concentration of Na_0^+ (140 mmole/liter). Consequently, the effect of the shift reagent on all of the sodium ions is mediated by a constant exchange between the ions bound to the reagent and the much larger pool of ions free in the aqueous solution. Since the rate at which this exchange occurs is fast compared with the relevant NMR time scale, averaging of all NMR parameters takes place between the two pools of sodium ions, the shift-reagent-bound and the free. The resulting observed values of chemical shift and relaxation rates then become the weighted averages of the corresponding intrinsic values in the two pools, and the mole fractions of sodium in these pools are the weighting factors. Nevertheless, although the mole fraction of the bound ions is typically quite small, because of the large magnitude of the paramagnetic effect its contribution is still significantly reflected in the weighted average values observed in the actual spectrum.

Because of the short-range effect of lanthanide paramagneticity, sodium ions that have no immediate access to the chelate, and are also in slow exchange with both the bound as well as the free extracellular pools, are not affected by the presence of a shift reagent. Such is the case with Na_i when the shift reagent is restricted to the extracellular compartment. The result in the NMR spectrum is the separation of the sodium signal into two resolved resonances: one, arising from Na_0, which is shifted from the original position, and a second resonance, arising from the intracellular pool, which remains unshifted. If there are several compartments which contain sodium ions in slow exchange with the extracellular, shift-reagent-affected, sodium pool, they all give rise to one superimposed resonance in the unshifted position. Therefore, the method of shift-reagent-aided spectroscopy does not differentiate, in principle, among the various possible intracellular pools of sodium ions.

3.1.2. Shift Reagents

3.1.2.1. General. Paramagnetic lanthanide metal ions with non-S-state electronic configuration, e.g., Dy^{3+}, can affect the chemical shift position of nuclei in their immediate proximity. Furthermore, the lanthanide ion can be chelated to ligands which will preferentially interact with a particular nucleus, such as Na^+ with the tripolyphosphate (PPP) ligand. Also, the choice of ligand can affect the properties of the complex which determine its clinical toxicity, such as a tendency to bind Ca^{2+}. Although a variety of lanthanide chelates have been offered as aqueous shift reagents for cationic NMR, reagents typically used are $Dy(PPP)_2$, DyTTHA, where TTHA is triethylenetetraminehexaacetic acid, and more recently DyDOTP and TmDOTP, where DOTP is tetraazacyclododecane-*N,N′,N″,N‴*-tetramethylenephosphonate (Fig. 4). Recently, polyoxa tetraaza macrocyclic complexes of dysprosium have been synthesized and characterized to serve as calcium-inert shift reagents for Na NMR (Sink *et al.*, 1990). One of them produces a dysprosium complex which is indeed more calcium inert than any other existing reagent, but its small induced shift makes it impractical as a shift reagent for most purposes.

3.1.2.2. $Dy(PPP)_2$. Dy chelated to two equivalents of the PPP ligand gives the largest shift known to date, but has a tendency to bind Ca^{2+} and in the presence of 4–6 mM of this reagent only low concentrations (0.1–0.7 mM) of free Ca^{2+} can be attained routinely (Fig. 5). Even to achieve these levels of free Ca^{2+}, about 6–8 mM of $CaCl_2$ needs to be added.

Lanthanide complexes containing PPP can be troublesome to use in buffer solutions because they bind Ca^{2+} and may precipitate. Therefore, a balance between added $CaCl_2$ and an acceptable shift must be obtained. The exact manner in which complexes are prepared can have a direct effect

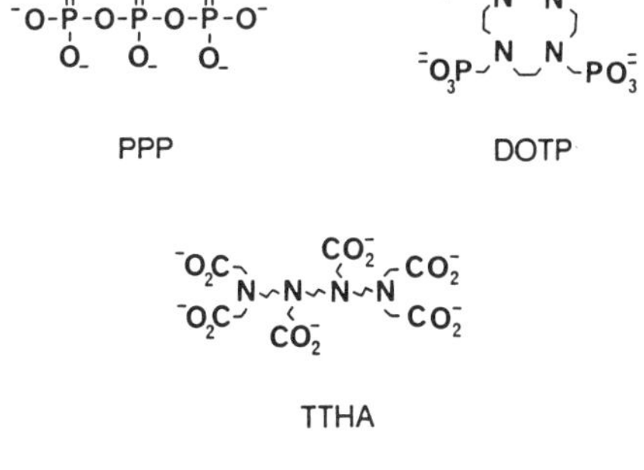

Figure 4. Chemical formulas of three shift reagent ligands: triphosphate (PPP), tetraazacyclododecane-*N,N′,N″,N‴*-tetramethylenephosphonate (DOTP), and triethylenetetraminehexaacetic acid (TTHA).

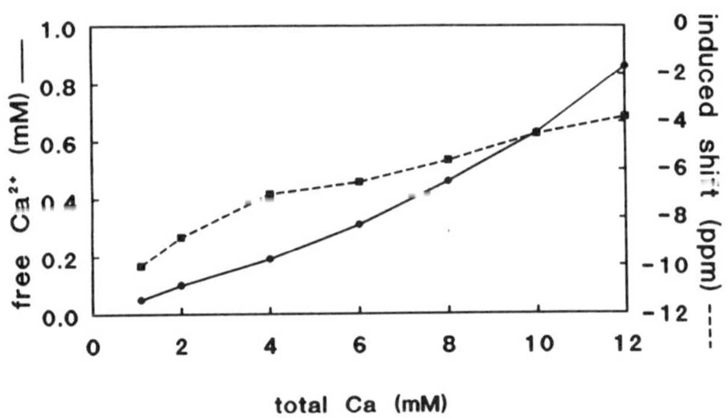

Figure 5. The induced ^{23}Na NMR shift (----) and the free Ca^{2+} (—) versus total Ca^{2+} in a Hepes buffer solution containing $Dy(PPP)_2$. The shift was varied by the addition of $CaCl_2$. Calcium addition causes removal of $Ca(PPP)^{3-}$ and $Dy(PPP)^{2-}$ from the solution, creating an inverse relationship between the induced shift and the concentration of free Ca^{2+}. (From Miller *et al.*, 1991.)

on the stability of the buffer solution. Our group has had success with preparing $Dy(PPP)_2$ using a modified procedure of Burstein and Fossel (1987c). Two solutions of $Na_5P_3O_{10}$(PPP) are prepared: the first, a 550 mM solution, is prepared ensuring complete dissolution by adjusting pH between 7 and 8; the second, a 250 mM solution, is prepared without pH adjustment. Dysprosium chloride ($DyCl_3 \cdot 6H_2O$) is dissolved to a concentration of about 250 mM with a pH of 4.8. The exact concentration is determined by colorimetric techniques. The dysprosium stock solution and the 550 mM PPP solution are then mixed in a 1:1 volume ratio. A precipitate forms from this mixture. The 250 mM PPP solution is added to this mixture until the precipitate dissolves. The pH is then adjusted to 7.0. We use the $Dy(PPP)_2$ stock mixed into buffer solutions, e.g., Hepes, that contain no carbonates. Because of the relatively low solubility of lanthanide carbonates, carbonate ions would cause sequestration and precipitation of lanthanide metal ions from the shift reagent chelate. In some studies, e.g., in perfused hearts, extra Ca^{2+} must be added to increase the free Ca^{2+} concentration. This causes some precipitation, and therefore the final solution is filtered through a micropore filter after the precipitate has settled.

The major problem with $Dy(PPP)_2$ is its high affinity for calcium and the upper limit of about 0.5–0.7 mM it sets to the free Ca^{2+} concentration. This in turn is particularly deleterious in perfused heart experiments because of the direct correlation between the extracellular free calcium concentration (Ca_0^{2+}) and developed contractile force (DF). Nevertheless, this limitation is not as severe as would be expected on the basis of typical DF versus Ca_0^{2+} curves. We routinely have been obtaining in isolated perfused rat hearts systolic pressures in the range of 50–80 mm Hg, and occasionally

100–120 mm Hg, while on the basis of the Ca_0^{2+} concentration as measured by a calcium-specific electrode a pressure of only about 10–20 mm Hg should be obtained. It seems that calcium is bound to the $Dy(PPP)_2$ chelate and that this bound calcium, not detected by the calcium-specific electrode, serves as "contractile calcium" in addition to Ca_0^{2+}. Indeed, approximately one equivalent of calcium ions has been found bound to $Dy(PPP)_2$ in our buffers prepared for heart perfusion (Chu and Elgavish, unpublished). Unfortunately, however, it seems that some yet unknown and thus experimentally uncontrolled mechanism, possibly an equilibrium between two different structures of the $Dy(PPP)_2{\cdot}Ca$ complex, makes difficult the routine preparation of perfusates that retain contractile function at the level of pre-shift-reagent control despite frequent, but random, success. The existence of such different $Dy(PPP)_2{\cdot}Ca$ complexes, most probably with stoichiometries of 1:1 and 2:1, has been shown by ^{43}Ca NMR (Vogel and Braunlin, 1985).

Despite the disadvantages discussed above, $Dy(PPP)_2$ has two major advantages: its large induced shift which produces excellent spectral resolution, and its simple spectrum, i.e., two resonances, a truly unshifted Na_i and an upfield shifted Na_0 signal. Figure 6 shows spectra from a rat heart

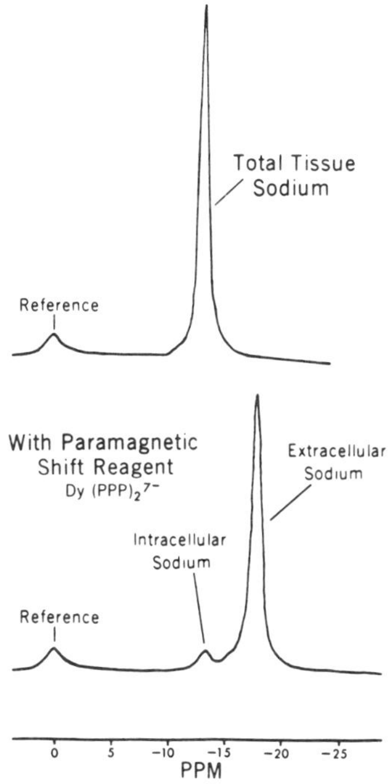

Figure 6. ^{23}Na NMR spectra of an isolated perfused rat heart in the absence (top) and presence (bottom) of the paramagnetic shift reagent $Dy(PPP)_2$. (From Elgavish, 1989.)

perfused with $Dy(PPP)_2$ (5.5 mM). A 5.0-ppm shift of the Na_0 peak is evident, exposing the Na_i resonance. The Na_0 resonance probably contains the contributions of all extracellular compartments, mainly interstitium, vasculature, and the bath surrounding the heart. This seems to indicate that $Dy(PPP)_2$ distributes quickly and homogeneously in all of these compartments without creating concentration gradients and complex BMS effects.

3.1.2.3. Dy(TTHA). Dy(TTHA) can tolerate large concentrations of Ca^{2+} with the addition of $CaCl_2$ to overcome calcium buffering by the reagent, but the shift is small and the signals are relatively broad and thus Na_i is poorly resolved from the extracellular resonances (Fig. 7). Furthermore, Na NMR spectra in organs in the presence of Dy(TTHA) display

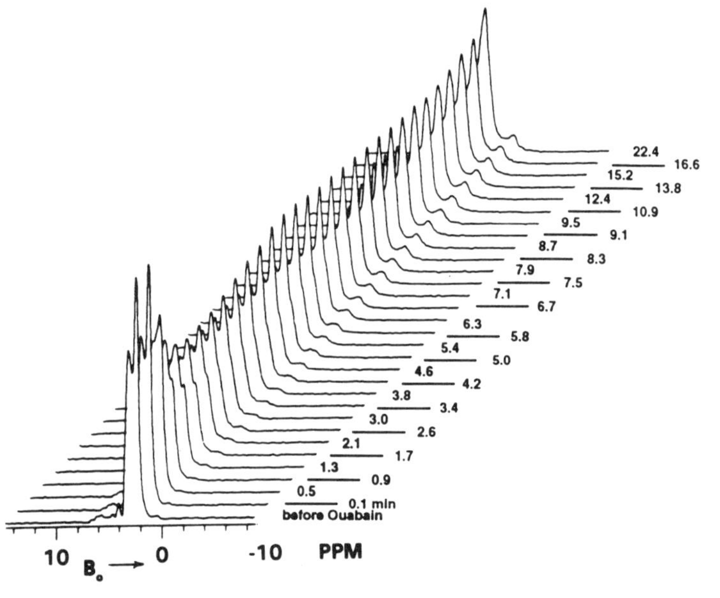

Figure 7. ^{23}Na NMR spectra (95.2 MHz, 8.5 T) of an isovolumic Langendorff-perfused beating rat heart. Before the acquisition of these spectra, the heart had been perfused for 20 min with the standard medium. The heart rate was 250 beats/min, the systolic pressure 175 mm Hg, and the coronary flow 24 ml/min. The first spectrum was obtained at the end of a subsequent 30-min period during which the heart was perfused with a solution containing 10 mM $Dy(TTHA)^{3-}$ and during which the heart rate was 204 beats/min and the systolic pressure was 175 mm Hg. Subsequent spectra are labeled with the time (midpoints of acquisitions) elapsed since a perfusion medium containing 10 mM $Dy(TTHA)^{3-}$ and 1 mM ouabain entered the heart. During the 23 min of this perfusion, the coronary flow was 20 ml/min. During all three perfusions, the heart was bathed with a flowing solution containing 10 mM $Dy(TTHA)^{3-}$. Each spectrum was the result of the acquisition of 96 transients except the last (22.4 min), for which only 48 transients were obtained. The vertical scale was adjusted to correct for this. The spectra displayed have been resolution enhanced with a DM value of 20. (From Pike *et al.*, 1985.)

more signals than with other shift reagents. While sodium spectra of isolated hearts perfused with $Dy(PPP)_2$ consist of only two signals, a relatively sharp, unshifted Na_i resonance and a somewhat broadened and upfield-shifted Na_0 resonance, spectra of hearts perfused with Dy(TTHA) have seven or eight poorly resolved peaks, with the one assigned as "unshifted" Na_i being in fact shifted 0.5–0.6 ppm downfield (Pike *et al.*, 1985; Kohler *et al.*, 1991). A valiant effort was carried out by Kohler *et al.* (1991) to resolve this problem. Through a series of modeling experiments and ischemia and reflow experiments in perfused hearts of various species, they identified the contributions to the sodium spectrum, in the presence of Dy(TTHA), of the four physical compartments in the perfused heart experiment. They also concluded that the Dy(TTHA)-aided ^{23}Na NMR spectrum is sensitive to effects of BMS and concentration gradients, and thus to heart geometry and positioning and to differences in animal species.

Although their work has led to a better understanding of the reasons for the complexity of the Dy(TTHA)-aided ^{23}Na NMR spectrum, the complexity and the typically poor spectral resolution induced by this reagent remain, and still hamper reliable Na_i quantitation. Furthermore, several questions remain unanswered: Why is Dy(TTHA) uniquely sensitive to the above-mentioned effects in contrast to other dysprosium chelates? After all, BMS effects, being dependent on the magnetic moment and concentration of the paramagnetic metal ion, should be similar for all dysprosium chelates at similar concentrations. Is it possible that, unlike other shift reagents, Dy(TTHA) distributes, and is retained, in tissue with a pattern that results in unique compartmentation and in local concentration gradients? Answers to these questions may be interesting *per se*, but these possibilities are hardly an encouragement for the use of this particular shift reagent.

The TTHA ligand is commercially available and forms a stable complex with Dy^{3+}. A slight excess of TTHA (1.1 : 1) should be coordinated with Dy^{3+} since free Dy^{3+} is quite toxic (Evans, 1990). The free Ca^{2+} concentration can be adjusted to a high (3 mM) level without precipitation problems. The Dy(TTHA) shift is unaffected by addition of Ca^{2+} and therefore it can be used with higher free Ca^{2+} concentrations.

3.1.2.4. Dy(DOTP) and Tm(DOTP). A new, stable ^{23}Na shift reagent, $Dy(DOTP)^{5-}$ (DOTP = 1,4,7,10-tetraazacyclododecane-*N*,*N*′,*N*″,*N*‴-tetramethylenephosphonate), has been offered by Sherry *et al.* (1988). With human erythrocytes it yielded a shift close in magnitude to that generated by $Dy(PPP)_2$, but allowing a somewhat higher free Ca^{2+} concentration in the perfusate.

Several recent studies used Tm(DOTP), a reagent developed by Sherry *et al.* (1988), as a shift reagent for ^{23}Na spectroscopy. Tm(DOTP) produces a significantly larger shift than Dy(TTHA) (Fig. 8) but not as large as $Dy(PPP)_2$, and can tolerate free Ca^{2+} levels up to about 1.0 mM while

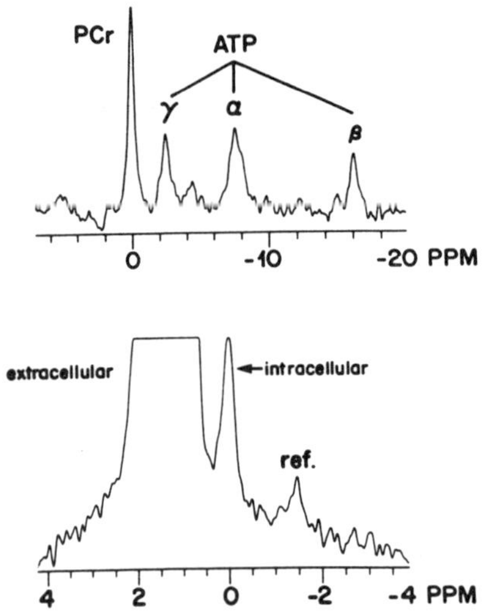

Figure 8. ^{31}P and ^{23}Na NMR spectra of a perfused rat heart. Each spectrum was acquired in 1 min; 3.5 mM $Tm(DOTP)^{5-}$ was present in the perfusate. The ^{31}P spectrum (upper panel) shows phosphocreatine (PCr) and ATP. Extracellular sodium, intracellular sodium, and sodium in the intraventricular balloon (ref.) are detected in the ^{23}Na spectrum (lower panel). (From Malloy *et al.*, 1990.)

retaining a reasonable shift (Buster *et al.*, 1990). Tm(DOTP), however, produces greater line broadening than Dy(TTHA), and thus a larger shift is needed for improved resolution between Na_i and Na_0. Tm(DOTP) binds Ca^{2+} but is stable with added Ca^{2+} up to a level of about 1 mM free Ca^{2+}, above which precipitation occurs. Fig. 9 shows the relationship between added Ca^{2+} and Tm(DOTP) concentration. BMS effects and hyperfine shifts were determined for Tm(DOTP) and compared to those of Dy(TTHA). It

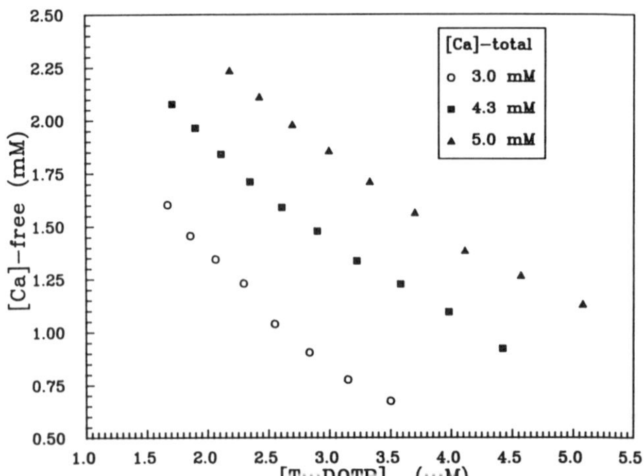

Figure 9. The effect of increasing $[Tm(DOTP)]^{5-}$ on the level of free calcium ions ([Ca]-free) measured by a Ca-sensitive electrode at 37°C from modified K-H buffer solutions prepared with total Ca^{2+} ion concentrations of 3.0, 4.3, and 5.0 mM. (From Buster *et al.*, 1990.)

was found that Tm(DOTP) had low BMS to hyperfine shift ratios indicating a dominance of the hyperfine interaction even in the presence of Ca^{2+} in a manner comparable to $Dy(PPP)_2$ (Kohler *et al.*, 1990). The DOTP ligand can be synthesized from the HBr salts of the corresponding cyclic amines via a Mannich-type reaction (Sherry *et al.*, 1988).

In *in vivo* studies, slightly better resolution of Na_i is achieved with Tm(DOTP) than with Dy(TTHA) but complete resolution of Na_i is not achieved and larger doses appear to be acutely toxic to the rat (Albert *et al.*, 1990b). The large affinity for Ca^{2+}, the high osmolarity at concentrations required for *in vivo* experiments, and the instability at high concentrations render $Tm(DOTP)^{5-}$ difficult, but effective, to use. Several studies with Tm(DOTP) in the heart have been recently presented (Kohler *et al.*, 1990; Pike *et al.*, 1990), as well as in the eye (Kolodny *et al.*, 1990) and in *in vivo* spectroscopy of the rat head (Albert *et al.*, 1990b).

3.2. Relaxation Method

3.2.1. T_1 Modification

Degani and Elgavish (1978) were the first to show that the Na_i and Na_0 resonances could be distinguished by paramagnetically modifying the T_1 of the Na_0 signal. The relaxation agent Gd(EDTA) was devised and added to a suspension of phosphatidylcholine vesicles which had been prepared to contain both intra- and extravesicular Na^+. The Gd(EDTA) complex did not cross the vesicular membrane and thus only Na_0 was appreciably affected by the paramagnetic chelate, shortening its T_1 predominantly by a hyperfine dipolar interaction but also partly by enhancing its quadrupolar relaxation rate (Elgavish, 1982). ^{23}Na NMR spectra of the vesicular suspension were acquired using an inversion-recovery pulse sequence with an interpulse delay time selected such that the sequence would null the Na_0 signal. Since the T_1 of the Na_0 signal was paramagnetically shortened, it would also null at a shorter delay time than would the Na_i signal with its now relatively longer T_1. The Na_i signal would now lag in recovery and remain inverted (Fig. 10). Since the Gd chelate did not introduce any appreciable shift in the chemical shift position, the nulled Na_0 signal and the inverted Na_i signal remained superimposed. This fact has somewhat hindered accurate quantitation of the area of the Na_i signal using this method.

3.2.2. T_2 Modification

Renshaw *et al.* (1986) exploited alteration of the T_2 value between the Na_0 and Na_i. They used dextran-magnetite as a relaxation enhancer because it has the effect of broadening the ^{23}Na signal and is too large to cross the

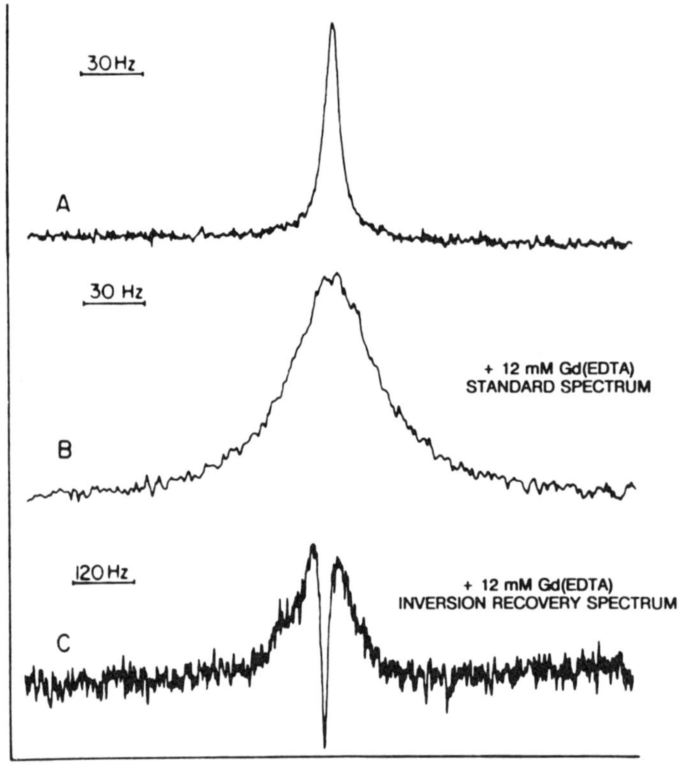

Figure 10. ^{23}Na NMR spectra of: (A) 150 mM NaCl dissolved in phosphatidylcholine vesicle suspension p^2H 9.2 at 28°C; (B) the same as A after addition of 11.6 mM Gd(EDTA); (C) the same as B using 180°–τ–90° pulse sequence in order to separate the signal due to the ions entrapped in the inner vesicular medium. Forty transients were accumulated in A and B and 4800 transients with τ = 4 msec in C, with a recycle time of 1.1 sec. (From Degani and Elgavish, 1978.).

cell membrane. The ^{23}Na resonance in a saline solution has a linewidth of approximately 20 Hz, whereas in the presence of dextran-magnetite it displayed a linewidth of 6900 Hz (Fig. 11). In *in vivo* measurements where dextran-magnetite was infused into a rat reaching a circulatory concentration of 0.1 μM, the following percentage reduction in Na_0 was observed: hind leg (14 ± 4), liver (19 ± 2), kidney (47 ± 4), and heart (53 ± 4). As Na_i is not well resolved in these spectra, this is not the best method for Na_i quantitation.

3.3. Shift-Relaxation Method

The shift-relaxation method, developed in the isolated, perfused heart system (Miller *et al.*, 1991), is a combination of the relaxation method and

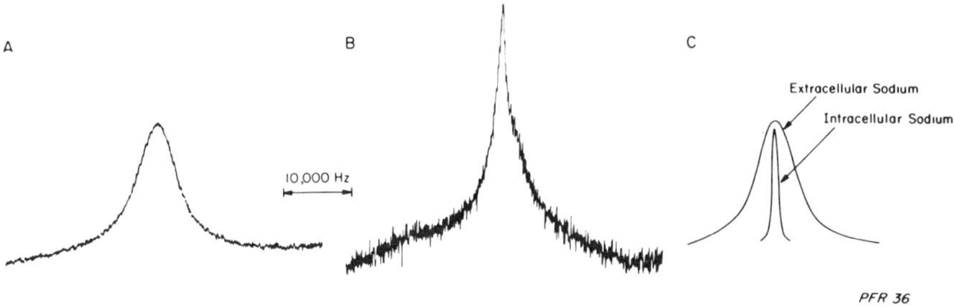

Figure 11. Sodium spectra (52.9 MHz). (A) Spectrum from 1.5-ml sample of 140 mM NaCl, 20 mM phosphate, 10 mM KCl, and 3.0 μM dextran-magnetite particles. Ten thousand averages were collected with a repetition rate of 0.1 sec and a 90° flip angle. (B) Spectrum of a 0.5-ml erythrocyte sample after washing and centrifugation. Thirty thousand averages were collected for this spectrum. (C) Spectrum in B is represented as the sum of two components, one of which has the lineshape shown in A and represents the extracellular sodium. The narrow component arises from the intracellular sodium. (From Renshaw *et al.*, 1986.)

the shift method. Previously, an approach utilizing the relaxation enhancement induced by a shift reagent was shown in solution, and a combined method proffered (Brown *et al.*, 1986). In our method, the sample is perfused with a lanthanide (Ln) agent such as $Ln(PPP)_2$ where Ln represents both a shift-inducing lanthanide ion, e.g., Dy^{3+}, and a lanthanide ion that strongly affects the relaxation rates, i.e., Gd^{3+}. As long as these lanthanide chelates do not cross the cell membrane, only the Na_0 resonance will experience the shift and relaxation effects.

The ^{23}Na spectra should be acquired using an inversion–recovery pulse sequence which consists of a 180° pulse, followed by a delay time τ, followed by a 90° pulse for acquisition. The signal intensity obtained using this pulse sequence is a function of T_1 and τ:

$$I_\tau = I_0(1 - 2e^{-\tau/T_1}) \tag{19}$$

where I_τ and I_0 are the signal intensity with a delay time τ and the equilibrium intensity, respectively. Based on the difference between the T_1 of the Na_i and the Na_0, it is possible to select a τ value which minimizes or nulls the shifted Na_0 signal leaving an Na_i signal which is a constant fraction of its original intensity as long as its T_1 remains constant. To illustrate the differential behavior of the Na_i and Na_0 signal intensities, induced by this method, the simulated signal intensities of these two resonances, generated with an inversion–recovery sequence, are shown in Fig. 12 as a function of the delay time selected. Using parameters typical for a perfused heart experiment, it is evident that optimal reduction of Na_0 would be obtained with $\tau = 6.2$ msec. Actual spectra, obtained with 100 scans from a rat heart perfused and submerged in $Dy(PPP)_2$:$Gd(PPP)_2$ at 2.5 mM:2.0 mM concentration, are shown in Fig. 13. Good resolution was

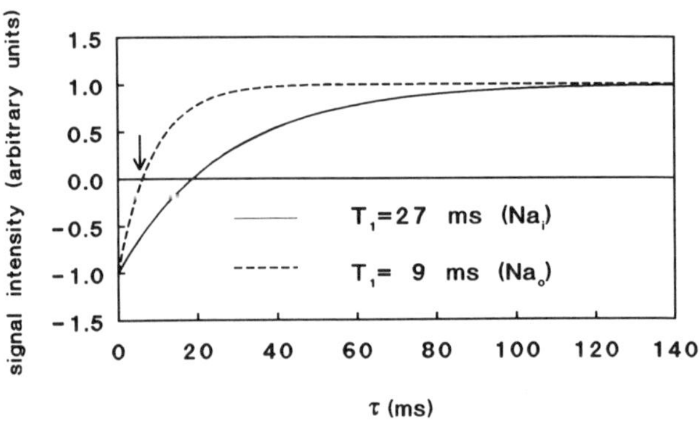

Figure 12. A simulation of signal intensity versus the delay time (τ) using an IR pulse sequence. Intensities were calculated with: $I = I_0\,[1\text{-}2\exp(-\tau/T_1)]$, $T_1 = 9$ msec, and $T_1 = 27$ msec, values which represent the T_1 of Na_0 in a $Dy(PPP)_2$:$Gd(PPP)_2$ (4 mM:2 mM) perfusate, and the T_1 of Na_i in rat hearts, respectively. (From Miller *et al.*, 1991.)

obtained with a 2.5-ppm shift between the Na_i and Na_0 resonances. When the spectrum was acquired using an inversion-recovery pulse sequence (with $\tau = 5.5$ msec) the Na_0 signal was reduced to $6.4 \pm 2.0\%$ of its original intensity, while $62 \pm 7\%$ of the equilibrium intensity of Na_i was retained. Since in this method the residual intensity of Na_i is a function of T_1, it is important that the latter remain constant during any experimental perturbations applied.

3.4. Selective Inversion–Recovery Method

A disadvantage of the shift-relaxation method discussed above is a reduction in the intensity of Na_i resulting from the nonselectivity of the inversion pulse. We have recently developed a selective inversion recovery (SIR) method to minimize the Na_0 signal without affecting the intensity of the Na_i resonance and have demonstrated its application in the isolated perfused rat heart model (Simor *et al.*, 1991).

^{23}Na spectra were obtained on a Bruker AM 360 wide-bore NMR spectrometer (8.5 T) equipped with a broadband probe accepting 20-mm sample tubes. The SIR pulse sequence consisted of a "soft" 180° preparation pulse followed by a delay time (τ), followed by a nonselective "hard" 90° sampling pulse and data acquisition. In spectra acquired in this manner, quantitation of Na_i could be carried out with good spectral resolution even in the presence of relatively small shifts (~2.5 ppm) between the Na_i and Na_0 signals. The ability to improve resolution with such small shifts should allow quantitation of Na_i with calcium-tolerant agents [e.g., Dy(TTHA)] which, however, induce relatively small shifts, or should allow a reduction

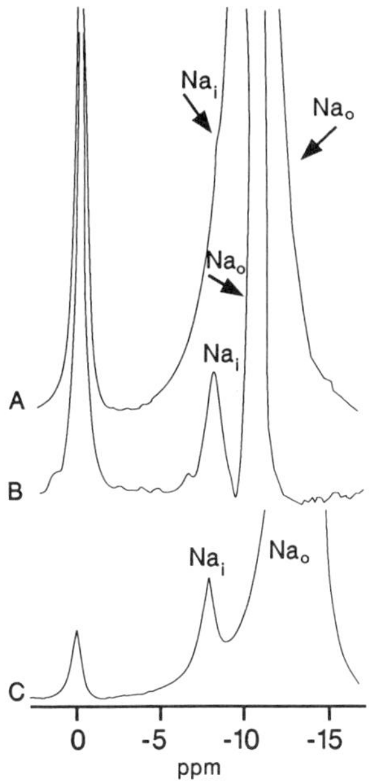

Figure 13. ^{23}Na NMR (95.132 MHz) spectra of the rat heart submerged in perfusate, and perfused with (A) Hepes plus the "shift-relaxation" reagent Dy(PPP)$_2$:Gd(PPP)$_2$ (2.5 mM:2 mM) using a 90° one-pulse experiment; (B) the same perfusate as in A with data acquisition by an inversion–recovery pulse sequence and phased upright with the scale adjusted such that the reference signal is of the same height as in spectrum A; (C) Hepes plus shift reagent Dy(PPP)$_2$ (4.5 mM) using a 90° one-pulse experiment and the scale adjusted such that tha Na$_i$ is similar in height as in B. Spectra were acquired with 100 scans, 4000 data points, and a 1-kHz spectral window. Note the good resolution between Na$_i$ and Na$_0$ in B, in contrast to the only partial resolution in C. (From Miller *et al.*, 1991.)

in the concentration of Dy(PPP)$_2$. Either way, the deleterious impact of these agents on contractile function can be substantially reduced.

To obtain, in perfused hearts, a ^{23}Na NMR spectrum with good resolution using conventional single-pulse acquisition, a shift of at least 5.5 ppm is necessary. This usually requires the use of Dy(PPP)$_2$, which due to its limiting effect on free Ca^{2+} concentration depresses (typically to about 50 mm Hg) the left ventricular developed pressure (LVDP). Using a lower concentration of this shift reagent (3.5 mM) or using 10 mM Dy(TTHA) improves LVDP (to 70–90 mm Hg). The disadvantage, however, is a shift reduced to about 2.5–3 ppm. Presaturating the Na$_0$ resonance in Dy(PPP)$_2$-aided spectra improves resolution, but may affect the Na$_i$ intensity due to saturation transfer.

Figure 39 (panel I; see p. 233) displays ^{23}Na NMR spectra that were acquired using 10 mM Dy(TTHA). LVDP was 108% of pre-shift-reagent control, and induced shift was 2.45 ppm. Spectrum A was acquired by a single pulse and was filtered by exponential multiplication. In B the same spectrum is shown after Gaussian filtering. Using either of these methods the typical poor resolution of Na_i obtained with Dy(TTHA) is evident. Spectrum C was obtained with the SIR method. Reasonable resolution allowing quantitative analysis of Na_i is demonstrated.

Figure 39 (panel II, p. 233) shows ^{23}Na NMR spectra of isolated rat hearts perfused with Hepes buffer solution containing high (6 mM) (A) or low (3.5 mM) (B, C) concentration of $Dy(PPP)_2$. LVDP in A and B was 50 and 80% of pre-shift-reagent control, respectively. Spectra A and B were obtained with single pulse acquisition. Spectrum B differentiated between Na_0 and Na_i, but total separation was not achieved. Spectrum C was obtained with the SIR method. Good spectral resolution is evident, allowing reliable quantitative analysis of Na_i.

3.5. Multiple Quantum Filter Method

Interactions between the quadrupole moment of the ^{23}Na nucleus and fluctuating electric field gradients at sodium-binding macromolecular sites may cause quadrupolar splitting and thus lift the degeneracy among the three NMR transitions of the sodium nucleus (see Section 2.1). This creates the possibility of generating double-quantum coherences between the $\pm 3/2$ and $\pm 1/2$ Zeeman levels. When the correlation time is long enough so that the condition for extreme narrowing does not hold, even when this is the case for only less than 1% of a pool of fast-exchanging nuclei, a biexponential transverse relaxation pattern is observed for all sodium nuclei in the pool. Such a ^{23}Na resonance which relaxes biexponentially can be passed through a state of double-quantum coherence, while monoexponentially relaxing Na with its three degenerate transitions cannot. Using this property, and assuming that while Na_i relaxes biexponentially Na_0 does not, it is theoretically possible to suppress the Na_0 resonance selectively. This would reveal the Na_i resonance without the use of a shift or relaxation reagent. Such double-quantum-filtered Na spectra were obtained in canine red blood cells (Fig. 14) (Pekar *et al.*, 1987) using the pulse sequence INADEQUATE (Bax, 1982)

$$90^0_x\text{-}\tau/2\text{-}180^0_y\text{-}\tau/2\text{-}90^0_x\text{-}\delta\text{-}90^0_x\text{-}t(ACQUISITION) \qquad (20)$$

with a 64-step phase cycle routine to eliminate other coherences. Jelicks and Gupta (1989a) utilized the same sequence with a 356-step phase cycle. The first part of the sequence, i.e., 90°-τ/2-180°-τ/2-90°, creates a single-quantum coherence, prepares part of it for transfer to double-quantum

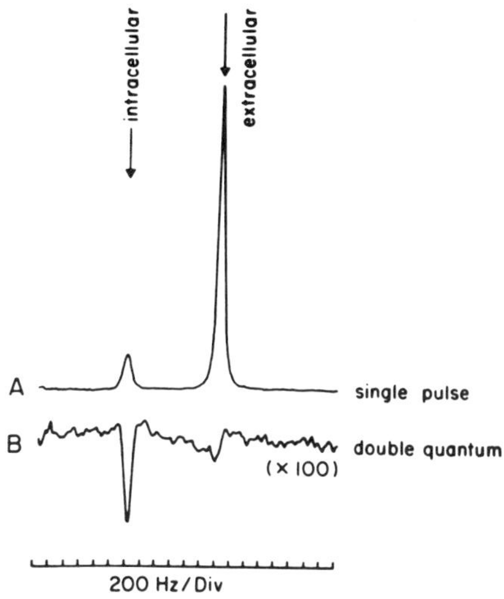

Figure 14. ^{23}Na NMR spectra of dog red cells suspended in isotonic medium containing 12 mM $Dy(P_3O_{10})_2^{7-}$. (A) Conventional spectrum; phased Fourier transform of averaged free precession signals following single 90° RF pulses. Sixty-four acquisitions in about 20 s. (B) Double-quantum-filtered spectrum; phased Fourier transform of averaged free precession signals using a phase-cycle pulse sequence. Thirty-two hundred acquisitions in about 17 min. (Vertical scale corrected for different number of acquisitions; intensity represents signal per scan.) (From Pekar *et al.*, 1987.)

coherence, and finally transfers it to double quantum with the second 90° pulse. The 180° pulse at the half-time of the preparation time τ eliminates transversal loss of coherence by refocusing the components of the magnetization dispersed by chemical shift. Not all magnetization is transferred into double quantum, but the phase cycle routine removes undesirable transitions. Finally, the last pulse, after a very short delay δ, reconverts the filtered double-quantum coherence to detectable single-quantum magnetization. The observed signal is then given by

$$f(t) = M_0(\tfrac{3}{4})(e^{-\tau/T_{2s}} - e^{-\tau/T_{2f}})\, e^{-\delta/T_{dq}} \cos(2\omega\delta)(e^{-t/T_{2s}} - e^{-t/T_{2f}})\, e^{i\omega t} \tag{21}$$

where T_{2f} is the T_2 of the more rapidly decaying outer, $-3/2 \rightarrow -1/2$ and $1/2 \rightarrow 3/2$ transitions, and T_{2s} is the T_2 of the slowly decaying, $-1/2 \rightarrow 1/2$ transition. T_{dq} is the T_2 of the double-quantum transition, τ is the double-quantum preparation time, δ is the short evolution time of the order of microseconds which serves to enable phase switching within the phase cycle routine, t is the variable time following the last pulse, ω is the resonance offset, and M_0 is the equilibrium longitudinal magnetization. The signal can be acquired as a function of the variable time t at a fixed value of τ which is selected to maximize the function $(e^{-\tau/T_{2s}} - e^{-\tau/T_{2f}})$ and thereby maximize the signal-to-noise ratio.

It should be pointed out that compared with single-quantum acquisition, the signal-to-noise ratio of corresponding spectra acquired by a double-quantum filter is reduced by an order of magnitude. The resulting penalty in total acquisition time to produce quantifiable Na_i resonances makes the practicality of this method in biological samples other than red blood cells somewhat difficult, although not impossible under particularly favorable experimental conditions.

A second potential problem of using the double-quantum method to obtain a differentiated Na_i without the need for a shift reagent was demonstrated by Jelicks and Gupta (1989a). Even in the relatively simple system of red blood cells, the extracellular sodium pool was found to contribute to the double-quantum signal. This fact was demonstrated by using $Dy(PPP)_2$ to shift the Na_0 signal while using the double-quantum method for acquisition (see Fig. 15). It is evident that the spectrum, although acquired with a double-quantum filter, is composed of an upfield-shifted signal in addition to the anticipated, unshifted signal due to Na_i. The shifted signal clearly should be attributed to a pool with fast-exchange access to $Dy(PPP)_2$, but no such access to the Na_i pool. Figure 15 also shows that with increasing $Dy(PPP)_2$ concentration, the double-quantum Na_0 component is progressively and selectively suppressed.

The paramagnetic quenching of the double-quantum extracellular component evident in the above experiment resulted most likely from the efficient dipolar relaxation of the sodium nucleus diverting it from the biexponential quadrupolar relaxation pathway necessary for the transfer to double-quantum coherence. This fact was further utilized by Jelicks and Gupta (1989b) using $Gd(PPP)_2$, a relaxation reagent which induced a substantially larger dipolar relaxation enhancement than its dysprosium analogue (Fig. 16). The nonquadrupolar, rapid, transverse relaxation induced by this exclusively extracellular gadolinium complex at 1 mM concentration allowed selective double-quantum observation of the intracellular component and thus solved the problem of signal differentiation, albeit not without the need for a paramagnetic reagent. This extracellular component which passes the double-quantum filter may be from interstitial sodium ions (Foy and Burstein, 1990).

Several studies that used double- and triple-quantum filters were recently presented on brain (Lyon *et al.*, 1990; Pekar *et al.*, 1990), muscle (Lyon *et al.*, 1990), liver (Lyon *et al.*, 1990), salivary gland (Seo *et al.*, 1990), and red blood cells (Seo *et al.*, 1990; Commodari *et al.*, 1990; Gullapalli *et al.*, 1990). Lyon *et al.* (1990) found that rates of Na^+ influx determined by the double-quantum method in muscle and liver were similar to those obtained with shift reagents. A study on cerebral ion transport suggested that multiple-quantum NMR might be more responsive than conventional NMR to Na^+ redistribution induced by physiological insults (Pekar *et al.*, 1990).

In general, obtaining a single ^{23}Na spectrum by using multiple-quantum methods takes at least 5 to 10 min compared with about 200 msec for a

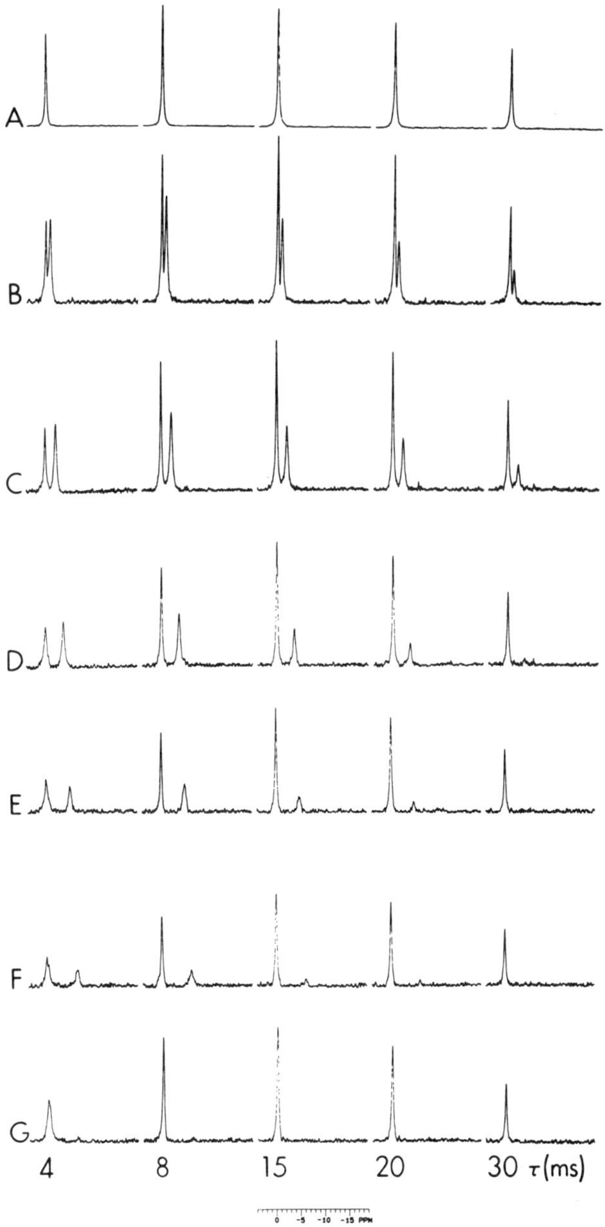

Figure 15. ^{23}Na double-quantum NMR spectra of human red cells washed with Hanks' balanced salt solution without shift reagent (A) (vertical scale reduced by a factor of four relative to other spectra), and with 1 mM (B), 2 mM (C), 3 mM (D), 4 mM (E), 5 mM (F), and 10 mM (G) $Dy(PPP)_2$. Spectra are presented in the absolute-value mode. In the presence of $Dy(PPP)_2$ the intra and extracellular ^{23}Na double-quantum resonances are resolved, the one to the right representing extracellular Na^+. Double-quantum preparation time was varied from 4 to 30 msec. (From Jelicks and Gupta, 1989a.)

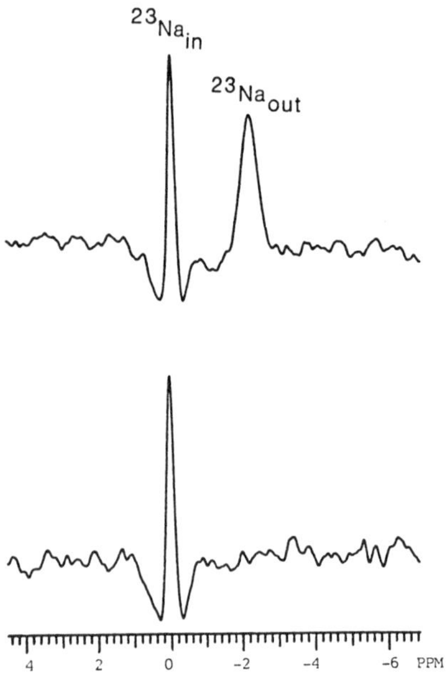

Figure 16. Double-quantum-filtered ^{23}Na NMR spectrum of red cells in the presence of 1.5 mM $Dy(PPP)_2$ (top). Addition of 1 mM $Gd(PPP)_2$ results in quenching of the shifted extracellular resonance (bottom). The number of transients was 2048. (From Jelicks and Gupta, 1989b.)

quantifiable shift-reagent-aided spectrum of Na_i produced in the presence of $Dy(PPP)_2$ (Elgavish *et al.*, unpublished). Therefore, the double-quantum method is most useful in experimental systems with very slow changes in the concentration of intracellular sodium, rather than in systems with a need for short time resolutions.

3.6. Measurement of Na_i Concentration

The determination of an absolute Na_i concentration from the corresponding NMR signal is not a trivial undertaking. Although the observed ^{23}Na NMR signal of Na_i is proportional to the concentration of the intracellular sodium ion pool, the proportionality factor contains several components not all of which can be easily determined in all experimental systems. Consequently, often in biological systems under study the *relative* Na_i signal intensity is presented, normalized to the intensity of the same signal under a control condition, e.g., before the onset of the studied perturbation. This manner of data presentation assumes no *absolute knowledge* of all components of the proportionality factor, only their *constancy* during the experiment. This approach may not provide Na_i concentrations in millimolar terms, comparable with values obtained by non-NMR techniques, but it also does not provide the often false sense of absolute knowledge not always really attainable.

There are several components that contribute to the proportionality factor: the NMR spectral visibility, still elusive in many biological systems (see Section 2.3); the filling factor, i.e., within what fractional volume inside the NMR receiver coil the spins contributing to the signal are distributed; strength of the H_1 RF field, which along with the duration of the sampling pulse determines the extent of signal excitation and which may be dependent on sample composition making calibration with an external standard sometimes misleading; and signal saturation, although due to the relatively short T_1 this is usually not a problem.

In relatively simple systems, reasonable steps can be taken for the accurate determination of some of the above factors, and for the elimination of others. In a suspension of red blood cells (RBC) (Gupta and Gupta, 1982) the Na_i concentration was determined by the following relationships:

$$S_0 = \frac{A_0}{A_c} \tag{22}$$

$$[Na_i] = \frac{A_i}{A_0} \cdot \frac{S_0}{(1 - S_0)} [Na_0] \tag{23}$$

where A_i and A_0 are the intensities of the Na_i and Na_0 resonances, respectively. A_c is the intensity of a noncellular control containing the same concentration of Na^+ as present in the extracellular medium (Na_0) in an identical sample geometry. S_0 is defined as the fractional extracellular space and is obtained by Eq. (22). This method for the determination of the fractional spaces assumes 100% visibility for Na_0 since the visibility of the control sample is clearly 100%. This may be a reasonable assumption in the RBC system, but generally it should be approached with caution. Also, while the H_1 field is likely to be homogeneous in the control solution, its deviations from homogeneity may not be predicted in a cell suspension, and even less in tissue and organ systems. H_1 inhomogeneity may cause less than optimal flip angles for parts of the sample and a decrease in overall A_0 which is not paralleled in A_c thus resulting in an underestimated S_0 and $[Na_i]$.

$[Na_i]$ is calculated with Eq. (23) from the intensities of the ^{23}Na resonances of the intracellular (A_i) and extracellular (A_0) ions and the independently determined $[Na_0]$. Clearly, A_i and A_0 usually should be normalized by their respective visibility factors for Eq. (23) to yield valid results assuming, of course, that these can be determined. In the RBC system the visibilities were clearly assumed to be 100%, although even in this system contradictory visibility results have been published to date (see Sections 2.3 and 4.1).

A different approach for the determination of fractional extra- and intracellular spaces in RBC suspension, in the presence of $Dy(PPP)_2$, is used by Cowan *et al.* (1985). They present a method for the determination of transmembrane sodium potential by ^{23}Na NMR. The fractions of intra- (α_i) and extracellular (α_0) water spaces are calculated from:

$$\alpha_0 = \frac{\delta_i - \delta_{obs}}{\delta_i - \delta_D}, \qquad \alpha_i = 1 - \alpha_0 \tag{24}$$

where δ_{obs} is the observed ^{1}H chemical shift of water in RBC suspended in $Dy(PPP)_2$ buffer relative to the frequency position of the ^{1}H signal from a coaxial reference tube containing deionized water. Using ^{1}H rather than ^{23}Na NMR, and chemical shift rather than signal intensity, circumvents the need for any assumptions about visibility and H_1 field homogeneity, respectively, in the determination of the extracellular space fraction (see above; also note the equivalence of α_0 and S_0). Equation (24) suffers, however, from a different set of assumptions which also may have only limited validity. This equation is obviously derived on the basis of the consideration that because of the fast exchange condition of water protons across the cell membranes between the extra- and intracellular compartments, the overall observed chemical shift position of water protons in the RBC suspension, δ_{obs}, will be the average of the extra- and intracellular shifts (δ_0 and δ_i) each weighted by the fractional size of the corresponding compartmental water space:

$$\delta_{obs} = \delta_0 \alpha_0 + \delta_i \alpha_i \tag{25}$$

Nevertheless, some assumptions do enter via the actual experimental setup for the determination of δ_0 and δ_i. By definition, these shifts should be identical to the overall experimental shift observed under the two limiting conditions, i.e., $\alpha_0 = 1$ or $\alpha_i = 1$, respectively, all other relevant experimental parameters remaining unchanged. In fact, Cowan *et al.* (1985) determined δ_0 from a solution of the $Dy(PPP)_2$ buffer, in the absence of cells, correcting for "effective" $Dy(PPP)_2$ concentration from a titration curve, and determined δ_i from an RBC sample with 100% hematocrit in the absence of shift reagent. Obviously, not all relevant parameters in these two limiting cases have necessarily remained unchanged compared with the RBC plus shift reagent sample. Particularly, it is likely that the BMS effect (see Section 2.2) will cause a different limiting intracellular chemical shift in RBC in the absence versus the presence of the paramagnetic shift reagent even for the same α_i. Thus δ_i, as determined, may not reflect the real parameter contributing to δ_{obs}. Again, in the RBC system this effect may not be large, but in tissue and organ systems with their more complicated compartment geometries it may become a significant problem.

The concentration of Na_i is then calculated from

$$[Na_i] = \frac{Na_i}{Na_0} \cdot \frac{1 - \alpha_i}{\alpha_i} [Na_0] \tag{26}$$

where Na_i and Na_0 are the intensity of the unshifted and shifted sodium peaks, respectively, and $[Na_0]$ is the sodium ion concentration in the suspension medium which needs to be determined separately. In a cell suspension this task is relatively simple. In a perfused organ the concentration in the perfusate may be used if one assumes no sodium gradients between the perfused vessels and the interstitium. A more serious issue is the implicit assumption in Eq. (26) that the extra- and intracellular visibilities are equal.

Using the above approach to obtain $[Na_i]$, Cowan *et al.* calculated the transmembrane sodium potential in RBC from a modified Nernst equation:

$$E_{Na^+} = \frac{RT}{zF} \ln \frac{[Na_0]}{[Na_i]} \tag{27}$$

where R, T, F, and z are the gas constant, absolute temperature, Faraday's constant, and electric charge, respectively, and $[Na_i]$ and $[Na_0]$ are the intra- and extracellular sodium concentrations. The values of the former, obtained by the NMR method, correlated well with those obtained from standard flame photometric measurements.

A detailed discussion of the problems encountered in the determination of an absolute Na_i concentration in the isolated perfused heart is given by Pike *et al.* (1985). Such a concentration c_i would be estimated from

$$c_i \sim \frac{a_i}{V_i \cdot F_i \cdot \nu_i} \tag{28}$$

where a_i, V_i, F_i, and ν_i are the intracellular sodium signal area, fractional volume, relaxation-flow factor, and NMR visibility, respectively. The authors conclude that under usual experimental conditions $F_i \approx 1$, and assume that the value of 0.54 for V_i in the heart obtained by Gonzales *et al.* (1980) is applicable (see next paragraph where $V_i = 0.25$, obtained by NMR). The most serious assumption made is the one relating to the value of the visibility, ν_i. After a lengthy discussion about the experimental variability of ν and the merit of determining it by "rigorous non-NMR analysis" and the empirical formula $\nu = n_{NMR}/n_{non\text{-}NMR}$, they end up admitting that such an analysis cannot be conducted on the perfused beating heart and thus the unsubstantiated assumption that $\nu_i = 1$ is made and further used.

Jelicks and Gupta (1989c) present a more rigorous treatment of Na_i in the isolated perfused heart. The concentration of Na_i, $[Na_i]$, is obtained from the ratio of the areas of the intra- and extracellular ^{23}Na NMR signals (A_i and A_0), the ratio of the fractional volumes (V_i and V_0), and $[Na_0]$

$$[Na_i] = \frac{A_i}{A_0} \cdot \frac{V_0}{V_i}[Na_0] \tag{29}$$

with

$$\frac{V_i}{V_0} = \frac{A_h^{(^2H)}}{A_s^{(^2H)}} \cdot \frac{A_s^{(x)}}{A_0^{(x)}} - 1 \tag{30}$$

where the ratio of the fractional volumes is determined by using the ratio of the areas of the water ^2H NMR signal in the heart perfused with 2% D_2O in the perfusion buffer ($A_h^{(^2H)}$) and the water ^2H signal in a sample of the same buffer solution filling up the entire sensitive volume of the NMR probe inside the receiver coil ($A_s^{(^2H)}$), as well as the ratio of the areas of the NMR signal of an x nucleus (^{23}Na, ^{35}Cl, or ^{59}Co) in the same solution sample versus the extracellular x NMR signal. The last is obtained with the aid of $Dy(PPP)_2$ in the case of x = ^{23}Na, and since Co^{3+} does not enter the cells and intracellular $^{35}Cl^-$ does not have an NMR-visible signal, the latter two yield only extracellular resonances.

The visibility of the Na_0 resonance is proven to equal 100% by obtaining from Eq. (30) virtually the same extracellular volume fraction, i.e., 0.75, using NMR data from any of the three x nuclei [cf. this value with $V_0 = 0.46$, obtained by Gonzales *et al.* (1980)]. A 100% visibility for Na_i as well is deduced from the fact that after 1 hr hypoxia the NMR-determined $[Na_i]$, assuming 100% visibility for the Na_i signal, was 158 mM, close to the perfusate sodium concentration. From this fact a visibility of 100% was extrapolated for the prehypoxic situation as well. The possibility that such prolonged hypoxia could cause physiological changes leading to a change in visibility was considered and rejected. Except for this leap of faith, this last approach comes closest to a reliable determination of an absolute concentration for Na_i in a complex organ system.

Malloy *et al.* (1990) determined $[Na^+]_i$ in a perfused heart by NMR spectroscopy using a solution enclosed in an intraventricular balloon as an internal concentration standard. Na_0 in the heart was differentiated by Tm(DOTP), while the intraballoon reference signal was shifted in the opposite direction using Dy(DOTP). The balloon volume was calibrated by the addition of 25- or 50-μl increments of a 140 mM Na^+ solution up to a volume of 150 or 250 μl (Fig. 17). The incremental area of this reference NMR signal (ΔA_r) can be plotted versus the increments of added amount of Na^+ (ΔNa), and from the slope of the linear relationship obtained and from the area of the Na_i peak (A_i) the total amount of intracellular sodium, in millimoles, can be calculated. The fractional intracellular volume (S_0) was determined by Co atomic absorption using $Co(EDTA)^-$ as a marker of extracellular space and was used to calculate $[Na]_{i,NMR}$. The following formula summarizes the procedure:

$$[Na]_{i,NMR} = \frac{1}{S_i} \cdot \frac{\Delta Na}{\Delta A_r} A_i \tag{31}$$

A similar calibration curve using an intraballoon reference signal had been employed in a perfused heart study by Fossel and Hoefeler (1986), with the difference that rather than using an intraventricular balloon, they had used a heart-sized balloon placed in the same position as the heart. In these two approaches, each solves one problem and creates another one: the intraventricular method retains the possible magnetic effects of the surrounding heart tissue, whereas the heart-sized balloon lacks these effects entirely; on the other hand, the intraventricular balloon samples only a

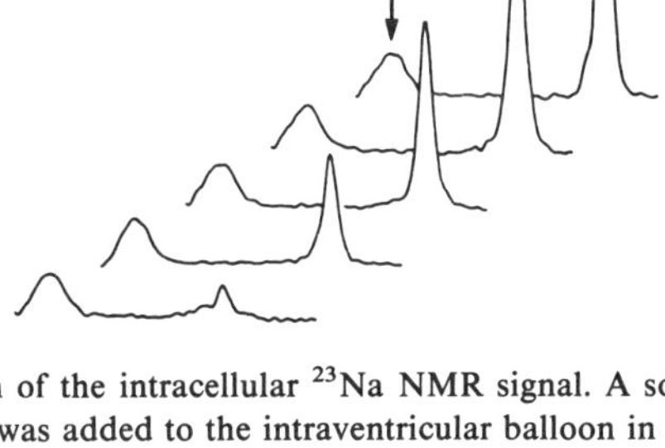

Figure 17. Calibration of the intracellular ^{23}Na NMR signal. A solution of 140 mM NaCl and 2 mM Dy(DOTP) was added to the intraventricular balloon in 50-μl increments. (From Malloy *et al.*, 1990.)

relatively small part of total sample volume and thus may not reflect faithfully inhomogeneity and BMS effects in the entire volume occupied by the heart tissue.

Note that Eq. (31) is equivalent to Eq. (29) if one replaces the calibration slope of the former by the term $[Na_0] \cdot V_0/A_0$ of the latter. In essence, Jelicks and Gupta (1989c) use the extracellular sodium signal and its concentration in the perfusate as reference. This, of course, assumes that there are no sodium ion gradients between the perfusate and tissue spaces unaccessible to the shift reagent. On the other hand, the Na_0 signal reflects better the magnetic conditions over the space occupied by the heart than a balloon reference. Furthermore, while the method of Jelicks and Gupta addresses, although with a mild assumption, the issue of NMR visibility, the method reflected in Eq. (31) does not. Rather, Malloy *et al.* present a separate intracellular sodium concentration, $[Na]_{i,AAS}$, determined by atomic absorption, which is about fivefold larger than $[Na]_{i,NMR}$ thus concluding a visibility of 20%, contrary to Jelicks and Gupta (1989c) who found 100% visibility for Na_i in the isolated, perfused rat heart. Interestingly, nevertheless, the absolute values for the concentration of Na_i, determined by these two groups, were quite similar: 6.2 versus 10 mM.

3.7. Shift and Relaxation Reagent Toxicity

The various shift reagents available to date have been used mainly for the determination of the levels of intracellular sodium in cells and perfused organs. Some of these systems are more sensitive than others to potential deleterious effects of the reagents. Such effects may be caused either by the toxic Dy^{3+} ion released from the reagent complex, or by the sequestration of physiologically essential ions, e.g., Ca^{2+}, by the reagent, or by some undesirable interactions of the intact reagent with tissue components.

Release of free Dy^{3+} may occur either because of insufficient binding affinity to a particular ligand used in a reagent or, more likely, because of decomposition of the ligand itself. $Dy(PPP)_2$, in particular, decomposes in certain systems; the triphosphate ligand of its diamagnetic analogue, $Lu(PPP)_2$, was found to be hydrolyzed in a suspension of rabbit renal brush border membrane with a half life of 20 min, possibly by membrane-associated phosphatases (Elgavish and Elgavish, unpublished; see Figs. 18 and 19). Matwiyoff *et al.* (1986) found a similar, although slower, decomposition in rat abdominal muscle, while no decomposition was reported from relatively long experiments in cell suspensions of erythrocytes, myocytes, etc. In our experience with isolated perfused hearts, no diminution of ^{23}Na shift took place when effluent buffer collected from a perfused heart was observed in comparison with the fresh, unused buffer (Fig. 20; Elgavish *et al.*,

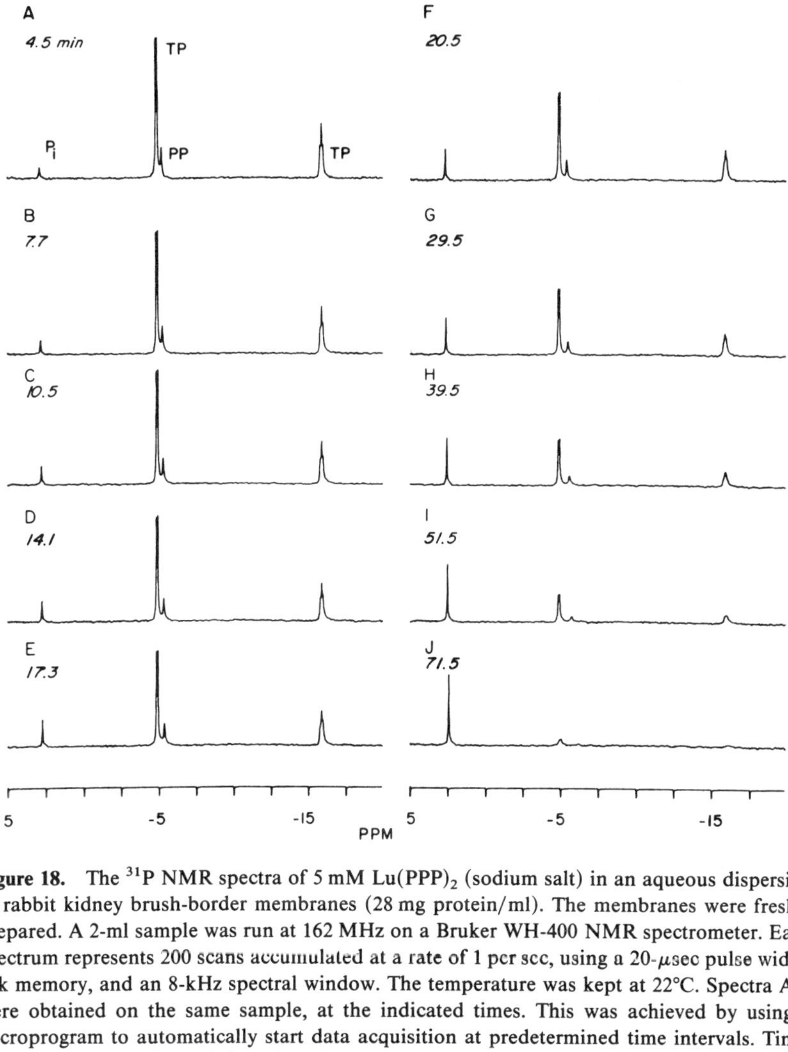

Figure 18. The ^{31}P NMR spectra of 5 mM Lu(PPP)$_2$ (sodium salt) in an aqueous dispersion of rabbit kidney brush-border membranes (28 mg protein/ml). The membranes were freshly prepared. A 2-ml sample was run at 162 MHz on a Bruker WH-400 NMR spectrometer. Each spectrum represents 200 scans accumulated at a rate of 1 per sec, using a 20-μsec pulse width, 16k memory, and an 8-kHz spectral window. The temperature was kept at 22°C. Spectra A–J were obtained on the same sample, at the indicated times. This was achieved by using a microprogram to automatically start data acquisition at predetermined time intervals. Time, in minutes, starting from mixing of membranes and metal chelate, is indicated on each spectrum. Signals: P$_i$, inorganic orthophosphate; TP, triphosphate; PP, pyrophosphate; The chemical shifts are referenced to the position of 85% phosphoric acid. (Elgavish and Elgavish, unpublished.)

unpublished). Also, no deleterious effects were observed on the left ventricular pressure trace of a rat heart perfused with 6 mM of the shift reagent, while perfusion with a buffer containing 0.1 μM free Dy^{3+} caused very conspicuous effects (Fig. 21; Elgavish *et al.*, unpublished). Both of these experiments strongly indicate that Dy(PPP)$_2$ does not decompose during

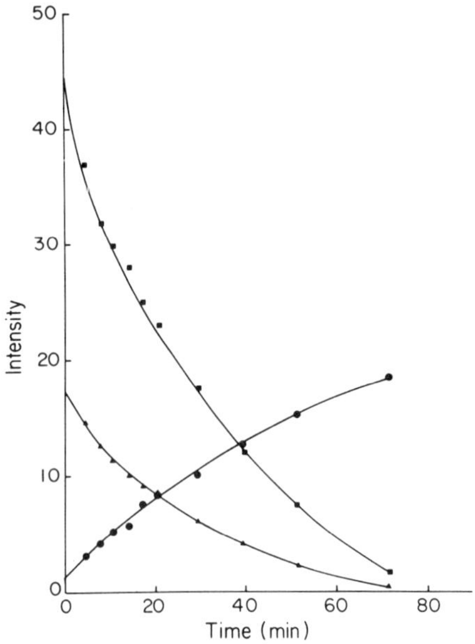

Figure 19. The time dependence of the ^{31}P NMR signal intensity of terminal P (squares) and middle P (triangles) of triphosphate in the $Lu(PPP)_2$ chelate, as well as of inorganic orthophosphate (circles). These intensities are derived from the spectra shown in Fig. 18. (Elgavish and Elgavish, unpublished.)

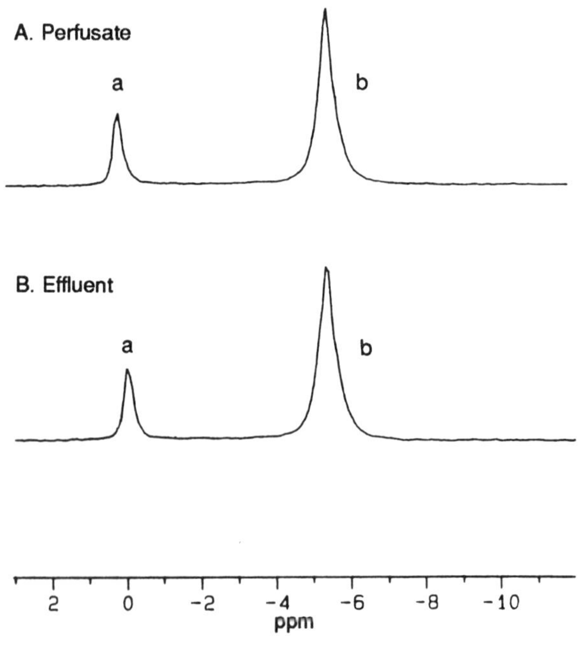

Figure 20. ^{23}Na NMR spectra of Hepes perfusate containing 6 mM $Dy(PPP)_2$ before its usage for perfusion of rat heart (A), and of the same perfusate collected from the effluent of 120 min of rat heart perfusion (B). (Elgavish *et al.*, unpublished.)

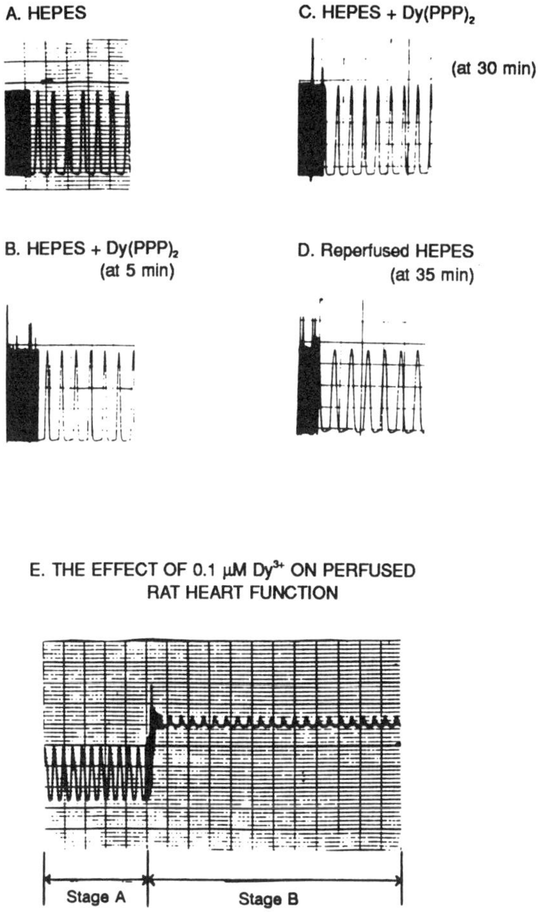

Figure 21. The left ventricular pressure trace of a rat heart perfused with Hepes buffer (A), 5 min with Hepes + 6 mM $Dy(PPP)_2$ (B), 30 min with Hepes + 6 mM $Dy(PPP)_2$ (C), same heart reperfused with Hepes buffer following 35 min of perfusion with shift reagent (D), and the effect of only 0.1 μM free, unchelated Dy^{3+} ions (stage A, control; stage B, perfusion with Dy^{3+}) (E). Note the substantial effect of unchelated dysprosium on pressure versus the absence of such effect in the $Dy(PPP)_2$-perfused heart. (Elgavish *et al.*, unpublished.)

perfusion of a rat heart. It is probable that, under the conditions of continuous flow in a perfused heart experiment, the shift reagent is not accessible to phosphatases, contrary to the situation in some other tissues, like skeletal muscle or renal membranes.

One of the goals of the development of shift-reagent-aided ^{23}Na NMR methods is their ultimate application to the observation of Na_i *in vivo*, in animals for the expansion of biochemical knowledge in intact living systems with all of their complexity, and in humans for early diagnosis of disease or for monitoring therapy. One agent, Gd(DTPA) (DTPA = diethylenetriaminepentaacetate), has been used extensively for several years in clinical MRI with good contrast enhancement of proton images and few adverse side effects. Unfortunately, the Dy^{3+} analogue of DTPA does not shift the Na_0 signal sufficiently for its effective use as a ^{23}Na shift reagent. Since most ^{23}Na NMR work has been carried out with Dy(TTHA) and $Dy(PPP)_2$, the following discussion will focus on these.

It is well known that the $Dy(PPP)_2$ complex limits the solution concentration of free Ca^{2+} ions, and thus it is not possible to achieve both a $Dy(PPP)_2$ concentration high enough for adequate separation of Na_0 and Na_i, and a free Ca^{2+} concentration reaching physiological level (1.25 mM). Perfusing, e.g., a rat heart with a typical free Ca^{2+} level of less than 0.5 mM in the presence of $Dy(PPP)_2$ on the average produces a left ventricular systolic pressure of about 50–80% of control, although occasionally pressures of 100–120% of control are also obtained (see Section 3.1.2.2).

Recently, the inhibition of lymphocyte stimulation by a number of shift reagents has been examined (Sze *et al.*, 1990). This work focused primarily on results obtained with Dy(TTHA) and $Dy(PPP)_2$. Human peripheral blood lymphocytes were prepared and mitogens of varying specifications and mechanisms were employed to trigger lymphocyte proliferation. Both Dy(TTHA) and $Dy(PPP)_2$ at commonly used concentrations inhibited thymidine incorporation in phytohemagglutinin (PHA)-stimulated cells up to 100-fold. Half-maximal levels of thymidine incorporation occurred at about 0.5 mM for Dy(TTHA) and at about 0.3 mM for $Dy(PPP)_2$, with inhibitory effects starting at about 0.2 mM for both. The time course of this inhibition was a continuous curve with the extent of toxicity depending on the length of time spent in contact with the shift reagent. The chelator ligands themselves, Na_5PPP and Na_6TTHA, were tolerated up to concentrations of 2 mM and 10 μM, respectively. Supplemental $CaCl_2$ added to the chelates, or chelator ligands alone, did not attenuate the toxicities. Stock solutions of Dy(TTHA) prepared in stoichiometric proportions of 1:1, 1:2, and 1:3 with $CaCl_2$ were not significantly less toxic than shift reagent prepared without $CaCl_2$. Also, supplemental $CaCl_2$ added to the culture medium had no significant effect. Therefore, the authors concluded that there was no clear evidence to support the hypothesis that disturbances in calcium regulation were the mechanism of inhibition. Other chelators such as GHA (glucoheptonic acid), DTPA, and aza-crown complexes were found to be nontoxic at 1 mM and did not reduce thymidine incorporation in PHA-treated cells. Unfortunately, however, dysprosium complexes of these ligands do not induce an adequate shift of Na_0 from Na_i. Some findings suggest that the toxicity may be reversible under specific conditions so that experiments requiring toxic levels may be performed in live animals or perfused organs if effective clearance or elution of the reagent can be ensured.

4. APPLICATIONS

4.1. Blood Cells

The majority of biological ^{23}Na NMR papers, to date, are on RBC. RBC were originally of interest in the examination of potential abnormalities in certain disease states. RBC are often used as a model system for ion fluxes and as a system to evaluate the effects of shift reagents on cells. Ogino *et al.* (1985) examined Na^+/K^+ activity in RBC in normal and diseased states using ^{23}Na and ^{3P}K NMR. In early work with RBC, evidence of Na^+ binding sites was obtained from detergent solubilization which freed Na^+ and K^+ binding sites that were not available for rapid chemical exchange with the bulk (Magnuson *et al.*, 1970). Sodium binding was proven by demonstrating increases in ^{23}Na linewidth in the presence of solubilized membrane and decreases in the presence of K^+ due to competition between the two ions for the binding sites.

Later, evidence that some fraction of the plasma and erythrocyte sodium was relatively immobilized, came from work by Pettegrew *et al.* (1984) which showed that there was a large difference between the T_1 and T_2 of this type of Na^+ compared with those in aqueous saline solution. Furthermore, their relaxation time values were comparable with the values for sodium with human albumin or bound to an ion-exchange resin.

Jelicks *et al.* (1989) examined the direction and magnitude of changes in levels of Mg^{2+}, Ca^{2+}, and Na^+ in reticulocytes and erythrocytes during reticulocyte maturation in guinea pigs. A sharp decrease in ATP was found in erythrocytes with 65% increase in free Mg^{2+}, and large increases in Na_i and Ca_i in erythrocytes compared to reticulocytes, suggesting that alterations in the kinetics of membrane ion transport systems accompany changes in phospholipid and cholesterol content. In contrast to dog erythrocytes, in guinea pigs no Na^+-Ca^{2+} exchange was found in either reticulocytes or erythrocytes. It was suggested that the lower Na_i in reticulocytes may arise from increased activity of Na^+-K^+-ATPase.

Post (1989) presents evidence on erythrocytes suggesting that Na^+ and Li^+ are free inside the cell. The addition of an electrostatic influence by fixed charges on the intracellular macromolecules may affect the quadrupolar NMR line shape, but this does not correlate well with a restriction of mobility. There is a contribution to the T_2 originating from cations diffusing through electrical field gradients imposed by the cytoskeleton which is reflected in the NMR line shape. The T_1 of Na_i is related to the viscosity of its environment and this can affect the line shape, whereas the T_2 is unaffected because the cytoskeleton's diffusional constraints yield a relatively long τ_{c2}.

Burstein and Fossel (1987a) examined the effect of temperature, buffer osmolarity, and shift reagent concentration on the Na_i and K_i relaxation times in RBC. Resonances shifted upfield with increasing temperature. This was expected due to the temperature dependence of the bulk susceptibility. The same effect was also tested in the presence of added D_2O, but the resonance position remained unchanged. The $1/T_1$ and $1/T_2$ rates increased with increasing temperature but not with increasing levels of Na_i. Na_i linewidth was unaffected by changes in buffer osmolarity. Relaxation rates tended to decrease with increasing osmolarity; however, the effect on T_2 was small. Changes in shift reagent concentration affected both position and linewidth of Na_i. The Na_i resonance was found to be 0.58, 0.74, and 0.82 ppm from the NaCl reference with 5, 7, and 9 mM $Dy(PPP)_2$, and had linewidths of 49, 66, and 77 Hz, respectively.

The effect of disease or disease states is examined with RBC. Erythrocytes from stroke-prone hypertensive (SHRSP) rats were compared with those from normotensive (WKY) rats. They were found to have no differences in sodium concentration during the measurement (Fig. 22) and no difference in osmotic fragility or total cell volume (Kwan *et al.*, 1987). These results do not support the hypothesis that altered Na^+ transport, resulting in an increase in erythrocyte sodium concentration, is associated with spontaneous hypertension. Ong and Cheung (1986) found a difference in binding affinity between RBC ghosts from normotensives and hypertensives as determined from plots of ΔR versus Na^+_{total}. Binding affinities were found to be 500–1000 and 50–150 M^{-1}, respectively.

The effect of sepsis, the systemic response of the body to serious infection, on RBC was investigated by ^{23}Na NMR and ^2H NMR (glucose metabolism) by Hotchkiss *et al.* (1990). In induced sepsis, mediators of sepsis such as cachectin, tumor necrosis factor, endotoxin, and interleukin 1 achieve significant concentrations in the bloodstream, and could lead to a systemic response in the body. It was concluded, based on the lack of a significant effect on Na_i or glucose metabolism, that sepsis would have no effect on RBC.

Anderson *et al.* (1988) studied Na^+ flux during volume regulation in *Amphiuma* RBC using ^{23}Na NMR with $Dy(PPP)_2$-containing perfusate and conventional chemical methods. They confirmed that the reagent neither entered the cells nor altered Na_i, K_i, and Cl_i so long as a Ca^{2+} concentration above 0.5 mM was maintained. They were able to titrate the 4 mM $Dy(PPP)_2$ solution up to 2 mM Ca^{2+}. They centrifuged the precipitate and thus worked with an unknown concentration of Dy. For volume-activated Na^+ in the

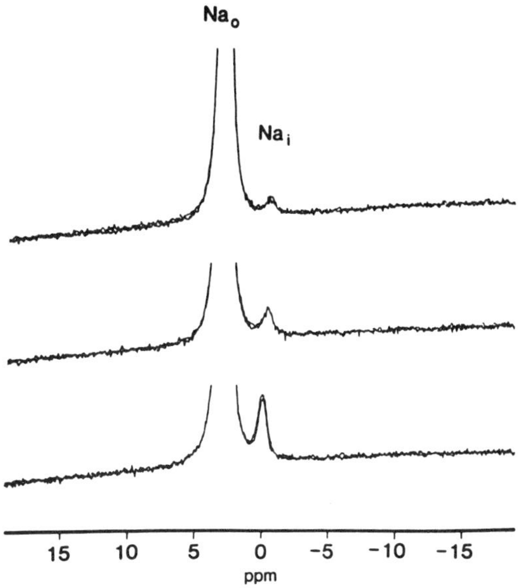

Figure 22. ^{23}Na NMR spectra (95.3 MHz) of blood samples (~50% hematocrit) from SHRSP and WKY at 0 hr (top), 2 hr (middle), and 6 hr (bottom) after incubation at 37°C in isotonic saline solution buffered with 10 mmole/liter Hepes at pH 7.4. In the absence of Dy(TTHA)$^{3-}$, a single, large peak at the position of 0 parts/10^6 (ppm) was observed (not shown). In the presence of 10 mmole/liter Dy(TTHA)$^{3-}$, two ^{23}Na NMR resonance signals representing the extracellular Na^+ space (Na_0) and intracellular Na^+ space (Na_i) were obtained. Note that the spectra of SHRSP and WKY samples were superimposable at all incubation periods. (From Kwan *et al.*, 1987.)

range 30–150 mM Na/kg dry cell solid (DCS), ^{23}Na NMR was within 4% of the results obtained by chemical methods and with 6 mM Na/kg DCS, 16% of Na_i was invisible. Transported Na^+ appeared to be NMR visible to the same extent, whether inside or outside the cell. Lysed cells were found to have altered the environment so as to increase the total NMR-visible sodium. They believed a methodical problem from magnetic interactions such as low S/N contributed to the invisible sodium.

Cala *et al.* (1988) reexamined the role of the Na^+–H^+ exchanger in the dual regulation role of cell volume and pH. Both RBC and rabbit hearts were examined. It was concluded that both functions can occur in a cell, yet it appears that when activated by increased $[H]_i$, the pathway is no longer subject to volume-dependent inactivation. As such, the H^+-activated Na^+–H^+ exchange will continue to swell even those cells with greater than normal volumes. These observations suggest a role for Na^+–H^+ exchange in pathologies associated with cytotoxic edema. The studies in rabbit hearts show that the increase of intracellular sodium levels during hypoxia is a result of increased dissipative (leak) entry of Na^+ and not decreased conservative (pump) extrusion.

A ^{23}Na NMR assay of Na^+-K^+ activity in RBC has been developed by Ouwerkerk *et al.* (1989). The method uses nystatin, which enhances sodium permeability in cells to an extent low enough for the pump to handle, but when ouabain is added the effect is doubled (Fig. 23). The rate of sodium increase can be used to calculate pump activity. They found 100% visibility of Na_i by causing an equilibrium between the Na_i and Na_0 pools and observing that the combined signal intensity did not change. This proved that there was no difference in visibility between the Na_0 and Na_i pools and, assuming 100% visibility for Na_0, one could conclude that Na_i visibility in RBC was also 100%. Also, they fit the Na_i signal to a broad component (~60%) and a narrow one (~40%). They also found that with a shift reagent of a Dy:PPP ratio of 1:2.5 they did not see leakage of Dy into the cells as evidenced by a lack of broadening of the ^{31}P NMR signals.

Knubovets *et al.* (1989) proposed a ^{23}Na NMR method to measure the maximal Na^+ efflux from RBC and found it to be 10.1 ± 1.0 mM/hr per liter cells (Fig. 24). Levels of Na_i and K_i were made to vary to decrease their concentration gradients across the erythrocyte membrane making active Na^+ efflux an ^{23}Na NMR-measurable process.

In a clinically relevant study by Monti *et al.* (1986), the relative contributions of the slow (s) and the fast (f) components of erythrocyte Na_i, as defined by the T_2 relaxation of these components, were compared in uremic

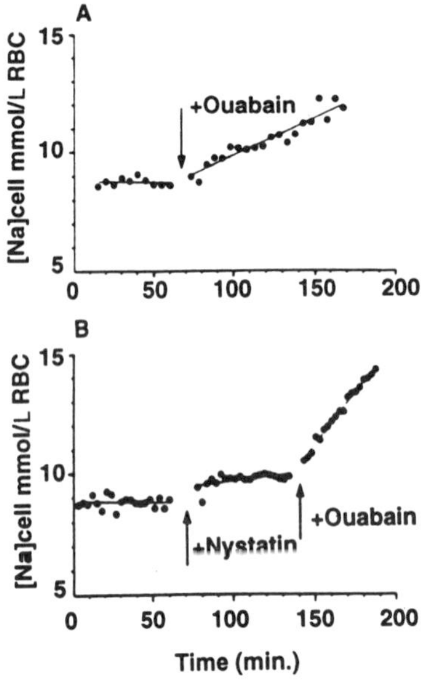

Figure 23. Intracellular sodium in erythrocytes as a function of time, measured by ^{23}Na NMR. (A) Na^+-K^+-ATPase assay on normal erythrocytes. The arrow indicates the addition of 1.0 mM ouabain. (B) Effect of nystatin. Arrows indicate the addition of nystatin (20 μg/ml packed cells) and of 1.0 mM ouabain, respectively. (From Ouwerkerk *et al.*, 1989.)

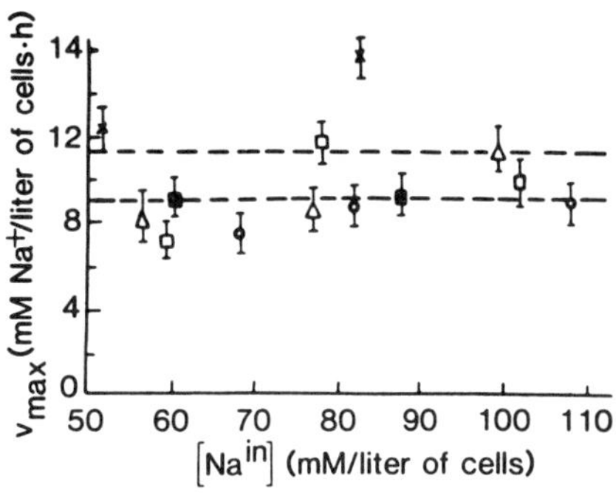

Figure 24. Dependence of active Na^+ efflux from RBC on intracellular Na^+ concentration. The different symbols used designate various donors. The bars reflect the accuracy of the determination. (From Knubovets *et al.*, 1989.) Note that the uppermost value on the x-axis should be 16 rather than 14.

patients versus normal volunteers. In 4 controls and 13 patients, values of 32 ± 10 and $68 \pm 10\%$ were found for the relative fractions of s and f, respectively. These fractions were close to the theoretically expected distribution of 40/60%. In some patients, however, the unusual fractions of 18 ± 3 and $82 \pm 3\%$ were found. The authors suggested that by some unknown mechanism the distribution of Na_i in some uremic patients was altered compared with normals.

Little is known about the mechanism of action of lithium which is used to treat manic-depressive disorders. A recent study showed quenching of the ^{23}Na double-quantum signal in erythrocytes by addition of Li^+ (Gullapalli *et al.*, 1990), confirming that Li^+ competes for Na^+ binding sites. The authors suggest the possibility of using this technique to monitor the presence and interactions of Li^+ in the brain *in vivo.*

4.2. Kidney and Renal Epithelial Cells

Studies using ^{23}Na NMR to investigate suspensions of kidney tubules began to appear in 1985 and 1986 (Rayson and Gupta, 1985; Kumar *et al.*, 1986). The size of the extracellular space penetrated by the shift reagent, $Dy(PPP)_2$, as well as the visibility of Na_i, were investigated by Rayson and Gupta (1985). The shift-reagent-accessible space was found to be similar to that which contained Cl_0^- ions. The latter was measured by ^{35}Cl NMR utilizing the fact that the intracellular ^{35}Cl NMR signal was broadened beyond detection (Fig. 25).

To evaluate its visibility, the Na_i level was raised artificially by simultaneously inhibiting Na^+ efflux and increasing the ion permeability of the plasma membrane. Thus, the NMR-observable intracellular Na^+ reached a level which was approximately 81% of that in the medium, as determined by chemical techniques, suggesting that the resonance of the Na_i pool was not subject to a 60% visibility reduction. The Na_i level measured in these cells was responsive to stimuli such as temperature, inhibitors of apical transport (furosemide and amiloride), and Na^+-K^+-ATPase inhibition

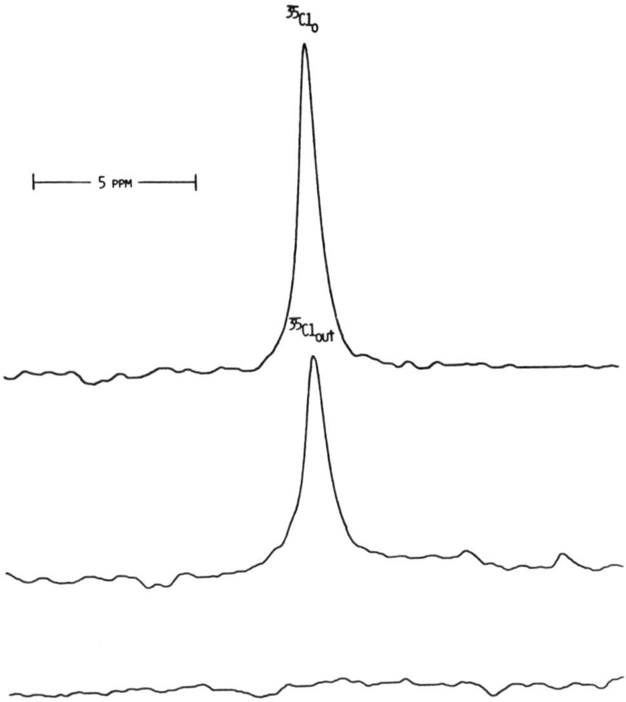

Figure 25. ^{35}Cl NMR spectra at 20 MHz (22°) of rat outer medullary kidney tubules in suspension with (lower trace) and without (middle trace) added Zn-HbO$_2$, demonstrating the invisibility of intracellular $^{35}Cl^-$ ions. The upper trace shows the ^{35}Cl NMR spectrum of the medium alone. (From Rayson and Gupta, 1985.)

which raised Na_i. Thus, the NMR method was found to be valuable for measuring Na_i in this system under basal conditions and to monitor, in the same preparation, the changes that occur under various experimental conditions without interfering with the morphological integrity of the cells.

Na^+-K^+-ATPase activity has been examined in proximal tubules (Avison *et al.*, 1987; Gullans *et al.*, 1985). In one study ^{23}Na NMR combined with K^+-electrode was used (Fig. 26) to measure Na^+–K^+ fluxes after abrupt stimulation of Na^+–K^+-ATPase (Avison *et al.*, 1987). Conclusions were drawn as follows: (1) initial Na^+ and K^+ movements after addition of K^+ (0.5 to 5 mM) to proximal tubules suspended in low $[K]_0$ medium are unidirectional fluxes through the Na^+–K^+-ATPase; (2) the Na^+–K^+ coupling ratio of the Na^+–K^+-ATPase was 1.54 ± 0.07; (3) this coupling ratio was fixed and did not vary; (4) at saturating values of $[K]_0$ the activity of the Na^+–K^+-ATPase increased with increasing $[Na]_i$ in the physiological range of Na_i. In another study, Na^+ efflux against a significant gradient was observed when sodium-loaded cells were warmed from 4 to 37°C (Fig. 27;

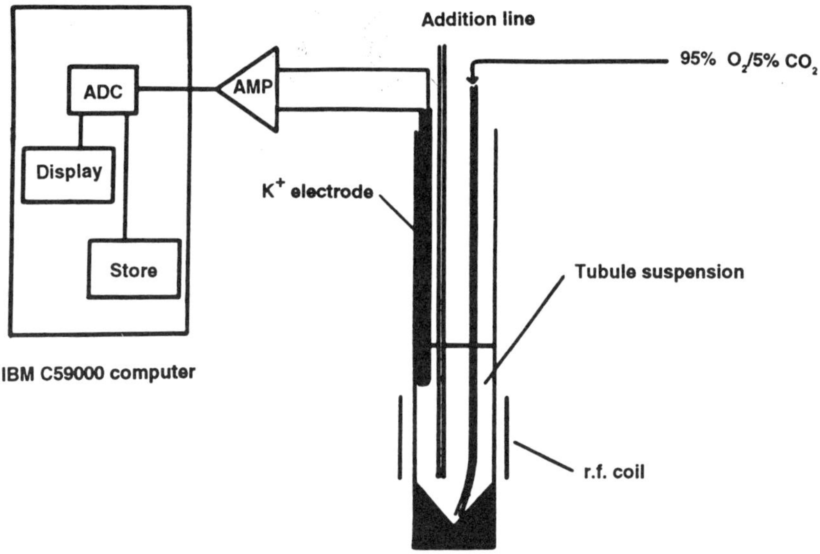

Figure 26. Schematic of apparatus used for simultaneous measurement of Na^+ and K^+ fluxes. NMR tube containing 8 ml of tubules suspension (10–20 mg protein/ml), sits in magnet of spectrometer, such that radio frequency coil of NMR probe surrounds tubule suspension. Temperature is regulated (25°C) by passing warm air around tube throughout experiment. Oxygen tension is maintained by O_2/CO_2 line, which also keeps tubule suspension well mixed. The addition line permits extracellular K^+ concentration step to be made without interrupting data acquisition. Output of K^+-selective electrode is measured by a high-impedance electrometer, then digitized and stored by an IBM C59000 computer for subsequent processing. (From Avison *et al.*, 1987.)

Gullans *et al.*, 1985). At 37°C, ouabain inhibited Na^+-K^+-ATPase activity in a dose-dependent, time-dependent manner. A biphasic time course of increased Na_i following ouabain (10^{3-} M) suggests that the Na^+ permeability of the plasma membrane may decrease following pump inhibition, thus limiting sodium influx. ^{23}Na visibility was assessed by several methods and was found to be 100%.

Proximal tubule suspensions have been used to study the effects of diabetes. Na_i was measured in proximal tubules of rats made diabetic by injection of streptozotocin in 5 mM citrate buffer (Kumar *et al.*, 1988). After 96 hr there was a 60% increase in Na_i in diabetic rats relative to controls with no more rise by 11 weeks (Fig. 28). Ouabain addition caused a 4-fold increase of Na_i in control samples and a 2.5-fold increase in samples from diabetic animals, but the tubules from diabetic animals still had a higher level of Na_i. Ouabain-inhibited Na^+-K^+-ATPase activity was substantially higher in the renal tubules of diabetic rats, the increase being proportional to that of Na_i. The effect of hyperglycemia on Na_i was tested by exposing

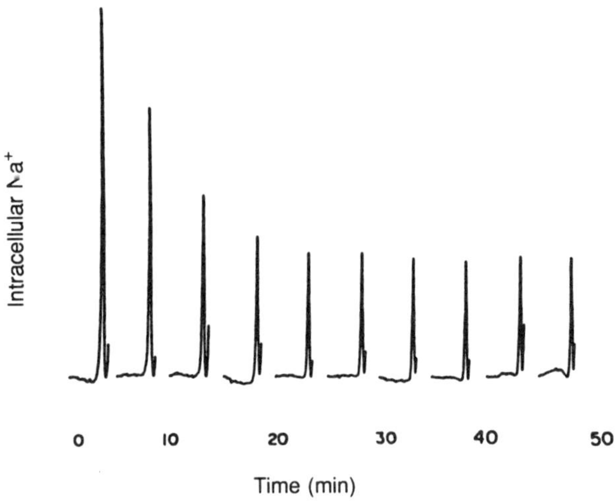

Figure 27. Typical experiment showing efflux of intracellular sodium as tubules were warmed to 37°C. Tubule suspensions were kept cold (4°C) during their preparation, resulting in an accumulation of intracellular sodium. When placed in the NMR, suspension reached 37°C within 10 min. Each spectrum is a 5-min accumulation of intracellular sodium only. Extracellular sodium peaks have been omitted. (From Gullans *et al.*, 1985.)

the tubules to low (5 mM) or high (25 mM) levels of glucose. There was a 52% rise in Na_i in normal rats, 11% in 96-hr diabetic animals, and 5% in 11-week diabetic rats. It was proposed that the rise in Na_i in the tubules of diabetic rats was due to the increased filtered load of glucose that stimulated the entry of glucose and sodium in the renal cells.

^{23}Na NMR visibility in the kidney was assessed for dog cortical tubules, rabbit cortical tubules, and dog thick ascending limbs. A difference was found between total sodium and transportable sodium (Boulanger *et al.*, 1987). Visibility was determined by flame photometry and by redistribution between intra- and extracellular compartments using nystatin (influx) and substitution of Na^+ by choline in extracellular fluid (efflux), and was found to be ~30% visible for total sodium and 58% for transportable sodium. Addition of sodium to sodium-depleted homogenates of dog renal cortex also showed a loss of visibility. T_1 and T_2 were measured, but did not establish a correlation with visibility.

Ammann *et al.* (1990) found a variation in the "apparent visibility" of Na_i induced by Na^+-substrate cotransport in dog cortical tubules. Expected Na_i changes due to addition of glutamine, glucose, and lactate were detectable by flame photometry but not by NMR. The authors proposed that compartmentation of Na_i in the tubules or inhomogeneous distribution of $Dy(PPP)_2$ within the extracellular space caused the observed variation in visibility.

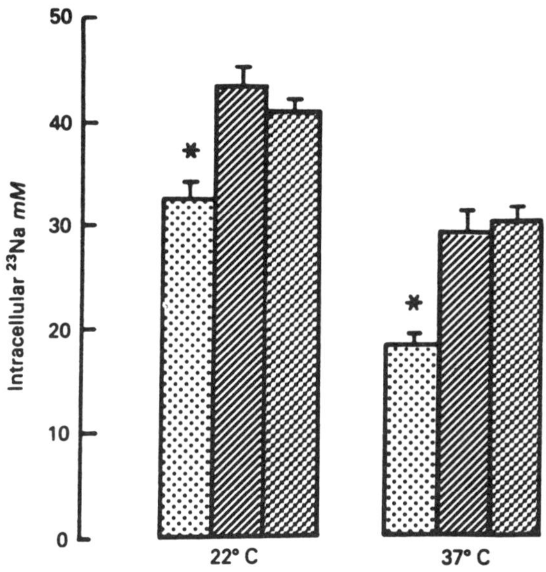

Figure 28. Intracellular Na^+ concentrations measured by NMR spectroscopy using suspensions of proximal tubules obtained from normal rats and rats with streptozotocin-induced diabetes. Spectra were obtained at 53 MHz with an acquisition time of 0.2 sec ($T_1 < 40$ msec) following 90° pulses over an interval of 3 min at room (22°C) and body (37°C) temperatures. Differences between control and diabetic rats are significant (*$p < 0.001$). Symbols: dotted bars–control; striped bars–diabetic, 96 hr; hatched bars–diabetic, 11 weeks. (From Kumar *et al.*, 1988.)

Bogusky *et al.* (1986) examined a rat kidney *in vivo* by spatially localized ^{31}P and ^{23}Na NMR using a surface-coil probe and rotating-frame methods. During altered physiologic states, changes in distribution of metabolites and sodium ions within the kidney were identified in one-dimensional metabolite maps (Figs. 29 and 30). No shift reagent was used in the ^{23}Na spectroscopy, and thus total tissue sodium was monitored. Localized ^{23}Na spectra of the cortex, outer medulla, renal pelvis, and abdominal wall are shown. Spectra illustrate how K^+ depletion results in less Na^+ in the outer medulla than in the untreated rat despite similar overall Na^+ uptake (Fig. 30).

Jans *et al.* (1988) examined renal epithelial LLC-PK1/C14 cell sodium fluxes during acidification by CO_2 and realkalinization. $Dy(PPP)_2$ in Ringer's solution was used. ^{23}Na visibility (V) was tested on these cells by the addition of nystatin and, using the following formula, was found to be 64 ± 4%:

$$V = (\Delta Na^+_{in}/\Delta Na^+_{out}) \cdot 100 \tag{32}$$

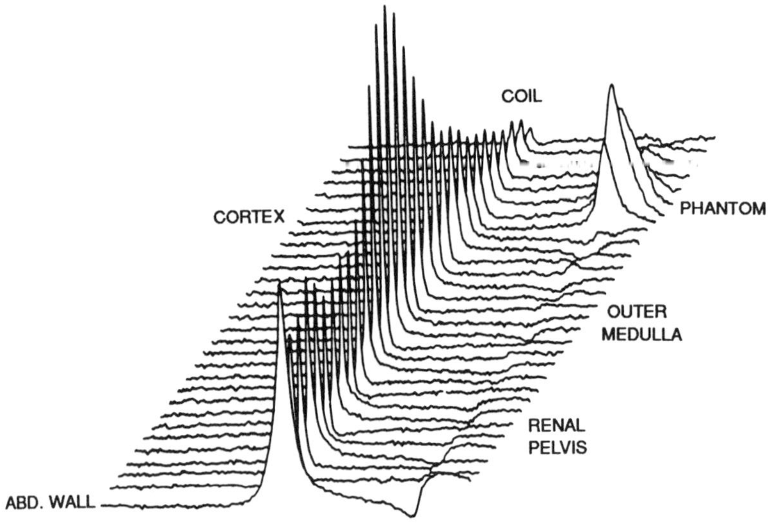

Figure 29. Distribution of sodium present in the *in vivo* rat kidney. The highest content of sodium in the kidney is found in the outer medulla. Identified in the figure is a resonance intensity standard containing 1 M NaCl plus dysprosium tripolyphosphate ("PHANTOM"). Also identified are the sodium resonances present in the cortex, outer medulla, renal pelvis, and abdominal wall. (From Bogusky *et al.*, 1986.)

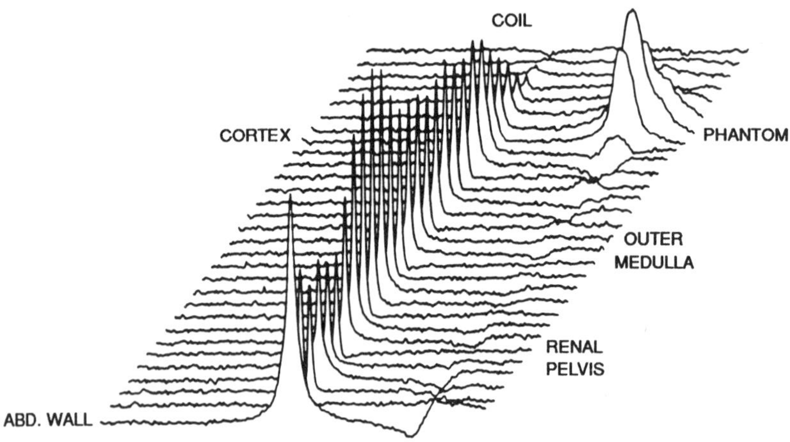

Figure 30. Sodium content of regions of kidney in potassium-depleted and DOCA-treated rats. This figure shows that the sodium content of the outer medulla of treated rats is less than in untreated controls despite similar sodium intakes at the time of experiment (compare to Fig. 29). Identified is the position of the coil and a phantom containing the same intensity standard as in Fig. 29. (From Bogusky *et al.*, 1986.)

where Δ indicates the change in sodium level. Bubbling with CO_2 had the effect of increasing Na_i to 120% which later settled at a new equilibrium level of 113%. In the presence of 1 mM amilorade, Na_i remained at ~100%. In the absence of amilorade, realkalinization was 0.057 pH unit/min versus 0.022 pH unit/min with amiloride. Proton efflux and sodium influx were measured and their ratio was found to be significantly smaller than 1.0, contrary to an expected 1:1 stoichiometry for the Na^+–H^+ exchanger. It was suggested that some of the sodium flux was utilized in the uptake of D-glucose with a stoichiometry of 1:1 for the sodium–D-glucose cotransport.

4.3. Yeast, Bacteria, and Other Cells

4.3.1. Yeast

Balschi *et al.* (1982) first reported a NMR study of Na_i in yeast. The transmembrane transport of Na^+ was monitored by examining the Na^+ efflux from Na^+-rich yeast cells (Fig. 31). Ogino *et al.* (1983) studied the relationships among fluxes of K^+, Na^+, and H^+ using ^{39}K, ^{23}Na, and ^{31}P NMR in yeast cells energized by oxygenation. When cells were deenergized after the aerobic period, K^+ efflux, H^+ influx, and Na^+ influx were calculated to be 1.6, 1.5, and 0.15 μmole. Sodium and potassium were found to be 40% visible.

Höfeler *et al.* (1987) have examined sodium transport and phosphorus metabolism in Na^+-loaded yeast by ^{23}Na and ^{31}P NMR. Yeast metabolism is interesting because of the formation of vacuolar polyphosphate species, P_n. There appears to be a hierarchical energy demand on the cells under different conditions. In particular, the energy for the Na^+–NH_4^+ exchange takes precedence over that required to produce and store polyphosphates. This conclusion is supported by the fact that when the cells are "forced" to exchange K^+, as well as NH_4^+, for Na^+ (by the addition of 5 times as much K^+ to the NH_4^+-containing medium), polyphosphates are never formed in significant amounts, and the initial linear phase of Na^+ efflux persists possibly 6 times as long. It is suggested that the ultimate consumption of polyphosphate species probably occurs because inorganic orthophosphate is required for the metabolic processes which produce the anions for the Na^+-anion coextrusion.

4.3.2. Bacteria

The gram-positive *Streptococcus faecalis* was used as a bacterial model to study the mode of action of monensin, a carboxylic polyether antibiotic, which is known for its cation-transporting ability across membranes (Delort *et al.*, 1989). The bacteria were studied in two states: (1) deenergized

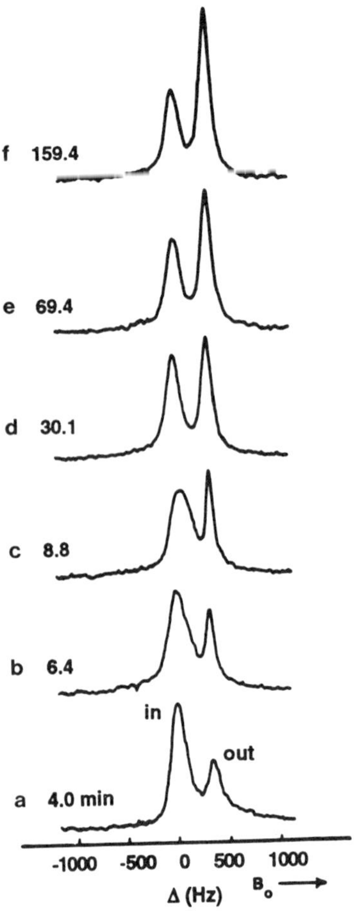

Figure 31. The time dependence of the ^{23}Na spectrum (132.3 MHz, 11.74 T) of a D_2O suspension [24% (wt/vol)] of Na^+-rich yeast cells. The spectra are labeled with the time elapsed since resuspension. Each spectrum is the result of the accumulation of 128 free-induction decays in 52 sec. The time labels correspond to the midpoints of the accumulation intervals. The suspension was occasionally stirred, to prevent settling, by a gentle pipetting. The temperature was 297 K. (From Balschi *et al.*, 1982.)

(depleted of ATP) and considered as an inert biological membrane on which cationic transport induced by the ionophore alone could be investigated; (2) energized mode, or active cell, where the ionophore-sensitive response of the living organism, particularly by the cation pumps and by glycolysis, was probed. The cationic movements were shown to be coupled to the existing ionic gradients of K^+ and Na^+. Also, ^{31}P and ^{13}C NMR spectra of intermediary metabolites detected in active cells showed that glycolysis was dramatically modified in the presence of monensin.

A study by Martin *et al.* (1987) showed that ^{23}Na NMR using $Dy(PPP)_2$ could be used to determine Na_i in packed pellets of intact *D. discoideum* amoebae with no apparent toxic effects of the shift reagent on the cells. Time-dependent changes in the ^{23}Na and ^{31}P NMR spectra showed a rapid degradation of $Dy(PPP)_2$. In particular, the reagent-induced shift in the position of the Na_0 resonance decreased a total of 7 ppm within 40 min due to hydrolysis of the reagent. This hydrolysis was suggested to be catalyzed by an acid phosphatase.

The cyanobacterium *Synechococcus* 6311, isolated from fresh water, was studied using ^{23}Na and ^{31}P NMR (Packer *et al.*, 1987). This species was interesting because of its tolerance for moderate salinity. The bioenergetic changes of the species were examined upon transition from growth medium of 0.01 M to 0.5 M NaCl. Na^+ was found to penetrate the cells rapidly under dark aerobic conditions; cells grown for several days in high-salt medium, however, reestablished a low internal sodium content, comparable to normal cells. Cells grown in high-salt medium for several generations exhibited a pattern of ^{31}P metabolites similar to control cells, except that they produced additional peaks, presumably of sugar phosphates, in the monoester phosphate region which also displayed increasing intensities.

Castle *et al.* (1986a, b) studied *E. coli* which, like other bacteria, were known to maintain an inwardly directed Na^+ gradient utilizable for driving various processes such as accumulation of nutrients or motility. It was hypothesized that the Na^+ transport was via the Na^+-H^+ exchanger. To test this hypothesis the authors examined the Na_i chemical potential compared with that of H^+. Sodium concentration measurements were done under aerobic and glycolytic conditions. Sodium extrusion and maintenance of a stable, low Na_i were found to correlate well with the development and maintenance of proton-motive force, a result that is consistent with proton-driven Na^+-H^+ exchanges, a means of Na^+ transport. Na_0 was found to be 100% visible while Na_i was 45%, as determined by extrusion methods.

The inwardly directed Na^+ gradient had a concentration ratio of about 25. The correlation between the steady-state sodium gradient (ΔpNa) and the proton-motive force developed by endogenously respiring *E. coli* was also examined. Although ΔpNa versus ΔpH displayed slopes close to 1, the value of ΔpNa consistently exceeded that of ΔpH by ~0.4 unit, indicating a partially electrogenic character for the putative H^+/Na^+ antiport. The H : Na stoichiometry was 1.13 ± 0.01 at external pH below 7.2. These results were discussed in the framework of a model in which two distinct, integral stoichiometries (H : Na, 1 : 1 and 2 : 1) were available with some relative probability; the model predicts futile cycling of Na^+ and a dissipative proton current. The magnitude of the pH gradient developed by cells was osmolarity-dependent, yielding steady-state pH_i values from 7.1 at 100 mOsm to 7.7 at 800 mOsm.

4.3.3. Other Cells

Greenfield *et al.* (1990) measured the Na_i content of a wall-less strain of *Neurospora crassa*, a slime variant of a filamentous fungus, and the effect on Na_i of the presence of insulin. This fungus is useful as an experimental organism for the study of certain aspects of fungal metabolism because it is easy to isolate its plasma membranes and vacuoles in high yields without the problems associated with removal of the cell wall. Several shift reagents were tested on this system. $Dy(PPP)_2$ was rapidly hydrolyzed by extracellular phosphatases. Of the reagents tested, Tm(DOTP) was the only agent found to provide consistently adequate resolution of Na_i without metabolic interference with the cells. Insulin-treated cells showed a slow, continued accumulation of sodium ($+1.14 \pm 0.39\%/min$) during the 3 to 12 min following pretreatment with insulin followed by glucose addition, while control cells lost sodium very slowly ($-0.63 \pm 0.29\%/min$).

Pilatus *et al.* (1990) used ^{31}P and ^{23}Na NMR spectroscopy to study resting and stimulated mast cells. Exocytosis induced by cross-linking the type I receptor for Fc_ε domains present on rat mucosal mast cells (RBL-2H3 line) required the influx of Ca^{2+} ions and was markedly influenced by the concentration of covalent cations (K^+, Na^+, and H^+) in the medium. RBL-2H3 cells were cultured on microcarrier beads to very high density ($3–4 \times 10^7$ cells/ml), suitable for NMR experiments. Approximately $6–8 \times 10^7$ cells, or 2 ml of beads, were transferred into a 10-mm-diameter NMR tube which was continuously perfused with growth medium equilibrated with 95% O_2 and 5% CO_2. $Dy(PPP)_2$ was added to the perfusate for Na_i measurements. In the absence of K^+ in the medium, ATP and pH_i decreased and Na_i increased. Following inhibition by ouabain of the Na^+–K^+-ATPase, the inward Na^+ current was determined to be 5×10^7 ions/cell per sec. This value was found to be constant, within 10%, during exocytosis.

4.4. Heart

Wittenberg and Gupta (1985) measured Na_i in cardiac myocytes from rats. The value obtained was consistent with values from ion-selective electrodes. Myocytes incubated for several hours in an insulin-deficient, Ca^{2+}-containing medium exhibited a significantly lower Na_i. Reversible sodium loading of the cells by manipulation of Ca_0 levels was observed. Incubation of myocytes in a Ca^{2+}-free, insulin-containing medium caused a threefold increase in Na_i. In contrast to cells with insulin, insulin-deficient myocytes exhibited a markedly lower level of Na_i in Ca^{2+}-free medium. These observations suggest that insulin may stimulate a pathway for Na^+ influx in heart cells.

The application of lanthanide shift reagents to the observation of Na_i in a perfused rat heart (Fig. 7) was first presented by Pike *et al.* (1985). Fossel and Hoefeler (1986) measured Na_i and K_i concentrations by NMR. Values for sodium levels were in agreement with values from ion-sensitive electrodes, but potassium values were not consistent with a large "invisible" fraction.

Burstein and Fossel (1987b, c) suppressed the Na_0 signal by presaturation and separated Na_i and Na_0 by mathematical means. Relaxation times for Na_i and Li_i in a frog heart were measured (Burstein and Fossel, 1987b). For ^{23}Na, T_1 was 23 msec and T_2 was biexponential with 2 and 17 msec. The latter were inconsistent with the expected from a homogeneous pool of nuclei undergoing quadrupolar interactions. Relaxation times were not changed by a fivefold increase in the Na_i level induced by ouabain perfusion. Addition of 10 μM ouabain to the perfusate plus zero potassium, low calcium, and replacement of 66% of the Na^+ with Li^+ resulted in the following percent changes in Na_i: ouabain, 460 ± 60%; K^+, 300 ± 30%; Li^+, 51 ± 6% (Burstein and Fossel, 1987c). A 45% increase in Na_i was observed with a change in pacing rate from 0 to 60 beats/min with proportional changes for intermediate pacing rates.

Jellicks and Gupta (1989c) explored the visibility issue in a multinuclear NMR study in Langendorff perfused rat hearts. Based on a comparison of the extracellular volumes, calculated using ^2H and ^{23}Na, ^{35}Cl, or ^{59}Co NMR, they concluded that resonances of Na_0 and Cl_0 (including those in interstitial spaces) were fully visible, contrary to previous assumptions in the literature. Furthermore, prolonged hypoxia or ischemia caused a dramatic increase in Na_i, such that it rose to approach that in the external medium indicating full visibility of Na_i. It was shown that contrary to assumptions in the literature that double-quantum ^{23}Na NMR detects only the intracellular resonance, both intra- and extracellular sodium ions contributed to the double-quantum-detected signal. This was clearly demonstrated by the presence of a double-quantum resonance shifted by an exclusively extracellular paramagnetic shift reagent. The authors used a combination of MQS and Gd-based paramagnetic relaxation reagents at relatively low concentrations to "quench" any Na_0 contribution and thus enable the specific quantitation of Na_i.

In our laboratory we have used the isolated perfused rat heart model in ^{23}Na NMR experiments using the shift reagent $Dy(PPP)_2$ for monitoring Na_i (Fig. 6), specifically in the study of the effects of hypoxia and ischemia, ouabain and amiloride action, and the staircase effect induced by increased pacing rate.

Changes in levels of Na_i and high-energy phosphates were monitored during varying durations of hypoxia (<25 mm HgO_2) or global ischemia, and upon subsequent 10 min of reoxygenation or reflow, in the isolated, perfused rat heart to determine possible correlations between Na_i and reversibility of function (Elgavish *et al.*, 1987). All of the hearts subjected to 20 min hypoxia, and three out of six hearts subjected to 30 min hypoxia, recovered left ventricular function upon reoxygenation. These hearts also displayed no increase in Na_i during hypoxia. Three out of six hearts subjected to 30 min hypoxia, and all hearts subjected to 40 min hypoxia (Fig. 32), failed to recover function upon reoxygenation. In all nonrecovering hearts a significant increase in Na_i was detected by 27 min into hypoxia,

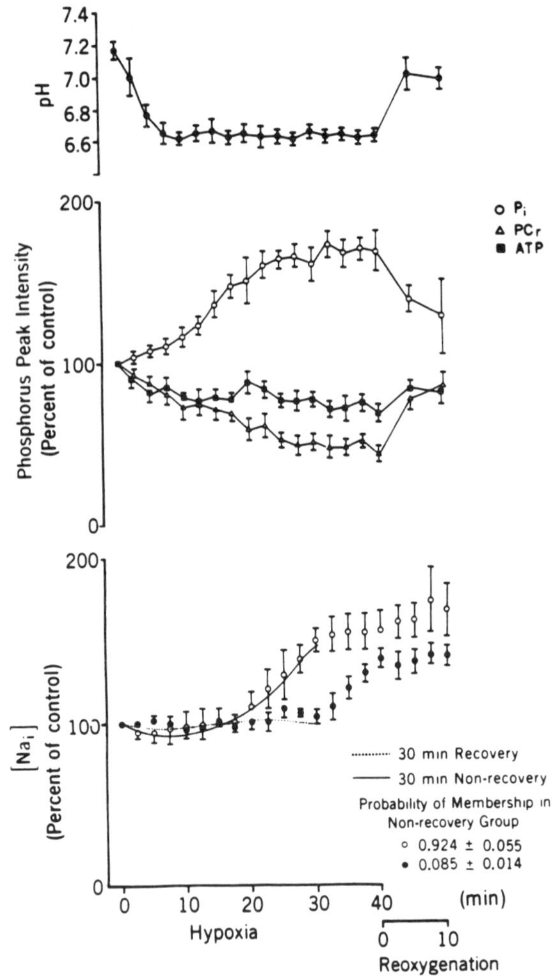

Figure 32. Hearts subjected to 40 min of hypoxia followed by 10 min of reoxygenation. None of these hearts regained function upon reoxygenation. The pH and phosphorus peak intensities are practically indistinguishable from those in hearts subjected to 20 and 30 min of hypoxia. The solid and dotted lines drawn through the sodium data points are the best-fit curves from the 30 min results of "with recovery" and "without recovery" hearts, respectively. Note that starting at 24 min the hearts subjected to 40 min of hypoxia could be separated into two groups. The probability of nonrecovery indicated in the figure was assessed at 30 min. (From Elgavish *et al.*, 1987.)

and this increased level remained elevated throughout reoxygenation. In hearts subjected to ischemia of 30 min or longer, a correlation was found between probability of recovery upon reperfusion, the extent of Na_i elevation, and duration of ischemia (Fig. 33). Contrary to Na_i, high-energy phosphate levels, measured by ^{31}P NMR interleaved with ^{23}Na NMR, were not found to be a predictor of reversibility of left ventricular function in hypoxic or ischemic hearts.

The increase in the level of Na_i, secondary to inhibition of the Na^+-K^+-ATPase by cardiac glycosides, is known to increase calcium influx via Na^+-Ca^{2+} exchange, and therefore to lead to an increase in contractility. This increase in Ca_i^{2+} has been related to the development of intracellular acidification and activation of Na^+-H^+ exchange as a measure to prevent further acidification. Thus, the efflux of intracellular H^+ results in an additional increase in Na_i. This may subsequently lead to an increased rate of Ca^{2+} influx and therefore to the potentiation of the effects of cardiac glycosides. To assess the role of Na^+-H^+ exchange in the mechanism of ouabain action, amiloride, a known inhibitor of the Na^+-H^+ exchanger, was used in a $Dy(PPP)_2$-aided ^{23}Na NMR and ^{31}P NMR study in isolated, perfused rat hearts (Lotan *et al.*, 1988, 1992).

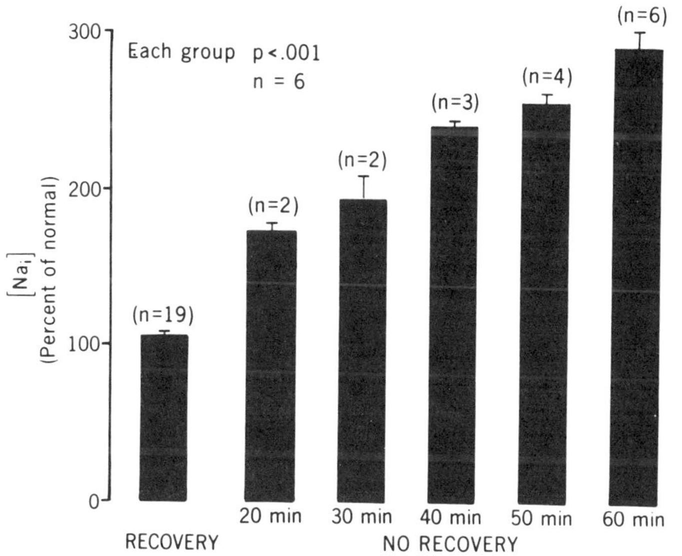

Figure 33. Bar graphs of NMR-determined post-reperfusion levels of Na_i in isolated perfused rat hearts subjected to varying duration (20, 30, 40, 50, 60 min) of total global ischemia followed by 10 min of reperfusion. Six hearts were measured in each of the five time groups, and the number of hearts in each group that did not recover function following reperfusion is indicated above each corresponding bar. (From Elgavish *et al.*, 1987.)

The results obtained in this study suggested that Na^+–H^+ exchange played a role in ouabain induced intracellular acidification, that amiloride displayed inhibitory effects beyond its effects on the Na^+–H^+ exchanger (Fig. 34), and that the arrhythmogenic effects of cardiac glycosides might be related to the rate of Na_i rise.

The staircase (treppe) phenomenon, or the positive interval-strength relationship, is the increase in myocardial force development resulting from an increase in stimulation frequency. It had been suggested that a rise in Na_i, attributed to a "sodium pump lag," may be the primary cause for the observed changes in developed force. ^{23}Na and ^{31}P NMR spectroscopy, with $Dy(PPP)_2$, were used to assess the relationship between stimulation frequency, level of Na_i, high-energy phosphate stores, intracellular pH, and left ventricular pressure, in the isolated, perfused, rat heart (Miller *et al.*, 1989). An increase in the level of Na_i, consistent with the "Na^+–K^+ pump lag" theory, was found with increasing stimulation frequency (Fig. 35). The ultimate pressure response, however, represented a delicate balance between this positive inotropic effect and a negative inotropic effect possibly resulting from the shortening of the time interval between beats (restitution effect).

^{23}Na NMR, aided by $Dy(PPP)_2$, has also been used to examine the intact thoracic aorta (Jelicks and Gupta, 1990). Multinuclear NMR spectroscopy was used to examine the relationship between primary hypertension and levels of Ca_i, Mg_i, and Na_i in intact thoracic aorta of spontaneously

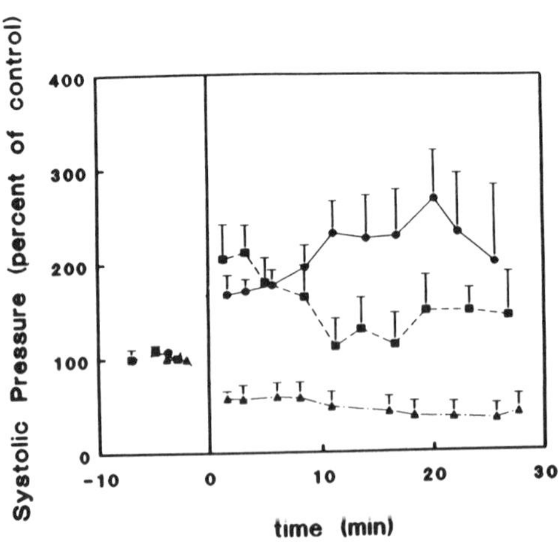

Figure 34. Systolic pressure (percent of control) versus time (min) for hearts perfused with 50 μM ouabain (Group I) (circles), with 50 μM ouabain plus 1.0 mM amiloride (Group II) (squares), and with 1.0 mM amiloride (Group III) (triangles). In Group I an initial increase is observed within the first 10 min followed by a plateau of 230 ± 30% of control in the second period of the experiment. In Group II an immediate increase in pressure was observed within the first 4 min reaching a level of 210 ± 30% of control. A sharp decrease followed, almost reaching control level by 11 min. Finally, pressure plateaued around 150% of control. In Group III, an immediate negative inotropic effect was observed with a decrease in pressure to around 60% of control, further drifting slowly toward 35% of control by the end of the experiment. (From Lotan *et al.*, 1992.)

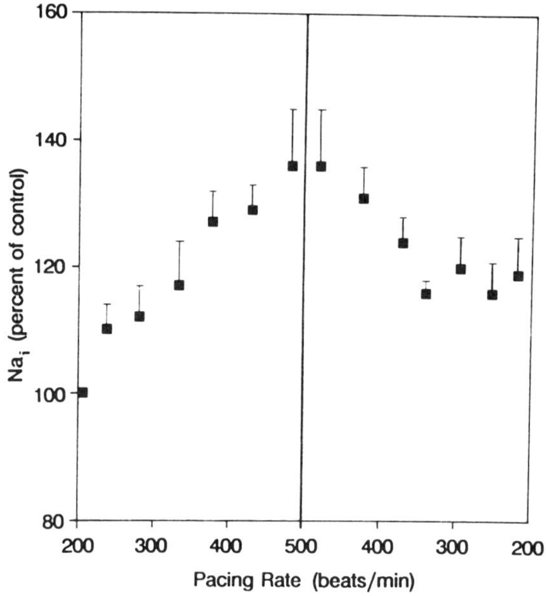

Figure 35. Na_i versus pacing rate (beats/min) for all hearts ($n = 7$). Levels of Na_i were determined from the area under the Na_i resonance in ^{23}Na NMR spectra, normalized to the area under the reference signal in the spectrum. Na_i values for each heart were expressed as percent of Na_i level in unpaced control, and then averaged for all seven hearts for each pacing rate. Vertical bars denote S.E.M. Note the almost linear increase of Na_i with upward pacing. A single heart, paced up to 500 beats/min, displayed an even higher Na_i of 154% (datum not shown). The decline of Na_i with downward pacing is slightly slower, and it does not fully return to control upon termination of pacing. (From Miller *et al.*, 1989.)

hypertensive rats. Values of intracellular ionic composition and systolic pressure found in this study are compiled in Table 1.

In vivo measurements of Na_i were recently reported by Balschi and Pohost (1990). Dy(TTHA)(meglumine)$_3$ with 4 mM $CaCl_2$ was infused into

TABLE 1
Intracellular Ionic Composition and Systolic Pressure in the Aorta of Hypertensive versus Control Rats[a]

	Control	Hypertensive
Systolic pressure	123 ± 8 mm Hg	159 ± 6 mm Hg
Ca^{2+}	151 ± 26 nM	277 ± 28 nM
Na^+	10.1 ± 1.8 mM	21.9 ± 6.4 mM
Mg^{2+}	246 ± 76 μM	180 ± 38 μM

[a] Adapted from Jelicks and Gupta (1990).

the femoral vein of an anesthetized pig and a 2-cm surface coil was sutured to the heart. ^{23}Na NMR spectroscopic measurements, using the surface coil and adiabatic excitation pulses, were performed before, during, and after ischemia was induced using an inflatable artery occluder around the left anterior coronary artery. Assuming equal intra- and extracellular visibilities, and using current literature values for relative intra- and extracellular volumes, a value of 10 ± 4.7 mM for the intracellular Na^+ was calculated from the Na_i peak area in the NMR spectrum.

During ischemia, Na_i increased twofold and returned to control level within 20 min from onset of reperfusion. Several animals underwent ventricular fibrillation (VF) in the course of ischemia. These animals also displayed a faster rise of Na_i compared with non-VF animals.

Changes in membrane permeability of the Na^+ ion are an important parameter in the action potential and thus in the cardiac cycle. Thus, a shift reagent-aided ^{23}Na NMR spectroscopic approach which enabled the monitoring of the cycle-dependent level of intracellular sodium in the isolated perfused heart would be desirable. We have developed such a method, and demonstrated its application to the isolated perfused rat heart paced at a constant rate of 250 beats/min by means of an external pacemaker (Elgavish *et al.*, 1991). ^{23}Na spectra were obtained on a Bruker AM 360 wide-bore NMR spectrometer. Prior to acquisition (AQ), a selective presaturation (P) of 40 msec duration was applied to reduce the intensity of the extracellular sodium signal. "In memory" signal acquisition and the acquisition pulse program shown in Fig. 40 (I) (see p. 234) were used allowing a time resolution of FID acquisition every 80 msec. Three hundred scans were averaged to obtain a spectrum for each of the six 40-msec intervals. Three consecutive cardiac cycles were used to provide one scan for each of six different intervals, denoted A through F, of the full cardiac cycle. The first two of each three-cycle set is shown in Fig. 40 (I). In the third cycle in each set, only interval A is used for NMR data collection. Each set of six spectra was collected from a total of 900 cardiac cycles, i.e., a total of 3.6 min/set.

The area under the intracellular Na (Na_i) peak was determined for each spectrum and expressed as percent of the Na_i level of the last 40-msec interval of the cardiac cycle [interval F in Fig. 40 (I)]. Figure 40 (II) (p. 234) depicts the means ($n = 5$) of the Na_i level during each of the six cycle intervals. An increase of 14% in Na_i is evident during intervals A and B which roughly correspond to the depolarization phase of the cardiac cycle. This Na_i level is significantly ($p < 0.05$) higher than those of intervals C–F. Thus, we have developed a method to obtain sodium NMR spectra with a differentiated Na_i resonance at every 40-mscc interval in the cardiac cycle. The levels of Na_i during each of six such intervals of the entire cycle show that a significant increase in Na_i takes place during the first 80 msec, as would be anticipated on the basis of the role of Na^+ in the cardiac depolarization phase.

A perfused heart study examining the calcium paradox was carried out using Dy(TTHA)-aided ^{23}Na NMR spectroscopy (van Echteld *et al.*, 1990). Accumulation of Na_i during ischemia or during Ca^{2+}-free perfusion had been suggested to play an important role in the development of damage from reperfusion or Ca^{2+} repletion caused by an increased Ca^{2+} influx via the Na^+-Ca^{2+} exchanger. In this study, hearts perfused with a Ca^{2+}-free solution did not display any rise in Na_i, although the calcium paradox did occur when reperfused with Ca^{2+}, and did show a rise in Na_i when subsequently made ischemic. In contrast, in the absence of both Ca^{2+} and Mg^{2+}, hearts showed a tenfold rise in Na_i which further increased when made ischemic. The authors concluded, therefore, that an increase in Na_i during Ca^{2+} depletion is not a prerequisite for the calcium paradox to occur, and that Mg^{2+} blocks the influx of Na^+ which during low-Ca^{2+} perfusion had been suggested to occur through the slow calcium channel.

Another study examined the mechanism of Na_i increase during ischemia by using amiloride to inhibit the Na^+–H^+ exchange and lidocaine to block the sodium channel in hearts perfused with Dy(TTHA) (Bernard and Ingwall, 1990). They found that with amiloride (1 mM) the Na_i increase was delayed and reduced by 50%, and with added lidocaine (150 μM) Na_i accumulation was the same as in controls. These results suggest that Na_i increase during ischemia is not caused by increased flux through the sodium channel. The authors suggest that Na^+–H^+ exchange is the mechanism for Na_i increase during ischemia but that the extent and rate of H^+ flux inhibition are inconsistent with amiloride activity unless alternative pathways for H^+ extrusion are masking the effect of amiloride on pH_i. There was evidence for other pathways of H^+ extrusion since the pH_i during ischemia was somewhat higher (pH_i 6.2 versus 6.0) when a HCO_3^--containing Krebs–Henseleit perfusate was used versus a HEPES perfusate that contained no HCO_3^-.

Our group (Chu *et al.*, 1990a) has studied the effect of mild hypoxia on isolated perfused rat hearts externally paced at a constant stimulation frequency of 260 beats/min. The hearts were perfused with a HEPES buffer that contained 6 mM $Dy(PPP)_2$. In this model of mild hypoxia, oxygen tension was maintained at pO_2 ~100 mm Hg for 60 min followed by 60 min of normoxic perfusion. In comparison with the nonrecovery group in the model of severe hypoxia of 30 min duration and no external pacing (see above, and Elgavish *et al.*, 1987), these hearts displayed more limited Na_i accumulation. It seems, therefore, that Na_i increase, and to a larger extent contractile function, seem to be sensitive to the extent of oxygen deprivation, in a manner such that a threshold level for possible posthypoxic recovery seems to exist below 100 mm Hg of pO_2 during hypoxia. The role of controlled stimulation frequency in determining cardiac response to hypoxic conditions, however, requires further clarification.

We also studied the effect of 30 min of hypoxia ($pO_2 \sim 25$ mm Hg) followed by 60 min of reoxygenation on the T_1 of the Na_i resonance in isolated perfused rat hearts (Prince *et al.*, 1992). The Na_i signal was differentiated by 6 mM $Dy(PPP)_2$, and T_1 measured by an inversion recovery sequence every 5 min. Prehypoxic control T_1 of Na_i was found to be 27 ± 1 msec ($n = 8$). Although during hypoxia the level of Na_i increased about 50% above prehypoxic control, no significant change in T_1 was found. No such change occurred during reoxygenation, and the overall mean T_1 for Na_i over the entire 120 min duration of perfusion was determined (27.7 ± 1.8 msec). This observed constancy of T_1 in spite of the increase in the level of Na_i was explained on the basis of the relatively low association constant for the binding of Na^+ to intracellular polyelectrolytes.

Membrane damage from hypothermic preservation of rat hearts was determined by ^{23}Na and ^{31}P NMR spectroscopy using Dy(TTHA) (Navon *et al.*, 1990). It was concluded that ^{23}Na NMR spectroscopy could distinguish between two types of membranal abnormalities that may be caused by cold storage: (1) a state of increased permeability to sodium ions (in such hearts Na_i increases gradually during cold storage, and they recover function to varying extents upon reperfusion) and (2) irreversible damage displaying regional, macroscopic infarcts that allow penetration of shift reagent and give rise to an additional ^{23}Na signal between the resonances of Na_i and Na_0.

Malloy *et al.* (1990) used Tm(DOTP) and atomic absorption to study the effect of global ischemia on Na_i in a perfused rat heart. Control values of Na_i were determined by atomic absorption to be 30.7 ± 3.3 mM versus 6.2 ± 0.5 mM by NMR. A reversible 54% increase in Na_i was seen with 10 min of ischemia (Fig. 36). Further accumulation of Na_i was produced by prolongation of ischemia to 30 min. The rate of Na_i increase was more than 4000 times greater than the increase in H^+ on a molar basis. There was an indication, however, that Na^+–H^+ exchange increased early in ischemia because the proportional change in H^+ was greater than that in Na_i. Their conclusions were (1) 20% of Na_i was visible by NMR, (2) intracellular acidosis and accumulation of Na_i developed simultaneously during global ischemia, (3) increased Na_i is not itself an indicator of irreversible injury, and (4) recovery of mechanical performance is associated with return of Na_i to baseline as measured by NMR.

Studies on the heart using multiple-quantum-filtered (MQF) ^{23}Na NMR spectroscopy have recently emerged. Using this technique, Payne *et al.* (1990) concluded that the MQF Na signal is predominantly from the extracellular space in normal heart and from the intracellular space in ischemic heart. However, the quenching effect of a shift reagent may hide some of the extracellular MQF signal, especially during ischemia.

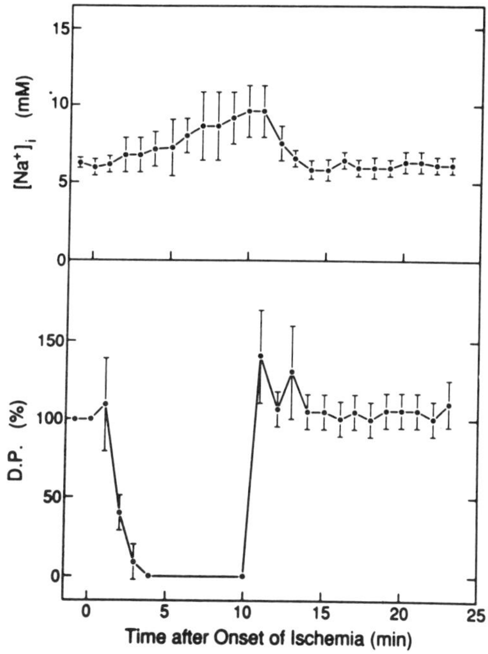

Figure 36. Relationship between $[Na^+]_i$ and developed pressure during 10 min of ischemia followed by reperfusion. Intracellular sodium in four hearts was measured by NMR at 1-min intervals; developed pressure was measured simultaneously. (From Malloy *et al.*, 1990.)

The issue of whether the interstitial sodium signal contributes to the multiple-quantum signal of Na_i is addressed by Foy and Burstein (1990). Multiple-quantum spectroscopy (MQS) relies on a compartment exhibiting biexponential T_2 decay as a result of quadrupolar interaction. Since the characteristics of interstitial sodium have not been determined, it is unknown whether interstitial sodium indeed contributes to the multiple-quantum signal. Table 2 shows relative magnitudes and T_1 of total cardiac, intracellular, and interstitial Na compartments for frog and rat hearts.

TABLE 2
Relative Magnitudes and T_1 for Cardiac Sodium Compartments[a]

	Relative magnitude (%)	T_1 (msec)
Frog hearts (n = 3)		
Total	100	29.4 ± 3.9
Intracellular	8.4 ± 4.9	22.4 ± 3.0
Interstitial	93.4 ± 6.1	30.7 ± 4.4
Rat hearts (n = 2)		
Total	100	31.0 ± 1.5
Intracellular	21.7 ± 9.3	23.0 ± 2.5
Interstitial	81.9 ± 1.4	34.2 ± 4.9

[a] Adapted from Foy and Burstein (1990).

4.5. Cancer

Liebling and Gupta (1987) examined Na_i concentrations of uterine leiomyoma and leiomyosarcoma, benign and malignant tumors of smooth muscle, respectively, in comparison with their nonneoplastic counterpart myometrium, as well as of colonic adenocarcinoma compared with its nonneoplastic counterpart colonic mucosa. Results ($[Na_i]$ in mmol/liter tissue) were: myometrium, 14.3 ± 1.2; uterine leiomyoma, 5.1 ± 1.2; uterine leiomyosarcoma, 34.6 ± 4.5; colonic mucosa, 8.0 ± 1.1; well-differentiated adenocarcinoma of the colon, 13.9 ± 1.5; poorly differentiated adenocarcinoma of the colon, 21.7 ± 0.6.

Conclusions: (1) there was an alteration in Na_i in neoplastic human tissue compared with its tissue of origin; (2) the degree of cellular proliferation activity and/or tumor invasiveness may have a direct correlation with the concentration of Na_i; (3) the benignity and malignancy may be distinguishable on the basis of the level of Na_i—i.e., once obtained, a measurement of Na_i may represent as good a diagnostic and/or prognostic indicator as histopathologic viewing; (4) the technique of ^{23}Na NMR spectroscopy and/or imaging has the potential of becoming a powerful, noninvasive diagnostic tool in cancer.

4.6. Frog Skin

Civan *et al.* (1983) observed two ^{23}Na signals in frog skin reflecting sodium pools in the extracellular space and a medium relatively inaccessible to $Dy(PPP)_2$. The reagent-insensitive signal could be increased by ouabain and decreased by amiloride. This signal may in part be Na_i and this suggests that NMR may become a useful complementary tool for monitoring the Na_i content of transporting epithelia.

4.7. Multiple Organs

Using ^{23}Na NMR aided by $Dy(PPP)_2$, Bárány and Venkatasubramanian (1986) measured Na_i and Na_0 concentrations in rat liver, kidney, and brain, and in frog muscle, skin, and liver. A capillary containing a NaCl solution of known concentration and shift reagent was inserted in the observed tissue in each case. The Na signal arising from this capillary was shifted 30 ppm away from the position of the tissue Na signal. The area under this shifted signal was used as a standard to calculate the concentrations of tissue Na_i and Na_0 from their corresponding peak areas. The concentration of total tissue sodium was also determined from the undifferentiated ^{23}Na peak area obtained from spectra acquired in the absence of shift reagent. All data are compiled in Table 3.

TABLE 3
Total, Intracellular, and Extracellular Sodium Content of Tissues[a]

Tissue	Sodium content[b]		
	Total	Intracellular	Extracellular
Rat liver	21.2 ± 2.3	11.8 ± 1.9	7.3 ± 0.8
Rat kidney	43.0 ± 5.1	30.7 ± 5.4	14.3 ± 4.0
Rat brain	35.9	20.3	19.8
Frog muscle	24.0 (20.9)	16.8	10.5
Frog skin	69.6 (73.5)	34.7	36.1
Frog liver	20.8 (21.0)	16.0	6.2

[a] Adapted from Bárány and Venkatasubramanian (1986).
[b] Values correspond to μmoles Na^+/g fresh weight. Data in parentheses refer to the total sodium content determined in frog tissues after trichloroacetic acid or Triton X-100 treatment.

In the frog, total tissue sodium concentrations were additionally measured by NMR in solutions obtained from each tissue following treatment by either trichloroacetic acid or Triton X-100 to release tissue-bound Na^+. These values are shown in Table 3 in parentheses.

The values for total tissue sodium concentration corresponded well, within experimental error, to the sum of Na_i and Na_0. Similarly, the values obtained for total sodium concentrations in the chemically treated frog tissues were not significantly different from those determined in the same tissues before treatment. These results, therefore, seem to indicate 100% visibility for both Na_i and Na_0. Overall, intracellular sodium concentration in the tested tissues covers the range between 12 and 35 mM.

Casu *et al.* (1986) carried out studies to measure the T_1 of Na_i differentiated by $Dy(PPP)_2$, in various rat tissues and found values within the range of 12–32 msec. In livers of CCl_4-treated rats a prolongation of T_1 was found, from 11.6 to 17.9 msec. The authors ascribe this effect to a CCl_4-induced reduction in the bound fraction of intracellular sodium, but not to any necrotic processes.

4.8. Brain and Cerebral Blood Flow

^{23}Na NMR has been used to measure cerebral blood flow in conjunction with brain metabolism in the assessment of diseases such as hypotension (Naritomi *et al.*, 1988) and seizures (Eleff *et al.*, 1988; Schnall *et al.*, 1988).

Naritomi *et al.* (1988) examined *in situ* gerbil brains for correlations among cerebral blood flow (CBF), energy metabolism, Na^+ pump activity, and electrocorticograms (ECoG) following graded hypotension. CBF in normotensive animals was 0.51 ± 0.07 ml per min. Following the onset of graded hypoxia, no changes in ^{31}P signals were observed until CBF fell to 0.21 ± 0.27 ml/g per min, at which level pH_i began to decrease in association with ECoG voltage reduction. At a CBF of 0.04–0.05 ml/g per min, PCr and ATP practically disappeared and the total tissue ^{23}Na signal suddenly decreased 8%. The authors ascribe the latter effect to a disruption in Na^+-K^+-ATPase activity due to the severe depletion of ATP. Such disruption would result in an increase of Na_i. The transfer of sodium spins from the 100% visible extracellular pool to the 40% visible intracellular pool would then be manifested in the NMR spectrum as a reduction in the narrow resonance normally observed for the undifferentiated tissue sodium signal. This demonstrates the thresholds for the development of tissue acidosis, changes in the PCr/P_i ratio, and ATP reduction. It appears that suppression of function occurs already prior to ATP changes, whereas sodium pump failure takes place only after severe ATP depletion.

Schnall *et al.* (1988) carried out ^{1}H, ^{31}P, and ^{23}Na measurements on cat brains to study cerebral metabolism during generalized seizures. No shift reagent was used but acute movement of Na^+ from outside to inside the cell was expected to be accompanied by a decrease in the intensity of the narrow ^{23}Na peak, because of spin transfer from a 100% visible to a 40% visible pool. Thus, the intensity of the narrow ^{23}Na peak corresponds to a particular extracellular-to-intracellular ^{23}Na content combination. To confirm that the change in ^{23}Na intensity was not an artifact of the increased cerebral blood flow and increased intracranial pressure, ^{23}Na NMR signals were observed from a cat brain with arterial CO_2 adjusted from 10 to 95 mm Hg. Under this range of CO_2 tensions, cerebral blood flow is known to increase by approximately 300% but, nevertheless, in this experiment no change in the ^{23}Na signal was observed, proving that the reduced ^{23}Na peak is indeed the result of an increase in the Na_i/Na_0 ratio.

Eleff *et al.* (1988) carried out concurrent measurements of cerebral blood flow, levels of sodium, lactate, and high-energy phosphates by ^{19}F, ^{23}Na, ^{1}H, and ^{31}P NMR on an *in situ* cat brain using a quadruply tuned coil. No shift reagent was used and, therefore, only total tissue sodium signal was measured, and intracellular changes were inferred from changes in intensity of this signal. Blood flow greatly increased during seizures with a 10% concurrent decrease in ^{23}Na signal intensity indicating cellular uptake of sodium, minimal lactate accumulation, no pH shift, and a change in the PCr/P_i ratio from 3.4 to 1.7. Changes in Na^+ distribution between intra- and extracellular pools grossly paralleled the time scale of the decline of ATP levels. Restoration of normal flow and ATP levels was associated with a return of the ^{23}Na NMR signal to a level about 9% above its control value. Such postseizure increases in the extracellular contribution to the tissue sodium signal may reflect an increase in extracellular volume secondary to interstitial edema and increased intravascular volume.

Naritomi *et al.* (1987) have carried out *in vivo* measurements of Na_i in gerbil brain and skeletal muscle using Dy(TTHA) by infusion. In both tissues, after 2 hr of shift reagent infusion the ^{23}Na spectrum is comprised of a single unshifted resonance and a broad component shifted about 5 ppm. In skeletal muscle, the shifted peak seems to consist of a single resonance, while in the brain it is split to two. The unshifted and shifted signals are assigned to intracellular, extracellular and intravascular sodium pools (Fig. 37). Under the same conditions, ^{1}H spectra were taken in both muscle and brain, and very similar spectra were obtained in terms of number of signals and their shifts. The authors concluded that these signals were representative of the intra- and extracellular and intravascular water spaces. They found the ratios of the ^{1}H signal areas to be close to known ratios of such spaces. Injection of ouabain increased both the signal assigned to intracellular sodium, as well as the ^{1}H signal representing the intracellular water space. Papaverine injection increased both ^{1}H and ^{23}Na signals ascribed to the intravascular compartment. The authors concluded that this method can provide useful physiological and biochemical information *in vivo.*

Burstein (1988) challenged several conclusions of Naritomi *et al.* (1987). First, the assignment of the observed ^{23}Na signals to pure intracellular and extracellular sodium pools was questioned because of the absence of a bulk susceptibility shift of the resonance assigned to the intracellular pool. Such a shift would be expected at the relatively high Dy(TTHA) concentration employed, and indeed had been observed in other studies. Second, the assignment of the ^{1}H resonances to intra- and extracellular water could not possibly be correct since water molecules would be in fast exchange between these compartments whereas the observed two separate signals, one shifted and one unshifted, clearly indicated a slow-exchange spectrum. Burstein (1988) suggested that the results in Naritomi *et al.* for both nuclei could be explained by the existence of tissue regions less accessible to the shift reagent.

Evidence implying that Dy(TTHA) crosses the blood–brain barrier (BBB) was presented recently (Albert *et al.*, 1990a). In the absence of BBB permeation by the shift reagent, the reagent-shifted ^{23}Na signal should be considerably smaller than in other tissues since it would contain only the intravascular, not the interstitial, pool. The conclusion about BBB permeability to Dy(TTHA) was drawn on the basis of the observation that the shifted signal was indeed larger than expected without the reagent crossing the barrier.

4.9. Skeletal Muscle

It has been found that in the presence of rat muscle, $Dy(PPP)_2$ decomposes much faster than in rat brain and heart (Matwiyoff *et al.*, 1986). If

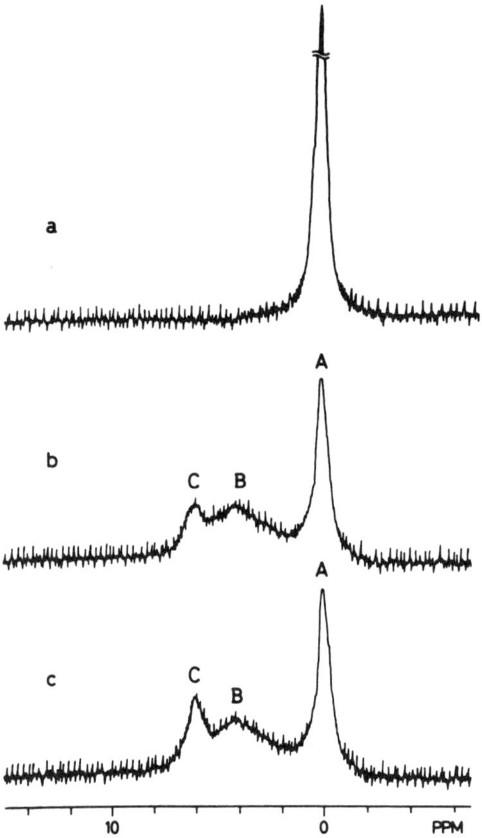

Figure 37. ^{23}Na NMR spectra of the brain: (a) before $Dy(TTHA)^{3-}$ infusion; (b) 2 hr after $Dy(TTHA)^{3-}$ infusion; (c) 5 min after papaverine hydrochloride injection. The peaks A, B, and C represent the intracellular, extravascular-extracellular, and intravascular Na^+, respectively. Five minutes after papaverine hydrochloride injection, the intravascular Na^+ (peak C) increased markedly. (From Naritomi *et al.*, 1987.)

only one extracellular compartment existed, then, as the $Dy(PPP)_2 : Na^+$ ratio decreased, the shift of the extracellular resonance would gradually decrease with its intensity unaltered until it merged with the ^{23}Na intracellular resonance to reconstitute the peak of undifferentiated, total tissue sodium. The spectra in this study show three or more Na_0 resonances and large changes in the intensity of the Na_0 peaks with time.

Blum *et al.* (1988a) examined the ischemic rat leg using $Dy(PPP)_2$ and Dy(TTHA), monitoring Na_i and high-energy phosphorus metabolites. With induced ischemia, the phosphorylation potential (λ) slowly fell. At a critical value of λ, associated with failure of the Na^+–K^+ antiport because of low availability of energy source, Na_i began to increase. At log λ of 3.12 ± 0.32 and pH_i of 6.86 ± 0.13, Na^+ influx, with and without ouabain, was 5.1 ± 4.3 and 4.0 ± 1.3 mole/liter per hr, respectively.

Balschi *et al.* (1989) investigated the magnetic-field dependence of shift-reagent-induced resolution. Dy(TTHA) was infused into a rat and a surface coil was used to observe Na_i in the gastrocnemius muscle at 8.4,

4.7, and 1.5 T. ^{23}Na linewidths increased almost linearly with increasing B_0 both in the absence and in the presence of shift reagent. The authors concluded from the results that both the shifted and unshifted sodium resonances were inhomogeneously broadened and that the observed linewidths were contributed primarily by bulk paramagnetic susceptibility effects. In the absence of shift reagent, these effects might have been caused by naturally present paramagnetic molecules, e.g., myoglobin.

As a consequence of the above, on a ppm scale, Na linewidths, both with and without shift reagent, showed little field dependence. Since the shift-reagent-induced shift difference between the resonances of Na_i and Na_0 also increased linearly with the external magnetic field, spectral resolution did not increase significantly with an increase in magnetic field strength. This, of course, meant that the efficacy of the shift-reagent method would not improve with the use of higher-field instruments. The authors' outlook, however, was rather more positively phrased, concluding that successful experiments would be possible at lower field strengths.

In an *in vivo* ^{31}P and ^{23}Na NMR study by Balschi *et al.* (1990) of a rat gastrocnemius muscle, using Dy(TTHA) infused into the jugular vein, the Na_i signal area gradually increased during 11 hr of total ischemia from 3-5 mM to 27 mM (Fig. 38). An Na_0 of 166 mM was found in the fast exchangeable extracellular space. Also determined was that the NMR visibility factors for Na_i and Na_0 were essentially the same. The authors concluded on the basis of this study that the shift reagent Dy(TTHA) was acutely nontoxic, and that sodium levels could be determined *in vivo.*

Excised dystrophic muscle from chicks showed a two-fold higher ^{23}Na NMR signal than normal muscle (Misra *et al.*, 1990). A similar ratio of sodium content between dystrophic and normal muscle was obtained by electron probe X-ray microanalysis. Also, it was found that the ^{23}Na T_1 of dystrophic muscle is longer than that of normal muscle, 26 ± 1 and 20 ± 1 msec, respectively.

Metabolic and ionic changes were examined in rat leg muscle before and during late decompensatory shock (LD) (Blum *et al.*, 1988b). Shock was induced by bleeding the animal until the mean blood pressure was 40 mm Hg. The removed blood was reinjected into the animal as needed to maintain this pressure. The LD phase occurred when all of the blood had been returned and this pressure could not be maintained. Until LD was reached, no increase in Na_i was observed during perfusion with Dy(TTHA) buffer. Although the high-energy metabolite PCr fell with time during shock, ATP did not decrease until the LD phase. When the phosphorylation ratio, $PR = [ATP]/[ADP][P_i]$, fell below $\log(PR) = 3.21 \pm 0.42$, Na_i increase commenced at a rate of 7.97 ± 0.60 meq/liter per hr. This rate corresponded to a twofold increase in Na^+ permeability compared with that of normal muscle.

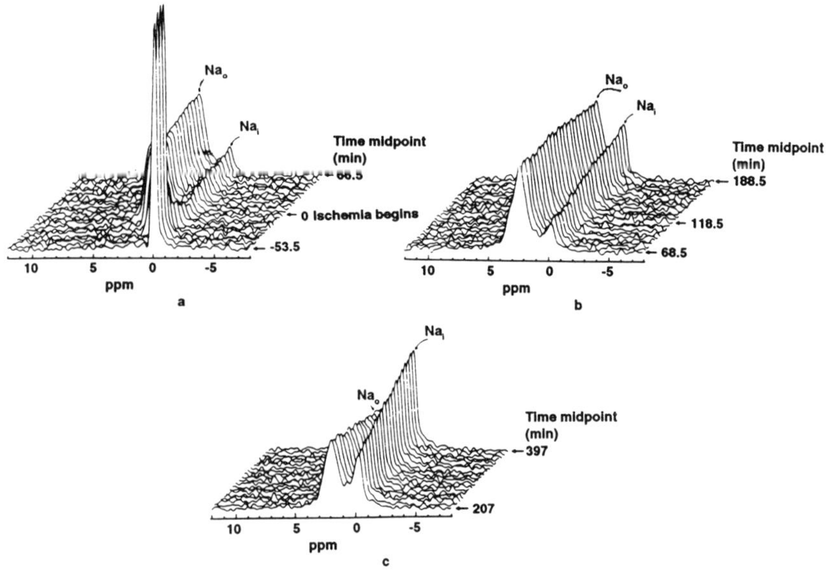

Figure 38. ^{23}Na NMR spectra obtained before and during ischemia in rat gastrocnemius muscle. (a) Spectra obtained during shift reagent infusion and the preischemic period ($t = -53.5$ to 0) and during the first hour of ischemia and after shift reagent infusion had ceased ($t = 0$ to 66.5 min). (b) ^{23}Na NMR spectra obtained during the second and third hour of ischemia. (c) ^{23}Na NMR spectra obtained after 3 hr of ischemia. The time between the start of each spectrum is 10 min. Each resolution-enhanced spectrum represents the sum of 600 scans obtained over 2 min; times indicated are the midpoints of the acquisitions. (From Balschi *et al.*, 1990.)

4.10. Oocytes

Visibility of sodium was found to depend on the stage of embryonic development in *Rana* oocytes, ovulated eggs, and early cleavage embryos (Gupta *et al.*, 1985). In prophase-arrested, follicle-enclosed oocytes, total oocyte sodium was about 17% visible (~40 mmole/kg cells). About 30% of total atomic-absorption-determined sodium in oocytes was associated with the yolk platelets isolated at room temperature by differential centrifugation. NMR analysis, however, did not yield a detectable ^{23}Na signal from these intact platelets. Thus, while yolk platelets were rich in Na^+, this sodium pool did not contribute to the NMR signal. Denuded oocytes, obtained by removing the follicular epithelium, gained about 13.6 mmole of total Na^+ per kilogram of cells, and exhibited a comparable increase in NMR-visible ^{23}Na, suggesting the existence of compartments with varying degree of NMR visibility within the oocyte.

Partially relaxed ^{23}Na NMR spectra revealed the existence of at least two major intracellular compartments of NMR-visible sodium with different magnetic environments and relaxation behavior in denuded oocytes. The

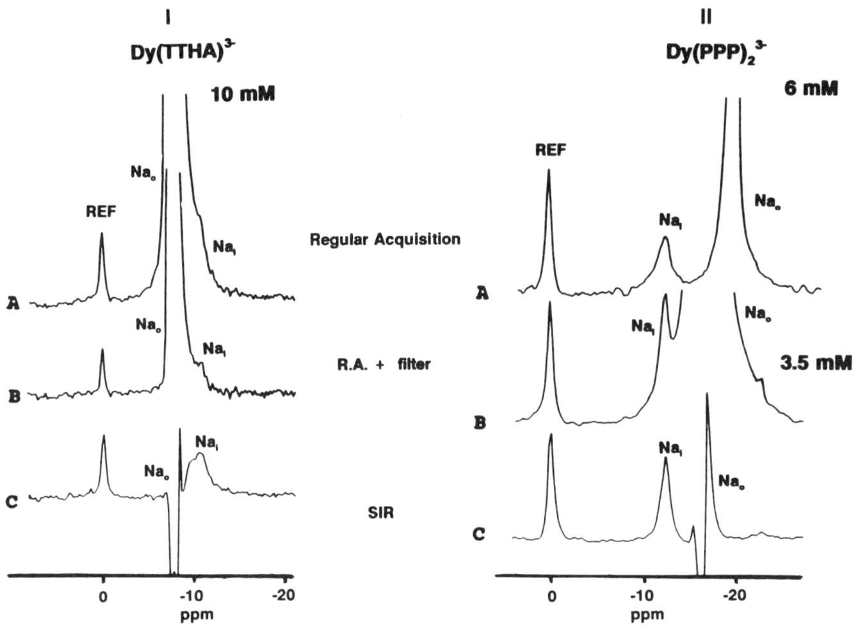

Figure 39. ^{23}Na NMR spectra of isolated rat hearts perfused with Hepes buffer containing as shift reagent either Dy(TTHA) (I) or Dy(PPP)$_2$ (II). Spectra A and B were acquired by a single pulse and were filtered by either exponential multiplication (A) or Gaussian filtering (B). In I, 10 mM Dy(TTHA) induced a 2.45-ppm downfield shift of Na$_0$ from Na$_i$ and using either of the two filters, the typical poor resolution of Na$_i$ obtained with 10 mM Dy(TTHA) is evident. In II, with 6 mM Dy(PPP)$_2$ (IIA), Na$_0$ is shifted ca. 6.5 ppm upfield, and the spectrum is well resolved. At the lower Dy(PPP)$_2$ concentration of 3.5 mM used to increase heart function, spectral resolution, however, is reduced below Na$_i$ quantifiability. Spectra C were obtained with the SIR method. Reasonable resolution allowing quantitative analysis of Na$_i$ is demonstrated both for 10 mM Dy(TTHA) (IC) and for 3.5 mM Dy(PPP)$_2$ (IIC). (From Simor *et al.*, 1991.)

authors surmised that since platelet Na^+ appeared to be NMR-invisible, one of the two observed compartments might be the nucleus. Progesterone action on the oocyte caused measurable changes in the level of NMR-visible Na^+. At ovulation (second metaphase), a gain in total egg sodium, as determined by atomic absorption, as well as in NMR-visible Na^+, was observed. Following fertilization, however, there was some loss of total cell sodium, but by the two- to four-cell stage about 70% of the total sodium became NMR-visible. These results indicated that a sizable fraction of the sodium in follicle-enclosed, prophase oocyte was sequestered and located in NMR-invisible compartments and that changes in NMR-visible Na$_i$ occurred following hormonal and developmental stimuli.

Morrill *et al.* (1985) examined insulin action on amphibian oocyte plasma membrane using NMR, electrophysiology, and ion flux techniques.

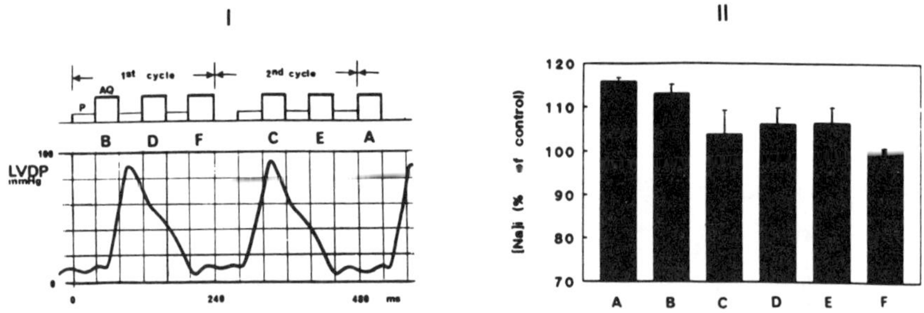

Figure 40. A shift-reagent-aided ^{23}Na NMR spectroscopic monitoring of the cycle-dependent level of intracellular sodium in isolated perfused rat hearts paced at a constant rate of 250 beats/min by means of an external pacemaker. ^{23}Na spectra were obtained on a Bruker AM 360 wide-bore NMR spectrometer. (I) The acquisition pulse program for monitoring by ^{23}Na NMR the heart-cycle-dependent level of Na_i is shown. Prior to acquisition (AQ), a selective presaturation (P) of 40 msec duration was applied to reduce the intensity of the extracellular sodium signal. "In memory" signal acquisition and the acquisition pulse program shown in I were used allowing a time resolution of FID acquisition every 80 msec. Three hundred scans were averaged to obtain a spectrum for each of the six 40-msec intervals. Three consecutive cardiac cycles were used to provide one scan for each of six different intervals, denoted A through F, of the full cardiac cycle. The first two of each three-cycle set is shown in (I). In the third cycle in each set, only interval A is used for NMR data collection. Each set of six spectra was collected from a total of 900 cardiac cycles, i.e., a total of 3.6 min per set. The area under the Na_i peak was determined for each spectrum and expressed as percent of the Na_i level of the last 40-msec interval of the cardiac cycle (interval F). The means ($n = 5$) of the Na_i level during each of the six cycle intervals are depicted by the bargraphs in (II). An increase of 14% in Na_i is evident during intervals A and B which roughly correspond to the depolarization phase of the cardiac cycle. This Na_i level is significantly ($p < 0.05$) higher than those of intervals C–F.

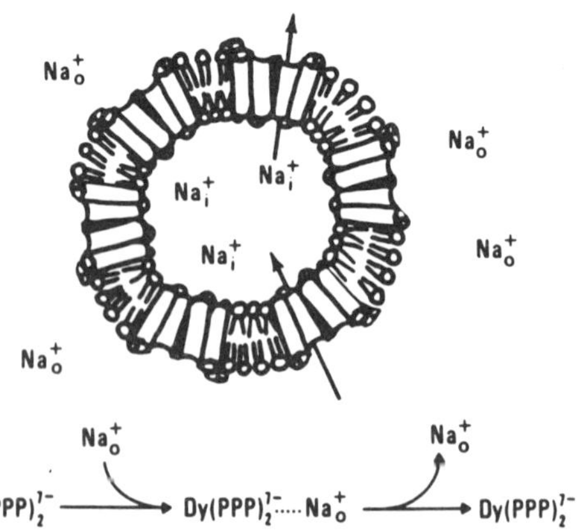

Figure 41. The exchange of sodium ions between intracellular (Na_i^+), free extracellular (Na_o^+), and extracellular shift-reagent-bound pools. (From Elgavish, 1989.)

The results suggested that insulin-induced mediator peptides stimulated Na^+–H^+ exchange resulting in an increase in pH_i and Na^+ uptake concomitant with an increase in receptor-mediated endocytosis and decrease in Na^+ conductance and associated membrane hyperpolarization. The net result appears to be a down-regulation of the Na^+ pump which together with a decrease in Na^+ conductance may direct high-energy phosphate compounds from cation regulation to anabolic processes of meiosis.

4.11. Salivary Glands

Na^+ influx has been measured in perfused rat mandibular gland (Seo *et al.*, 1987; Murakami *et al.*, 1986). Levels of Na^+, Li^+, and Cl^- in perfused salivary gland were monitored using Dy(TTHA) (Murakami *et al.*, 1986). The Na^+ concentration in the perfusate was varied by replacement with Li^+. The time constant for Na^+ exchange was slower following complete removal of Na_0 than following replacement of half of it, suggesting that Na^+ extrusion by Na^+–K^+-ATPase was reduced by lowering the Na_0 level. The time constant for Li^+ exchange was slower than that for Na^+ exchange. At rest, sodium concentration was not changed by inhibiting the Na^+–K^+-ATPase with ouabain and atropine, implying that in the absence of stimulation the spontaneous Na^+ influx across the plasma membrane must have been negligibly small.

By affecting the onset of stimulation with acetylcholine (Seo *et al.*, 1987), Na_i increased by 9.1 ± 1.5 mmole/liter intracellular fluid. The observed Na^+ influx rate was about 50% of the sodium secretory rate at the steady-state phase of secretion, estimated from the secretory rate by assuming an isotonic primary secretion. The authors found this fully compatible with the operation of a Na^+–K^+–$2Cl^-$ cotransport system for which they would expect a Na^+ influx rate exactly half of the rate of Na^+ and Cl^- secretion.

More recently, Seo *et al.* (1990) measured ^{23}Na in the perfused rat mandibular salivary gland using spin-echo double-quantum-filtered ^{23}Na NMR spectroscopy. The resonance due to Na_i consisted of two Lorentzian signals one stemming from the sharp $1/2$ to $-1/2$ transition with a T_2 of $95 \pm 4\ \text{sec}^{-1}$ and another, broad resonance due to the $-1/2$ to $-3/2$ and $3/2$ to $1/2$ transitions with a T_2 of $1360 \pm 75\ \text{sec}^{-1}$. A signal due to interstitial Na^+ was also observed, had longer T_2's, and disappeared upon administration of Dy(TTHA). This shift reagent was also used for the assignment of the intracellular signals.

4.12. Liver and Pancreas

Banni *et al.* (1987) have carried out a ^{23}Na relaxation study on rats acutely intoxicated with CCl_4. An increase in T_1 was found in CCl_4-treated rats, independently of necrosis. Neither fat accumulation nor proliferative processes affected the observed prolongation of T_1. When T_1 relaxation times were measured in the liver of vitamin E-treated rats subsequently intoxicated with CCl_4, a slight shortening of T_1 compared with CCl_4-intoxicated rats was observed. These results were discussed in terms of the antioxidant action exerted by vitamin E, taking into account that peroxidation of microsomal lipids was the key factor in the process of CCl_4-induced liver injury. Also, the observed changes in T_1 were discussed in terms of the interactions of Na^+ with cell membranes and/or the occurrence of changes in viscosity. ^{23}Na spectra of rat liver acquired in the presence of $Dy(PPP)_2$ were shown, as well as the changes in the ^{23}Na signals with time as the $Dy(PPP)_2$ complex became degraded.

A ^{23}Na and ^{31}P NMR study of a mouse liver perfused with Dy(TTHA) during nitrogen-induced hypoxia showed an apparent lag in Na_i increase compared with the time scale of decrease in ATP (Bowers *et al.*, 1990). The authors suggested that either the Na^+-ATPase in the hepatocytes was saturated with ATP under normoxic conditions, or that ATP supply for ion transport was preferentially maintained at the expense of other cellular metabolic functions during oxygen deprivation.

Membrane damage was assessed by the time dependence of ^{23}Na NMR spectroscopic line-shape changes in excised pancreas from rats after pancreatitis was induced by injection of sodium taurocholate (Kushnir *et al.*, 1990). Using 10 mM $Dy(PPP)_2$, it was concluded that the temporal changes in the spectrum of the affected pancreas were the result of the penetration with time of the intracellular space of membrane-damaged cells by the shift reagent.

ACKNOWLEDGMENTS. Partial support by NHLBI (P50 HL17667/IV, G.A.E.) and (5-F32 HL07851, S.K.M.) is gratefully acknowledged. S.K.M. also acknowledges support by The Frederik Philips Magnetic Resonance Research Center at Emory University, Atlanta, Georgia.

REFERENCES

Abragam, A., 1961, *The Principles of Nuclear Magnetism*, pp. 257–261, Oxford University Press (Clarendon), London.

Abragam, A., and Bleaney, B., 1970, *Electron Paramagnetic Resonance of Transition Ions*, Chapter 5, Oxford University Press (Clarendon), London.

Albert, M. S., Lee, J.-H., Patlak, C. S., Fenstermacher, J. D., and Springer, C. S., Jr., 1990a, *Soc. Magn. Reson. Med. 9th Annu. Meet. Abstr.* **1**:36.

Albert, M. S., Lee, J.-H., and Springer, C. S., Jr., 1990b, *Soc. Magn. Reson. Med. 9th Annu. Meet. Abstr.* p. 1269.

Ammann, H., Boulanger, Y., and Vinay, P., 1990, *Magn. Reson. Med.* **16**:368.

Anderson, S. E., Adorante, J. S., and Cala, P. M., 1988, *Am. J. Physiol.* **254**:C466.

Avison, M. J., Gullans, S. R., Ogino, T., Giebisch, G., and Shulman, R. G., 1987, *Am. J. Physiol.* **253**:C126.

Balschi, J. A., and Pohost, G. M., 1990, *Soc. Magn. Reson. Med. 9th Annu. Meet. Abstr.* **2**:798.

Balschi, J. A., Cirillo, V. P., and Springer, C. S., Jr., 1982, *Biophys. J.* **38**:323.

Balschi, J. A., Kohler, S. J., Bittl, J. A., Springer, C. S., Jr., and Ingwall, J. S., 1989, *J. Magn. Reson.* **83**:138.

Balschi, J. A., Bittl, J. A., Springer, C. S., Jr., and Ingwall, J. S., 1990, *NMR Biomed.* **3**:47.

Banni, S., Casu, M., Corongiu, F. P., Dessi, M. A., Lai, A., and Meloni, C., 1987, *Chem. Biol. Int.* **63**:207.

Bárány, M., and Venkatasubramanian, P. N., 1986, *Physiol. Chem. Phys. Med. NMR,* **18**:233.

Bax, A., 1982, *Two-Dimensional Nuclear Magnetic Resonance in Liquids,* Delft University Press/Reidel, Dordrecht, pp. 157–161.

Berendsen, H. J. C., and Edzes, H. T., 1973, *Ann. N.Y. Acad. Sci.* **204**:459.

Bernard, M., and Ingwall, J. S., 1990, *Soc. Magn. Reson. Med. 9th Annu. Meet. Abstr.* **2**:801.

Bleaney, B., 1972, *J. Magn. Reson.* **8**:91.

Blum, H., Schnall, M. D., Chance, B. and Buzby, G. P., 1988a, *Am. J. Physiol.* **255**:C377.

Blum, H., Schnall, M. D., Renshaw, P. F., and Buzby, G. P., 1988b, *Circ. Shock* **26**:341.

Bogusky, R. T., Garwood, M., Matson, G. B., Acosta, G., Cowgill, L. D., and Schleich, T., 1986, *Magn. Reson. Med.* **3**:251.

Boulanger, Y., and Vinay, P., 1989, *Can. J. Physiol. Pharmacol.* **67**:820.

Boulanger, Y., Vinay, P., and Desroches, M., 1985, *Biophys. J.* **47**:553.

Boulanger, Y., Vinay, P., and Boulanger, M., 1987, *Am. J. Physiol.* **253**:F904.

Bowers, J. L., Metz, K. R., Kruskal, J. B., Lee, R. G. L., Lanir, A., and Clouse, M. E., 1990, *Soc. Magn. Reson. Med. 9th Annu. Meet. Abstr.* **2**:807.

Brown, M. A., Stenzel, T. T., Ribeiro, A. A., Drayer, B. P., and Spicer, L. D., 1986, *Magn. Reson. Med.* **3**:289.

Bull, T. E., 1972, *J. Magn. Reson.* **8**:344.

Burstein, D., 1988, *Biophys. J.* **54**:191.

Burstein, D., and Fossel, E. T., 1987a, *J. Magn. Reson.* **73**:150.

Burstein, D., and Fossel, E. T., 1987b, *Magn. Reson. Med.* **4**:261.

Burstein, D., and Fossel, E. T., 1987c, *Am. J. Physiol.* **252**:H1138.

Buster, D. C., Castro, M. M. C. A., Geraldes, C. F. G. C., Malloy, C. R., Sherry, A. D., and Siemers, T. C., 1990, *Magn. Reson. Med.* **15**:25.

Cala, P. M., Anderson, S. E., and Cragoe, E. J., Jr., 1988, *Comp. Biochem. Physiol.* **90A**:551.

Castle, A. M., Macnab, R. M., and Shulman, R. G., 1986a, *J. Biol. Chem.* **261**:3288.

Castle, A. M., Macnab, R. M., and Shulman, R. G., 1986b, *J. Biol. Chem.* **261**:7797.

Casu, M., Corongiu, F. P., Dessi, M. A., Frau, M., and Lai, A., 1986, *Biochem. Biophys. Res. Commun.* **134**:1079.

Chu, W.-J., Kim, S. K., Pohost, G. M., and Elgavish, G. A., 1990a, *9th Annu. Meet. Soc. Magn. Reson. Med. Abstr.* **2**:916.

Chu, S. C.-K., Xu, Y., Balschi, J. A., and Springer, C. S., Jr., 1990b, *Magn. Reson. Med.* **13**:239.

Civan, M. M., and Shporer, M., 1978, in *Biological Magnetic Resonance,* Vol. 1 (L. J. Berliner and J. Reuben, eds.), Plenum Press, New York, p. 1.

Civan, M. M., Degani, H., Margalit, Y., and Shporer, M., 1983, *Am. J. Physiol.* **245**:C213.

Commodari, F., Tauskela, J. S., Sanctuary, B. C., and Shoubridge, E. A., 1990, *9th Annu. Meet. Soc. Magn. Reson. Abstr.* p. 1101.

Cope, F. W., 1965, *Proc. Natl. Acad. Sci. USA* **54**:225.

Cope, F. W., 1967, *J. Gen. Physiol.* **50**:1353.

Cope, F. W., 1970, *Biophys. J.* **10**:843.

Cowan, B. E., Sze, D. Y., Mai, M. T., and Jardetzky, O., 1985, *FEBS Lett.* **184**:130.

Degani, H., and Elgavish, G. A., 1978, *FEBS Lett.* **90**:357.

Delort, A.-M., Dauphin, G., Guyot, J., and Jeminet, G., 1989, *Biochim. Biophys. Acta* **1013**:11.

Eleff, S. M., Schnall, M. D., Ligetti, L., Osbakken, M., Subramanian, V. H., Chance, B., and Leigh, J. S., Jr., 1988, *Magn. Reson. Med.* 7:412.

Elgavish, G. A., 1982, in: *The Rare Earth In Modern Science and Technology* (McCarthy, Silber, and Rhyne, eds.), Vol. 3, Plenum Press, New York, p. 193.

Elgavish, G. A., 1989, *Invest. Radiol.* **24**:1028.

Elgavish, G. A., and Reuben, J., 1977, *J. Am. Chem. Soc.* **99**:1762.

Elgavish, G. A., Foster, R. E., Canby, R. C., and Pohost, G. M., 1987, *6th Annu. Meet. Soc. Magn. Reson. Med. Abstr.* p. 512.

Elgavish, G. A., Simor, T., Kim, S. K., Chu, W. J., and Pohost, G. M., 1991, *15th Int. Congr. Biochem.* p. 192.

Evans, C. H., 1990, *Biochemistry of the Lanthanides*, Plenum Press, New York, p. 339.

Fernandez, E., Grandjean, J., and Laszlo, P., 1987, *Eur. J. Biochem.* **167**:353.

Fossel, E. T., and Hoefeler, H., 1986, *Magn. Reson. Med.* **3**:534.

Foy, B. D., and Burstein, D., 1990, *Biophys. J.* **58**:127.

Fukushima, E., and Roeder, S. B. W., 1981, *Experimental Pulse NMR*, Addison-Wesley, Reading, Mass., p. 108.

Golding, R. M., and Halton, M. P., 1972, *Aust. J. Chem.* **25**:2577.

Golding, R. M., and Pyykkö, P., 1973, *Mol. Phys.* **26**:1389.

Gonzales, F., Harris, C., and Bassingthwaighte, J. B., 1980, *Physiologist* **23**:78.

Greenfield, N. J., Cherapak, C. N., Adebodun, F., Jordan, F., and Lenard, J., 1990, *Biochim. Biophys. Acta* **1025**:15.

Gueron, M., 1975, *J. Magn. Reson.* **19**:58.

Gullans, S. R., Avison, M. J., Ogino, T., Giebisch, G., and Shulman, R. G., 1985, *Am. J. Physiol.* **249**:F160.

Gullapalli, R. P., Hawk, R. M., and Komoroski, R. A., 1990, *Soc. Magn. Reson. Med. 9th Annu. Meet. Abstr.* p. 1271.

Gupta, R. K., 1987, in *NMR Spectroscopy of Cells and Organs*, Vol. 2 (Gupta, R. K., ed.), CRC Press, Boca Raton, Fla., p. 1.

Gupta, R. K., and Gupta, P., 1982, *J. Magn. Reson.* **47**:344.

Gupta, R. K., Kostellow, A. B., and Morrill, G. A., 1985, *J. Biol. Chem.* **260**:9203.

Höfeler, H., Jensen, D., Pike, M. M., Delayre, J. L., Cirillo, V. P., Springer, C. S., Jr., Fossel, E. T., and Balschi, J. A., 1987, *Biochemistry* **26**:4953.

Hotchkiss, R. S., Song, S.-K., Ling, C. S., Ackerman, J. J. H., and Karl, I. E., 1990, *Am. J. Physiol.* **258**:R21.

Hubbard, P. S., 1970, *J. Chem. Phys.* **53**:985.

Jans, A. W. H., William, R., Kellenbach, E. R., and Kinne, R. K., 1988, *Magn. Reson. Med.* 7:292.

Jelicks, L. A., and Gupta, R. K., 1989a, *J. Magn. Reson.* **81**:586.

Jelicks, L. A., and Gupta, R. K., 1989b, *J. Magn. Reson.* **83**:146.

Jelicks, L. A., and Gupta, R. K., 1989c, *J. Biol. Chem.* **264**:15230.

Jelicks, L. A., and Gupta, R. K., 1990, *J. Biol. Chem.* **265**:1394.

Jelicks, L. A., Weaver, J., Pollack, S., and Gupta, R. K., 1989, *Biochim. Biophys. Acta* **1012**:261.

Joseph, P. M., and Summers, R. M., 1987, *Magn. Reson. Med.* **4**:67.

Knubovets, T. L., Revazov, A. V., Sibeldina, L. A., and Eichhoff, U., 1989, *Magn. Reson. Med.* **9**:261.

Kohler, S. J., Ingwall, J. S., and Kolodny, N. H., 1990, *9th Annu. Meet. Soc. Magn. Reson. Med. Abstr.* **1**:39.

Kohler, S. J., Perry, S. B., Stuart, L. C., Atkinson, D. E., Clarke, K., and Ingwall, J. S., 1991, *Magn. Reson. Med.* **18**:15.

Kolodny, N. H., Kohler, S., Weinberg, D., Burr, T., Celi, A., D'Amico, D., and Gragoudas, E., 1990, *9th Annu. Meet. Soc. Magn. Reson. Med. Abstr.* **2**:666.

Kumar, A. M., Spitzer, A., and Gupta, R. K., 1986, *Kidney Int.* **29**:747.

Kumar, A. M., Gupta, R. K., and Spitzer, A., 1988, *Kidney Int.* **33**:792.
Kushnir, T., Kaplan, O., Askenasy, N., and Navon, G., 1990, *9th Annu. Meet. Soc. Magn. Reson. Med. Abstr.* **2**:982.
Kwan, C.-Y., Seo, Y., Ito, H., Murakami, M., and Watari, H., 1987, *J. Hypertens.* **5**:359.
Lewis, W. B., Jackson, J. A., Lemons, J. F., and Taube, J., 1962, *J. Chem. Phys.* **36**:694.
Liebling, M. S., and Gupta, R. K., 1987, *Ann. N.Y. Acad. Sci.* **508**:149.
Lotan, C., Miller, S. K., Pohost, G. M., and Elgavish, G. A., 1988, *7th Annu. Meet. Soc. Magn. Reson. Med. Abstr.* p. 37.
Lotan, C., Miller, S. K., Pohost, G. M., and Elgavish, G. A., 1992, *J. Mol. Cell. Cardiol.* **24**:243.
Lyon, R. C., Pekar, J., Moonen, C. T. W., and McLaughlin, A. C., 1990, *9th Annu. Meet. Soc. Magn. Reson. Med. Abstr.* **1**:37.
Magnuson, J. A., Shelton, D. S., and Magnuson, N. S., 1970, *Biochem. Biophys. Res. Commun.* **39**:279.
Malloy, C. R., Buster, D. C., Castro, M. M. C. A., Geraldes, C. F. G. C., Jeffrey, F. M. H., and Sherry, A. D., 1990, *Magn. Reson. Med.* **15**:33.
Martin, J.-B., Klein, G., and Satre, M., 1987, *Arch. Biochem. Biophys.* **254**:559.
Matwiyoff, N. A., Gasparovic, C., Wenk, R., Wicks, J. D., and Rath, A., 1986, *Magn. Reson. Med.* **3**:164.
Miller, S. K., Lotan, C., Hefner, L. L., Pohost, G. M., and Elgavish, G. A., 1989, *8th Annu. Meet. Soc. Magn. Reson. Med. Abstr.* p. 497.
Miller, S. K., Chu, W.-J., Pohost, G. M., and Elgavish, G. A., 1991, *Magn. Reson. Med.* **20**:184.
Misra, L. K., Hazle, J. D., Kim, E. E., and Dennis, L. W., 1990, *9th Annu. Meet. Soc. Magn. Reson. Med. Abstr.* **2**:803.
Monti, J.-P., Gailice, P., Crevat, A., Mehdi, M. E., Durand, C., and Murisasco, A., (1986), *Clin. Chem.* **32**:104.
Morrill, G. A., Weinstein, S. P., Kostellow, A. B., and Gupta, R. K., 1985, *Biochim. Biophys. Acta* **844**:377.
Mukidjam, E., Elgavish, G. A., and Barnes, S., 1987, *Biochemistry* **26**:6785.
Murakami, M., Seo, Y., Matsumoto, T., Ichikawa, O., Ikeda, A., and Watari, H., 1986, *Jpn. J. Physiol.* **36**:1267.
Naritomi, H., Kanashiro, M., Sasaki, M., Kuribayashi, Y., and Sawada, T., 1987, *Biophys. J.* **52**:611.
Naritomi, H., Sasaki, M., Kanashiro, M., Kitani, M., and Sawada, T., 1988, *J. Cereb. Blood Flow Metab.* **8**:16.
Navon, G., Askenasy, N., Vivi, A., and Tassini, M., 1990, *9th Annu. Meet. Soc. Magn. Reson. Med. Abstr.* **2**:918.
Nissen, H., Jacobsen, J. P., and Hørder, M., 1989, *Magn. Reson. Med.* **10**:388.
Ogino, T., den Hollander, J. A., and Shulman, R. G., 1983, *Proc. Natl. Acad. Sci. USA* **80**:5185.
Ogino, T., Shulman, G. I., Avison, M. J., Gullans, S. R., den Hollander, J. A., and Shulman, R. G., 1985, *Proc. Natl. Acad. Sci. USA* **82**:1099.
Ong, R. L., and Cheung, H. C., 1986, *Biophys. Chem.* **23**:237.
Ouwerkerk, R., van Echteld, C. J. A., Staal, G. E., and Rijksen, G., 1989, *Magn. Reson. Med.* **12**:164.
Packer, L., Spath, S., Martin, J.-B., Roby, C., and Blighy, R., 1987, *Arch. Biochem. Biophys.* **256**:354.
Payne, G. S., Seymour, A.-M. L., Styles, P., and Radda, G. K., 1990, *NMR Biomed* **3**:139.
Pekar, J., Lyon, R. C., Moonen, C. T. W., and McLaughlin, A. C., 1990, *9th Annu. Meet. Soc. Magn. Reson. Med. Abstr.* **2**:802.
Pekar, J., Renshaw, P. F., and Leigh, J. S., Jr., 1987, *J. Magn. Reson.* **72**:159.
Pettegrew, J. W., Woessner, D. E., Minshew, N. J., and Gloner, T., 1984, *J. Magn. Reson.* **57**:185.
Pike, M. M., and Springer, C. S., Jr., 1982, *J. Magn. Reson.* **46**:348.
Pike, M. M., Fossel, E. T., Smith, T. W., and Springer, C. S., Jr., 1984, *Am. J. Physiol.* **246**:C528.

Pike, M. M., Frazier, J. C., Dedrick, D. F., Ingwall, J. S., Allen, P. D., Springer, C. S., Jr., and Smith, T. W., 1985, *Biophys. J.* **48**:159.
Pike, M., Clark, D., Kitakaze, M., Madden, M., Golden, L., Cragoe, E., and Pohost, G. M., 1990, *9th Annu. Meet. Soc. Magn. Reson. Med. Abstr.* **1**:40.
Pilatus, U., Degani, H., and Pecht, I., 1990, *FEBS Lett.* **269**:292.
Post, J., 1989, *Scanning Microsc.* **3**:877.
Prince, L. S., Miller, S. K., Pohost, G. M., and Elgavish, G. A., 1992, *Magn. Reson. Med.* **23**:376.
Rayson, B. M., and Gupta, R. K., 1985, *J. Biol. Chem.* **260**:7276.
Renshaw, P. F., Blum, H., and Leigh, J. S., Jr., 1986, *J. Magn. Reson.* **69**:523.
Reuben, J., and Elgavish, G. A., 1979, in *Handbook on the Physics and Chemistry of Rare Earths* (K. A. Gschneider, Jr., and L. Eyring, Eds.), North-Holland, Amsterdam.
Reuben, J., and Fiat, D., 1967a, *J. Chem. Soc. Chem. Commun.* **1967**:729.
Reuben, J., and Fiat, D., 1967b, *J. Chem. Phys.* **47**:5440.
Reuben, J., and Fiat, D., 1969a, *J. Chem. Phys.* **51**:4909.
Reuben, J., and Fiat, D., 1969b, *J. Chem. Phys.* **51**:4918.
Schmidt, V. H., 1972, in *Proc. Ampère Int. Summer School II* (R. Blinc, ed.), University of Ljubljana, Ljubljana, Yugoslavia, p. 75.
Schnall, M. D., Yoshizaki, K., Chance, B., and Leigh, J. S., Jr., *Magn. Reson. Med.* **6**:15.
Seo, Y., Murakami, M., Matsumoto, T., Nishikawa, H., and Watari, H., 1987, *Pflugers Arch.* **409**:343.
Seo, Y., Murakami, M., and Watari, H., 1990, *Biochim. Biophys. Acta* **1034**:142.
Sherry, A. D., Malloy, C. R., Jeffrey, F. M. H., Cacheris, W. P., and Geraldes, C. F. G. C., 1988, *J. Magn. Reson.* **76**:528.
Shinar, H., and Navon, G., 1984, *Biophys. Chem.* **20**:275.
Shporer, M., and Civan, M. M., 1972, *Biophys. J.* **12**:114.
Shporer, M., and Civan, M. M., 1974, *Biochim. Biophys. Acta* **354**:291.
Simor, T., Kim, S. K., Chu, W. J., Pohost, G. M., and Elgavish, G. A., 1991, *10th Annu. Meet. Soc. Magn. Reson. Med. Abstr.* p. 1189.
Sink, R. M., Buster, D. C., and Sherry, A. D., 1990, *Inorg. Chem.* **29**:3645.
Springer, C. S., Jr., 1987, *Annu. Rev. Biophys. Biophys. Chem.* **16**:375.
Sze, D. Y., Corbelletta, N. L., Shochat, S. J., and Jardetzky, O., 1990, *Magn. Reson. Med.* **13**:14.
van Echteld, C. J. A., van Nieuwenhoven, M. A., van der Meer, P., and Ruigrok, T. J. C., 1990, *9th Annu. Meet. Soc. Magn. Reson. Med. Abstr.* **2**:799.
Vega, A. J., and Fiat, D., 1976, *Mol. Phys.* **11**:347.
Vogel, H. J., and Braunlin, W. H., 1985, *J. Magn. Reson.* **62**:42.
Wittenberg, B. A., and Gupta, R. K., 1985, *J. Biol. Chem.* **260**:2031.

5

In Vivo ^{19}F NMR

Barry S. Selinsky and C. Tyler Burt

1. INTRODUCTION

The field of *in vivo* NMR, made possible by the development of FT NMR techniques and large-bore stable superconducting magnets, has grown steadily in the last two decades. The increasing number of papers in this field has spawned several journals dedicated exclusively to the topic. New techniques and improved instrumentation allow an ever-increasing variety of NMR experiments which address significant *in vivo* questions.

When the topic of *in vivo* NMR is mentioned, thoughts automatically turn to ^{31}P NMR studies of energy metabolism. This topic has been reviewed here by Hollis (1980) and Elgavish (1987). NMR studies in living systems of ^{23}Na and ^{39}K, ions of great biological importance, have been addressed also (Civan and Shporer, 1978). Additionally, the use of isotopically labeled compounds in ^{13}C (Baxter *et al.*, 1983; London, 1988) and ^{15}N (Blomberg and Ruterjans, 1983) NMR studies of cell and organ metabolism are reviewed.

We will add to the *in vivo* NMR periodic table by discussing ^{19}F NMR and its application to living systems. *In vitro* analysis has been reviewed in Volume 1 by Gerig (1978). Historically, *in vivo* ^{19}F NMR is a relatively new technique. The first ^{19}F NMR study on living tissue was carried out in 1979 using isolated platelets (Costa *et al.*, 1979) and the first ^{19}F surface coil study of an intact animal was reported in 1983 (Wyrwicz *et al.*, 1983). Fluorine images in phantoms were first obtained in 1977 (Holland *et al.*,

Barry S. Selinsky • Department of Chemistry, Villanova University, Villanova, Pennsylvania 19085. **C. Tyler Burt** • Magnetic Resonance Center, University of Illinois at Chicago, Chicago, Illinois 60680.

Biological Magnetic Resonance, Volume 11: In Vivo Spectroscopy, edited by Lawrence J. Berliner and Jacques Reuben. Plenum Press, New York, 1992.

1977) and *in vivo* images by 1982 (Heidelberger and Lauterbur, 1982). Applications of ^{19}F NMR to living systems have blossomed in recent years, primarily due to the inherent advantages of fluorine over other nuclei of biological interest. Fluorine nuclei on an atom-for-atom basis are inherently better than ^{31}P nuclei for observation by NMR because the NMR sensitivity of fluorine is greater than twice that of phosphorus. The chemical shift range of interest of fluorine for biologic compounds is greater than four times the range for phosphorus. As ^{19}F is 100% naturally abundant, synthesis of compounds of interest need not depend on the purchase of expensive labeled compounds as one must with ^{13}C or ^{15}N. Fluorine-19 is a spin-1/2 nucleus and therefore has no quadrupolar contributions to its relaxation to complicate spectral interpretation. Since there are no endogenous compounds in animals which contain NMR-observable fluorine, interference from background resonances (such as water in *in vivo* ^{1}H NMR) is never a problem.

Several physiologically relevant questions are particularly well suited for exploration by ^{19}F NMR. The uptake, elimination, and metabolism of a drug or chemical which contains fluorine can be followed. In addition, several fluorinated probes for biological ions exist, and ^{19}F NMR can be used to monitor variations in cell ion concentrations in response to physiological stress or cell injury. In the study of specific tissues (such as tumors) the uptake of fluorinated drugs relative to the rest of the body can be exploited to study either the metabolic function of the tissue of interest or to identify the tissue by ^{19}F NMR.

These experiments depend upon low toxicity of fluorinated compounds which are taken up into cells (or whole animals) at NMR-observable concentrations. Fortunately, many fluorinated drugs and other substances in common use meet these requirements. The toxicity of fluorinated compounds varies with the chemical structure of the compound. Some fluorinated compounds, such as fluoroacetate, are potent metabolic inhibitors, and cause cell death at very low doses. Other compounds such as trifluoroacetate are metabolically stable and harmless to cells at micromolar to low millimolar concentrations. In general, compounds developed as drugs or metabolic probes either contain a trifluoromethyl group or a fluorinated aromatic moiety since the trifluoromethyl group is not metabolized and fluorine on an aromatic ring is either not metabolized or is completely degraded to a fluoride ion.

Thanks to advances in NMR electronics and hardware, doubly or triply tuned surface coils (Schnall *et al.*, 1985; Chang *et al.*, 1987) can be used to study the metabolism of a fluorinated compound while simultaneously determining the effect of the drug on phosphorus energy metabolism. This technique provides new ways to uncover the mechanism of drug action relative to its cellular toxicity.

For this review, we will define as *in vivo* those experiments actually carried out in the body using localized spectroscopy or exposed organs, or done with perfused organs. Studies of urinary metabolites of a fluorinated compound will be included only when they complement a similar *in vivo* study, or when they identify potentially useful *in vivo* experiments. This criterion will hold for isolated tissue or cells in culture.

As a reference, the ^{19}F chemical shifts of biological fluorinated compounds are provided in Table 1.

2. Metabolized Probes

2.1. Fluorodeoxyglucose

The technique of positron emission tomography (PET) has been extremely useful as a noninvasive measure of brain activity (Phelps *et al.*, 1979). The technique is important in identifying regions of brain affected by stroke, epilepsy, and other diseases of the central nervous system. PET most often involves the administration of a radioactive, nonmetabolized analogue of glucose. Radiolabeled glucose is transported to actively metabolizing regions of the brain. Once into the cell, the radioactive emission from the glucose analogue is recorded and the location of emission determined by image reconstruction (Phelps *et al.*, 1979). The most commonly used glucose analogue is 2-[^{18}F]fluoro-2-deoxy-D-glucose (2-^{18}FDG) (Reivich *et al.*, 1979; Phelps *et al.*, 1979), which is taken up in the brain and phosphorylated by hexokinase to [2-^{18}F]fluoro-2-deoxy-D-glucose 6-phosphate (2-^{18}FDG-6-PO_4). This metabolite was until recently believed to be metabolically stable with a half-life in the brain of approximately 8 hr (Reivich *et al.*, 1979). However, the validity of the metabolic stability of 2-^{18}FDG-6-PO_4 has been questioned (Fox, 1984).

Several groups have undertaken a study of 2-fluoro-2-deoxy-D-glucose (2-FDG; see Table 2 for abbreviations of fluorosugar metabolites) metabolism in brain and other tissues using ^{19}F NMR and the corresponding nonradioactive analogues of compounds used as PET tracers. The first metabolic study of 2-FDG metabolism in rat brain using ^{19}F NMR was reported by Nakada *et al.* (1986). An NMR resonance resulting from both 2-FDG and 2-FDG-6-PO_4 was identified in rat brain based on extrapolation from the ^{19}F chemical shifts of these compounds in aqueous solution (Nakada *et al.*, 1985). Two additional resonances were observed but not positively identified. ^{19}F NMR analysis of rat urine following 2-FDG administration demonstrated that fluoride ion and either 2-fluoro-2-deoxy-6-phosphogluconate or 2-fluoro-2-deoxy-δ-phosphogluconolactone were present as 2-FDG metabolites (Kwee *et al.*, 1986).

TABLE 1
^{19}F Chemical Shifts of Fluorine Compounds of Biological Interest and Their Metabolites

Standards	δ (ppm)
50 mM trifluoroacetate in water	0.0
50 mM trifluoroacetate in saline	0.0
Neat trifluoroacetic acid	−2.4
6-Trifluorotryptophan	−46.2
Fluorotrichloroethane	−76.1
1,2-Dibromo-1,1,2,2-tetrafluoroethane	−12.7

Compound	Solvent/buffer/tissue	δ (ppm)	Reference
Halothane	Neat	−1.0	Wyrwicz *et al.* (1983)
	Hexane	−1.2	Wyrwicz *et al.* (1983)
		−1.4	Trudell and Hubbell (1976)
	Water	0.5	Wyrwicz *et al.* (1983)
		0.4	Trudell and Hubbell (1976)
	Plasma	0.6	Wyrwicz *et al.* (1983)
	Erythrocytes	0.5–1.7	Wyrwicz *et al.* (1983)
	Normal kidney	0.2	Burt *et al.* (1983)
	Kidney tumor	0.0, 0.3	Burt *et al.* (1983)
	Muscle	0.0	Burt *et al.* (1983)
	Multilamellar vesicles	0.2	Trudell and Hubbell (1976)
	Brain	−2.2	Wyrwicz *et al.* (1983)
	Liver	−0.6	Selinsky *et al.* (1987)
Halothane metabolites			
Trifluorochloroethane	Liver homogenate	4.0	Selinsky *et al.* (1986)
Difluorochloroethane	Liver homogenate	−11.4, −15.6	Selinsky *et al.* (1986)
Trifluoroethanolamine	Liver homogenate	−0.3	Selinsky *et al.* (1986)
Fluoride ion	Liver homogenate	−43.8	Selinsky *et al.* (1986)

Methoxyflurane	Neat	−11.0	Wyrwicz *et al.* (1983)
	Hexane	−11.9	Wyrwicz *et al.* (1983)
	Water	−9.4	Wyrwicz *et al.* (1983)
	Plasma	−9.6–−9.7	Wyrwicz *et al.* (1983)
	Erythrocytes	−6.6–−12.6	Wyrwicz *et al.* (1983)
	Brain	−12.1	Wyrwicz *et al.* (1987b)
	Liver	−11.0	Selinsky *et al.* (1988a)
	Liver microsomes	−10.7	
Methoxyflurane metabolites			
Methoxydifluoroacetate	Liver (*in vivo*)	−7.1	Selinsky *et al.* (1988a)
	Liver microsomes	−7.6	Selinsky *et al.* (1988a)
Enflurane	Neat	−11.2 (-CF_2-)	Selinsky *et al.*
		−10.9 (CF_2H-)	(unpublished data)
		−81.1 (CHClF-)	(unpublished data)
	0.1 M buffer pH 7.4	−9.4 (-CF_2)	(unpublished data)
		−9.8 (CF_2H)	(unpublished data)
		−80.2 (CHClF-)	(unpublished data)
Enflurane metabolites			
2,2-Difluoro-2-difluoro-methoxyacetate	Urine	−3.3 (-CF_2-)	(unpublished data)
		−9.6 (CF_2H-)	
Isoflurane			
	Brain	−3.4 (CF_3-)	Wyrwicz *et al.*
		−9.7 (CF_2H-)	(1983)
Fluoropyrimidines			
5-Deoxy-5-fluorouridine	Water	−88.4	Malet-Martino *et al.* (1985)
	Blood	−89.5	Malet-Martino *et al.* (1985)
	Urine	−90.4	Malet-Martino *et al.* (1983)
5-Fluorouracil	Water	−93.8	Malet-Martino *et al.* (1985)
	Blood	−93.9	Malet-Martino *et al.* (1985)
	Urine	−93.9	Malet-Martino *et al.* (1983)

(continued)

TABLE 1. *Continued.*

Compound	Solvent/buffer/tissue	δ (ppm)	Reference
FBAL	Water	−112.9	Malet-Martino *et al.* (1985)
	Blood	−112.6	Malet-Martino *et al.* (1985)
	Urine	−112.9	Malet-Martino *et al.* (1983)
FUH_2	Water	−126.5	Malet-Martino *et al.* (1985)
	Blood	−126.4	Malet-Martino *et al.* (1985)
	Urine	−126.7	Malet-Martino *et al.* (1983)
FUPA	Urine	−111.2	Malet-Martino *et al.* (1983)
Fluoride	Urine	−43.1	Malet-Martino *et al.* (1983)
Fluorosugars and metabolites			
2-FDG			
α anomer	Water	−123.8	Nakada *et al.* (1985)
β anomer	Water	−123.7	Nakada *et al.* (1985)
2-FDG-6-PO_4			
α anomer	Water	−123.9	Nakada *et al.* (1985)
β anomer	Water	−123.8	Nakada *et al.* (1985)
2-Fluoro-2-deoxy-6-phosphogluconate	Water	−120.7	Nakada *et al.* (1985)
2-Fluoro-2-deoxy-D-sorbital	Water	−125.0	Nakada *et al.* (1986)
5-Fluoro-5-deoxy-L-sorbose (α anomer)	Water	−123.4	Nakada *et al.* (1986)
5-Fluoro-5-deoxy-L-sorbose-1-phosphate (α anomer)	Water	−122.9	Nakada *et al.* (1986)
2-Fluoro-2-deoxy-L-glyceraldehyde	Water	−127.5	Nakada *et al.* (1986)
2-Fluoro-2-deoxyglycerol	Water	−128.85	Nakada *et al.* (1988)
FDM			
α anomer	Water	−129.1	Kojima *et al.* (1988)
β anomer	Water	−147.5	Kojima *et al.* (1988)
3-FDG			
α anomer	Water	−124.5	Kwee *et al.* (1987)
β anomer	Water	−119.5	Kwee *et al.* (1987)
3-Fluoro-3-deoxy-D-sorbital	Water	−137.7	Kwee *et al.* (1987)
3-Fluoro-3-deoxy-D-fructose	Water	−133.1	Kwee *et al.* (1987)

Ion chelators			
BAPTA type[a]			
5-Fluoro BAPTA	Isotonic buffer		
Uncomplexed		−43.9	Smith *et al.* (1986)
Ca^{2+} bound		−38.2	Levy *et al.* (1987)
Zn^{2+} bound		−41.7	Metcalfe *et al.* (1985)
4-Methyl-5-fluoro BAPTA	Isotonic buffer		
Uncomplexed		−47.7	Levy *et al.* (1987)
Ca^{2+} bound		−41.8	Levy *et al.* (1987)
Quin type[b]			
QuinF (uncomplexed)	Isotonic buffer	−44.3	Levy *et al.* (1987)
QuinMF	Isotonic buffer		
Uncomplexed		−47.8	Levy *et al.* (1987)
Ca^{2+} bound		−39.5	Levy *et al.* (1987)
APTRA type[c]			
4F-APTRA	Isotonic buffer		
Uncomplexed		−46.8	Levy *et al.* (1988)
Mg^{2+} bound		−44.9	Levy *et al.* (1988)
5F-APTRA	Isotonic buffer		
Uncomplexed		−45.5	Levy *et al.* (1988)
Mg^{2+} bound		−41.2	Levy *et al.* (1988)
MF-APTRA	Isotonic buffer		
Uncomplexed		−49.3	Levy *et al.* (1988)
Mg^{2+} bound		−37.3	Levy *et al.* (1988)
F-cryp-1[d]	Buffer, pH 7		
Uncomplexed		−38.5	Smith *et al.* (1986)
Na^{+} bound		−36.5	Smith *et al.* (1986)
Trifluoromethylalanine			
$-H^{+}$	Hanks–Hepes buffer	−0.3	Taylor and Deutsch (1983)
$+H^{+}$		−2.0	Taylor and Deutsch (1983)

(continued)

TABLE 1. *Continued.*

Compound	Solvent/buffer/tissue	δ (ppm)	Reference
Difluoromethylalanine methyl ester	Hanks–Hepes buffer	−54.0	Taylor and Deutsch (1985)
Perfluorotripropylamine	Neat	−8.4, −10.8, −48.0	Taylor and Deutsch (1988)
Other fluorinated indicators			
Mefloquine	PBS	−8.1, −10.0	San George *et al.* (1984)
Tetrafluoroputrescine	d_6-acetone	−40.5	Diginis *et al.* (1986)
Trifluoromethylalanine			
$-H^+$	Hanks–Hepes buffer	−0.3	Taylor and Deutsch (1983)
$+H^+$	Hanks–Hepes buffer	−2.0	Taylor and Deutsch (1983)

[a] BAPTA, 1,2-bis-(2-aminophenoxy)ethane-*N,N,N′,N′*-tetraacetic acid.
[b] Quin, 2-(2-aminophenoxymethyl)-8-aminoquinoline-*N,N,N′,N′*-tetraacetic acid.
[c] APTRA, *o*-aminophenol-*N,N,O*-triacetic acid.
[d] Sodium ion indicator.

TABLE 2
Fluorosugar and Fluoropyrimidine Metabolite Abbreviations

Compound	Abbreviation
2-Fluorodeoxyglucose	2-FDG
3-Fluorodeoxyglucose	3-FDG
2-Fluorodeoxyglucose-6-phosphate	2-FDG-6-PO_4
Fluorodeoxymannose	FDM
5-Fluorouracil	5-FU
5′-Deoxy-5-fluorouridine	5′dFUrd
Fluorodeoxyuridine monophosphate	FdUMP
Fluoro-β-alanine	FBAL
Dihydrofluorouracil	FUH_2
Fluoro-β-ureidopropionic acid	FUPA
5-Fluorocytosine	5-FC

Further studies by Nakada and Kwee (1986) indicated significant metabolism of 2-FDG via the aldose reductase sorbitol pathway to the metabolites 2-fluoro-2-deoxy-D-sorbitol, 5-fluoro-5-deoxy-L-sorbose, 5-fluoro-5-deoxy-L-sorbose-1-phosphate, 2-fluoro-2-deoxy-L-glyceraldehyde, and 2-fluoro-2-deoxyglycerol. These assignments were based on ^{19}F NMR resonances in both *in vivo* and urine studies with chemical shifts similar to the chemical shifts for the above compounds. The most prominent metabolite resonance was assigned to 2-fluoro-2-deoxyglycerol, the proposed end product of the aldose reductase sorbitol pathway. These assignments were later strengthened by the observation that coadministration of sorbinil, an inhibitor of the aldose reductase sorbitol pathway, resulted in both a decrease in 2-FDG metabolism and a decrease in the resonance intensity for 2-fluoro-2-deoxyglycerol (Nakada and Kwee, 1987).

In conflict with the above assignments, Kanazawa *et al.* (1986) noted the presence of 2-deoxy-2-fluoro-D-mannose (FDM) as a contaminant in their preparation of synthetic 2-FDG. They also observed ^{19}F NMR resonances for FDM in mouse tissues after highly purified 2-FDG was administered. Kanazawa *et al.* attributed the presence of FDM to isomerization of 2-FDG *in vivo* by phosphoglucose isomerase. While other groups have not reported the presence of FDM in tissue or urine after 2-FDG administration, close examination of the *in vivo* spectra of Nakada and Kwee (1986) and Berkowitz and Ackerman (1987) reveals resonances with similar chemical shifts to those attributed to the α and β anomers of FDM. Resonances for FDM are clearly observed in studies of urine metabolites of 2-FDG (Kwee *et al.*, 1986); the relative intensities of these resonances are consistent with the equilibrium concentrations of α and β mannose in solution.

Thus, the data compiled here indicate three possible mechanisms for the metabolism of 2-FDG in brain:

1. Phosphorylation to 2-FDG-6-PO_4 by hexokinase, followed by oxidation to either 2-fluoro-2-deoxy-δ-phosphogluconolactone or 2-fluoro-2-deoxy-6-phosphogluconate by pentose phosphate pathway enzymes.
2. Phosphorylation to 2-FDG-6-PO_4 by hexokinase, followed by isomerization to FDM, possibly by the enzyme phosphoglucose isomerase.
3. Metabolism of 2-FDG using the aldose reductase sorbitol pathway to 2-fluoro-2-deoxy-L-glyceraldehyde and possibly 2-fluoro-2-deoxyglycerol.

Returning to the assumptions in PET scanning, one notes that regardless of metabolism the PET scan requires only that the radioactive label be retained by the cells for the duration of the procedure. While metabolism of 2-FDG increases the possibility of its elimination, studies which specifically measure the time-dependent elimination of the fluorine label from brain are necessary. ^{19}F NMR is particularly suited for measuring the time course of fluorine elimination from the brain. Wyrwicz *et al.* (1985) have determined the half-life of 2-FDG in brain using ^{19}F NMR to be ~80 min in fed rats. This result differs from the half-life of 8 hr estimated by Reivich *et al.* (1979), but is not significantly different from the half-life of 70 min determined by Hawkins and Miller (1978) for 2-deoxyglucose-6-phosphate. A shorter retention time for 2-FDG in brain should caution clinicians using PET to reconsider the quantitative aspects of their images. Some discussion of the effect that high levels of 2-FDG used in NMR experiments might have on overall glucose metabolism has been put forth. A recent report by Nakada and Kwee (1987) demonstrated that if sorbinil, an inhibitor of the aldose reductase sorbinol pathway, was coadministered with 2-FDG, the metabolism of 2-FDG-6-PO_4 was greatly reduced. The use of sorbinil along with 2-FDG may improve PET images by increasing the tissue retention time of the radiolabel.

Berkowitz and Ackerman (1987) have carried out the first *in vivo* proton-decoupled fluorine experiment in studying the metabolism of 2-FDG in rat brain using orthogonal decoupling/observe coils. While proton decoupling improved the resolution of the ^{19}F NMR spectra of brain shown in this report, the advantage of decoupling in light of *in vivo* linewidths is marginal. Furthermore, if such experiments were scaled to larger animals (such as humans), higher decoupling power would be necessary and sample heating and patient discomfort may become a problem.

In a recent study, Kwee *et al.* (1987) introduced 3-fluoro-3-deoxy-D-glucose (3-FDG) as a probe of aldose reductase activity in brain. 3-FDG is apparently a poor substrate for the hexokinase reaction (Bessel *et al.*, 1972); also, the binding affinity of phosphohexose isomerase is low relative to D-glucose (Bessel and Thomas, 1973). Kwee *et al.* demonstrated that 3-FDG is metabolized in the brain, and that is metabolism is inhibited by cotreatment with sorbinil. In addition, the elimination of 3-FDG from brain is slow relative to 2-FDG (Kwee *et al.*, 1987). Also, 3-FDG causes less perturbation of electroencephalograph readings from rat brain relative to equal doses of 2-FDG, and is apparently nontoxic to rats at high dosage (Halama *et al.*, 1984). These criteria suggest that 3-FDG might be suitable for use as a contrast agent for metabolic imaging in brain. Pilot studies in rats have supported the feasibility of this idea (Nakada *et al.*, 1988).

2.2. Anesthetics

2.2.1. Elimination from Brain and Mechanism of Action

A second class of fluorinated compounds of interest for NMR study are the inhalation anesthetics. Several fluorinated anesthetics are commonly used clinically, including halothane ($CF_3CHClBr$), isoflurane (CF_3CHCl-O-CF_2H), and enflurane ($CHClFCF_2$-O-CF_2H). These anesthetics are rapidly absorbed by the body in millimolar concentrations and hence are ideal for study by NMR. The first *in vivo* studies exploiting the anesthetics were carried out by Burt and Wyrwicz (Burt *et al.*, 1982; Wyrwicz *et al.*, 1982). In these, the anesthetic methoxyflurane ($CHCl_2CF_2$-O-CH_3) was observed in circulating dog blood brought to the NMR magnet by an arteriovenous bypass shunt. Soon afterwards, Wyrwicz *et al.* (1983) used a surface coil to demonstrate that fluorinated anesthetics could be observed in brain tissue and that anesthetic clearance could be monitored. Wyrwicz *et al.* (1983) also made the surprising discoveries that halothane clearance from the brain was relatively slow in comparison to recovery from anesthesia and that a halothane metabolite or adduct remained in brain at measurable levels for as long as 98 hr after a 30-min halothane exposure. Later, more refined studies of halothane clearance from rabbit brain demonstrated that after 120 min exposure to 1.0% halothane, the anesthetic existed in two different cerebral environments characterized by different half-lives and different ^{19}F chemical shifts (Fig. 1; Wyrwicz *et al.*, 1987a). The half-life of one environment correlates well with the recovery of the animal after anesthesia, indicating that halothane found in this environment is directly responsible for anesthesia. Wyrwicz *et al.* (1987b) also reported the presence of a nonvolatile halothane metabolite in brain with a half-life of several days, consistent with earlier work by her group (Wyrwicz *et al.*, 1983) and others (Selinsky and Burt, unpublished data).

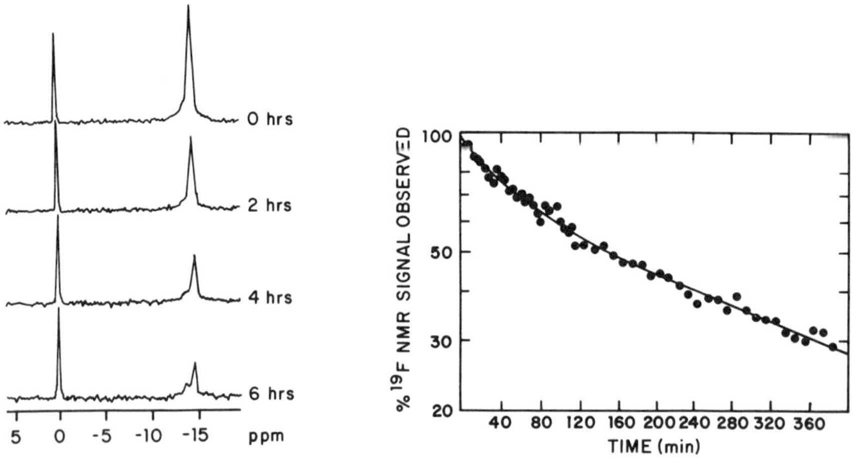

Figure 1. Halothane elimination from rabbit brain. The spectra on the left are consecutive *in vivo* surface coil ^{19}F NMR spectra of rabbit brain following the administration of 1% halothane for 120 min. The resonance at 0 ppm is $C_2Br_2F_4$ in olive oil used as an external chemical shift standard. The graph on the right describes the elimination of halothane from rabbit brain as a function of time, which is biphasic under these experimental conditions. (Reproduced with permission from Wyrwicz *et al.*, 1987a.)

A recent report by Litt *et al.* (1987) of halothane elimination from rat brain gives quantitative support to the above clearance data. The elimination of halothane from rat brain after a 60-min exposure to 1.0% halothane anesthetic could be described by two exponential decay curves. The rate constants for halothane clearance compared favorably with those determined by Wyrwicz *et al.* (1987a) for rabbits, when differences in animal mass were taken into account. Note that both of the above studies were done using 1% halothane. No ^{19}F resonance for a halothane metabolite was reported by Litt *et al.* in rat brain, probably due to the large linewidths seen in their spectra. No extracts of brain tissue were reported which would have demonstrated whether significant concentration of halothane metabolites was present (Wyrwicz *et al.*, 1987a).

Two environments of halothane in excised rat brain as characterized by T_2 relaxation were also observed by Evers *et al.* (1987). One environment is characterized by a T_2 of 3.6 msec. This environment is saturable at halothane levels in brain which are necessary for the onset of anesthesia. The second environment is characterized by a T_2 of 43 msec and is not saturable at physiological concentrations (Evers *et al.*, 1987). Interestingly, the two T_2 environments observed by Evers *et al.* apparently do not correspond to the two environments determined by the rates of halothane clearance from brain. When the dose dependence of the population of fluorine nuclei in the two T_2 environments is examined, a 0.95% dose of halothane administered for 60 min results in an experimental T_2 curve consisting of 85% of the short (3.6 msec) T_2 component. Litt *et al.* (1987) did not measure T_2 of halothane in rat brain but in excised rabbit brain. Wyrwicz *et al.* (1987a) reported a single T_2 of 3.8 msec after a 120-min exposure to 1% halothane. The long T_2 environment reported by Evers *et al.* only noticeably contributes to the experimental T_2 curve when the administered halothane dose exceeds 2% in 60 min.

Therefore, there appear to be three environments of halothane in brain, as described in Table 3. The different environments are defined by T_2 and by elimination rates. The long T_2 environment, which only becomes evident after extended halothane exposures at high concentrations, resembles halothane in synthetic liposomes (Evers *et al.*, 1987; Koehler *et al.*, 1977). Of the two environments characterized by a short T_2, the clearance rate of only one environment is fast enough to account for the recovery of the animal after anesthesia (Litt *et al.*, 1987). Dose-dependent studies of halothane clearance from brain are necessary to determine the relationship between these two environments. However, the existence of these environments, of which only one is consistent with anesthetic action, indicates that a single saturable halothane binding site in the brain is responsible for anesthesia. The short T_2 of this environment is consistent with motional restriction of the anesthetic molecule and is likely due to binding to an intermembrane protein (Evers *et al.*, 1987).

A recent report by Evers *et al.* (1988) provides further evidence for a saturable protein site for anesthetic action. They determined that the T_2 of fluorinated anesthetics in excised brain tissue correlated with the potency of the anesthetic. Hexafluoroethane, structurally similar to halothane but with minimal anesthetic behavior, had a T_2 greater than five times that for the least potent anesthetic measured. The next step necessary in these studies is to identify and characterize the environment in brain membranes (presumably a protein) which is producing the motional restriction of the anesthetic molecules.

Studies of the elimination of isoflurane, a more clinically relevant anesthetic, from rabbit and rat brain have also been performed. Like

TABLE 3
Halothane Environments in Rat Brain

	k	T_2	Reference
1	0.030 min^{-1}	3.6 msec	Evers *et al.* (1987), Litt *et al.* (1987)
2	0.12 min^{-1}	3.6 msec	Evers *et al.* (1987), Litt *et al.* (1987)
3	—	43 msec	Evers *et al.* (1987)

halothane, isoflurane elimination can be characterized as a two-component decay with half-lives of 26 and 174 min (Wyrwicz *et al.*, 1987b). The shorter decay time is approximately equal to that seen for halothane (25 min; Wyrwicz *et al.*, 1987a), but the longer half-life ($t_{1/2}$) for isoflurane is significantly less than the longer half-life for halothane. These data again suggest that the cellular environment characterized by the shorter half-life includes the anesthetic molecules causing anesthesia. The second component is likely due to anesthetic nonspecifically bound to membranes or cellular lipids. This work has been corroborated by Mills *et al.* (1987), who reported a slightly longer $t_{1/2}$ (36 min) for the rapidly decaying component of the ^{19}F isoflurane resonances. No estimate of error was given by Mills *et al.*; therefore, the statistical significance of difference between 26 and 36 min cannot be determined.

2.2.2. Hepatic Anesthetic Metabolism

The use of halothane as a surgical anesthetic occasionally results in severe liver damage, termed halothane hepatitis (Stock and Strunin, 1985). Halothane hepatitis may be caused by the formation of a harmful halothane metabolite in liver (Plummer *et al.*, 1982). Preliminary studies of halothane metabolism in rat (Burt *et al.*, 1984) and rabbit liver (Brown *et al.*, 1984) indicated that hepatic halothane metabolism could be successfully monitored. Selinsky *et al.* (1987) recently published an *in vivo* study of halothane metabolism in the liver of intact rats. The only metabolite visible under normal anesthesia conditions was trifluoroacetic acid. However, ^{19}F spectra were not recorded until at least 90 min after ending halothane exposure. Studies of anaerobic halothane metabolism in rat liver homogenates demonstrated that ^{19}F NMR can be used to monitor the complete metabolism of halothane to trifluoroacetic acid, chlorotrifluoroethane, chlorodifluoroethane, and inorganic fluoride (Selinsky *et al.*, 1986). An ^{19}F NMR spectrum of halothane metabolites generated by a rat liver homogenate is shown in Fig. 2. Resonances B, C, F, and G correspond to chlorotrifluoroethane, halothane, chlorodifluoroethene, and fluoride ion, respectively. Resonance A has not yet been identified. Resonance D also has not been positively identified, but its chemical shift relative to halothane suggests that it corresponds to halothane in a different membrane environment, as observed in brain (Wyrwicz *et al.*, 1987a). Resonance E is not observed in purified microsomes, or in perfused lives prepared from glutathione-depleted rats, and may therefore correspond to the glutathione adduct of halothane (Cohen *et al.*, 1975).

^{19}F NMR has also been used to answer questions about the hepatic metabolism of methoxyfluorane (Selinsky *et al.*, 1988a, b). The dose-dependent elimination of methoxyflurane from rat liver was reported as well as possible mechanisms of hepatic methyoxyflurane metabolism.

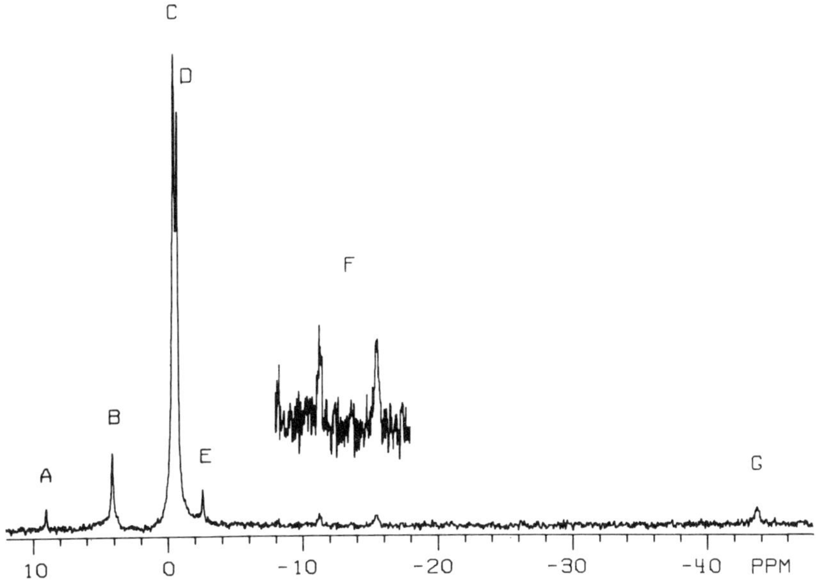

Figure 2. Halothane metabolism in a homogenate of rat liver. Halothane was added to a final concentration of 2 mM to 5 ml of rat liver homogenate which was made 5 mM in NADPH. After incubation for 1 hr at 37°C, an ^{19}F NMR spectrum was recorded. Identification of A–G is described in the text.

2.2.3. Anesthetics as Lipophilic Probes

Malignant cells contain different types and amounts of various lipids than do nonmalignant cells. Fluorinated anesthetics may be capable of distinguishing between normal and cancerous tissues. Burt and co-workers (Burt *et al.*, 1983; Wyrwicz *et al.*, 1983b) have demonstrated that the ^{19}F chemical shift of both halothane and methoxyfluorane are sensitive to the tissue environment. Later work by Burt *et al.* (1984) and Moore and Roberts (1985) showed that both the chemical shift and linewidth of halothane were sensitive to the presence or thermally induced regression of the tumor. The sensitivity of ^{19}F NMR parameters of halothane to malignancy indicates the potential use of fluorinated anesthetics as tumor contrast agents.

2.3. Fluoropyrimidines

Nucleoside analogues have proven to be effective chemotherapeutic agents in the treatment of various tumors (Chabner, 1981). Of particular interest are the fluoropyrimidines, particularly 5-fluorouracil (5-FU; see Table 2 for abbreviations of fluoropyrimidines and metabolites) and 5′-deoxy-5-fluorouridine (5′dFUrd). Both of these drugs are metabolized in the tumors to the active compound, fluorodeoxyuridine monophosphate

(FdUMP), which inhibits DNA synthesis by inhibition of deoxythymidine monophosphate (FdUMP), which inhibits DNA synthesis by inhibition of deoxythymidine monophosphate synthetase (Heidelberger, 1974). A second effect may be caused when 5-FU is extensively incorporated into RNA, altering the structure and function of some cellular RNAs (Heidelberger, 1974). The relative contributions of these two effects to the antitumor properties of fluoropyridines are not clear. However, ^{19}F NMR can be used to determine tumor metabolism of fluoropyrimidines *in vivo* and to provide insight into their mode of action.

The first *in vivo* studies of 5-FU metabolism were carried out by Stevens *et al.* (1984), who studied drug metabolism in mouse liver and in implanted Lewis lung carcinomas. In liver, 5-FU was converted to fluoro-β-alanine (FBAL), and to an unidentified second compound. Similar ^{19}F NMR spectra of rabbit liver after 5-FU treatment were observed by Nunnally *et al.* (1985a). The NMR resonance of the second compound was attributed to dihydro-fluorouracil (FUH_2) by Stevens *et al.* (1984). Studies of human blood after FdUrd treatment showed that the resonance originally attributed to FUH_2 was assigned incorrectly (Malet-Martino *et al.*, 1983). Subsequent studies of human urine using synthesized 5-FU metabolites identified the resonance as fluoro-β-ureidopropionic acid (FUPA; Malet-Martino *et al.*, 1985; Bernadou *et al.*, 1985). In examining tumor metabolism of 5-FU, Stevens *et al.* (1984) found that when high doses of the drug are administered, an ^{19}F NMR resonance characteristic of the active compound, 5FdUMP, is observed. As expected, conversion of 5-FU to fluoronucleotides in tumor cells is inhibited by allopurinol (Prior *et al.*, 1987).

5-FU administered to cancer patients at a dose of approximately 20 mg/kg was eliminated by the liver with a hepatic half-life of about 20 min (Wolf *et al.*, 1987). The only fluorinated metabolites clearly observable were 5-FU and FBAL. The inability to observe the other metabolite observed in mice, FIPA, likely is due to the ninefold lower dose in the human study than in the mouse study (Stevens *et al.*, 1984). All known metabolites can be observed in the urine of a treated patient within several hours of treatment (Bernadou *et al.*, 1985), including 5-FU, FUH_2, FUPA, FBAL, and inorganic fluoride. Most 5-FU metabolites are excreted in the urine, with less than 1% excreted in the bile (Martino *et al.*, 1987). Human biliary metabolites of 5′dFUrd are primarily bile acid conjugates of FBAL, with lesser amounts of free FBAL and inorganic fluoride (Malet-Martino *et al.*, 1988).

A different fluoropyrimidine, 5-fluorocytosine (5-FC), is a commonly used antifungal agent. Two separate studies of the metabolism of 5-FC in sensitive and resistant strains of *Candida albicans* have been carried out (DiVito *et al.*, 1986; Vialaneix *et al.*, 1986). In both studies, conversion of 5-FC to FdUMP was necessary to produce cell sensitivity to the drug. Urinanalysis of a patient given 5-FC to treat cryptococcal meningitis has been carried out (Vialaneix *et al.*, 1987).

The studies described above demonstrate the utility of ^{19}F NMR to answer questions about fluoropyrimidine metabolism in cell cultures, laboratory animals, and humans. A recent report by Rand *et al.* (1987) describes a possible use of *in vivo* ^{19}F NMR to diagnose herpes simplex virus infection in humans. Herpes simplex infection in cells causes induction of the cellular enzyme thymidine kinase, which will phosphorylate thymidine and similar compounds thus causing their accumulation in virus-infected cells. In herpes-infected mice, Rand *et al.* found that infected livers preferentially accumulated trifluorothymidine in sufficient concentration to be observed by ^{19}F NMR (Fig. 3). If this technique could be modified for human use, it would allow both the clinical diagnosis of herpes infection and the assessment of potential drug treatments to fight the infection.

2.4. Tumor Probes

One particularly exciting application of *in vivo* ^{19}F NMR is the use of fluorinated tumor-specific agents, which localize in cancerous tissue, allowing for their detection using either spectroscopic or imaging techniques. Two delivery strategies are possible for tumor identification. A fluorinated

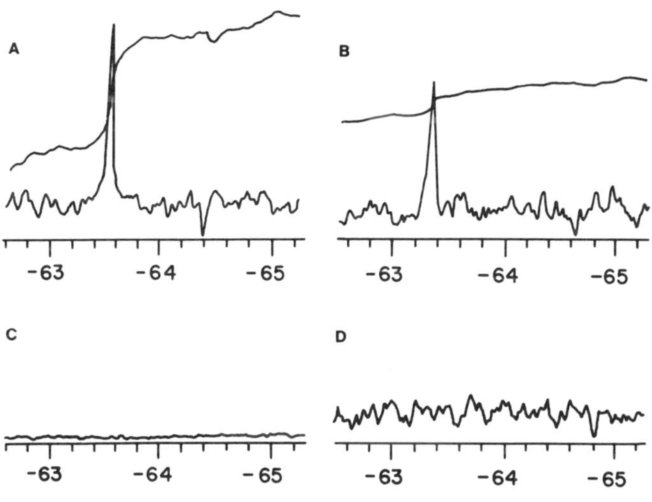

Figure 3. Trifluorothymidine uptake into mouse tissues. (A) Trifluorothymidine uptake standard in water. The chemical shift scale is arbitrary. (B) Homogenate prepared from the liver of a mouse 3 days after infection with herpes simplex virus 2. The mouse was given 100 mg/kg trifluorothymidine i.v. 3.5 hr before sacrifice. (C) Homogenate prepared from the liver of a carbon tetrachloride-treated mouse. The mouse was given 100 mg/kg trifluorothymidine I.V. 3.5 hr before sacrifice. (D) Homogenate prepared from the liver of an uninfected mouse 3.5 hr after I.V. administration of 100 mg/kg trifluorothymidine. (Reprinted with permission from Rand *et al.*, 1987.)

drug could be absorbed and retained in tumors in higher concentrations than in surrounding tissues or a drug could be designed which specifically binds to the cell surface of tumor cells but not to surrounding cells. Both approaches have been addressed in preliminary studies.

Joseph *et al.* (1987) found that the amino acid analogue difluoromethylornithine is preferentially distributed in prostate carcinomas implanted in rats relative to surrounding muscle. Difluoromethylornithine is also taken up into small intestine and kidney and therefore one must be careful in identifying prostate tumors *in situ.* Another potential tumor-imaging agent is tetrafluoroputrescine. A preliminary study by Diginis *et al.* (1986) demonstrated that tetrafluoroputrescine can be taken up into red blood cells. A more complete study by Sarhan *et al.* (1987) showed that in rats bearing Lewis lung carcinomas or hepatomas, 2,2-difluoroputrescine was selectively taken up into tumors and small intestine. Cellular difluoroputrescine was metabolically transformed in the tumors to difluorospermidine and difluorospermine. This transformation was enhanced by depletion of tumor putrescine and spermidine by difluoromethylornithine (Sarhan *et al.*, 1987).

A recent report by Hull *et al.* (1988) demonstrated that treatment of rats containing implanted tumors with difluoromethylornithine followed by 2,2-difluoroputrescine results in incorporation of the latter compound into cellular 6,6-difluorospermidine and 6,6-difluorospermine. Since polyamines are produced in higher concentrations in tumor cells than in nonmalignant cells of the same type, difluoroputrescine could potentially be used to identify cancerous or precancerous regions in apparently normal tissue.

With respect to preferential cell surface binding, Raleigh *et al.* (1986) have reported the reaction of a hexafluorinated 2-nitroimidizole with tumor tissues. While the results appear promising, there does not at present appear to be enough bound label to allow for practical *in vivo* detection. Recently, fluorinated carbohydrates have been synthesized as tumor plasma membrane markers which may prove to be useful as contrast agents (Sharma *et al.*, 1987).

2.5. Miscellaneous

Kawashima *et al.* (1985, 1986) point out how fluorinated antibiotics can be synthesized and their interaction with simple systems followed.

3. NONMETABOLIZED PROBES

Fluorinated compounds have been identified whose ^{19}F chemical shifts are sensitive to such variables as pH, oxygen tension, and free metal ion concentrations. These compounds in general are metabolized slowly by cells. Therefore, they are capable of probing metabolically important processes in living cells.

3.1. General Strategy for Introduction of Fluorinated Probes into Cells

Several strategies exist to introduce fluorinated indicators into cells. The chosen method depends somewhat on the environment and variable chosen to be monitored. The potential user of these methods should be aware of their relative merits and possible pitfalls.

Fluorinated compounds which probe membrane environments are as a rule nonpolar, hydrophobic ones which preferentially partition into cell membranes. The probe molecule can be introduced without chemical modification, although in animal or human studies the compound may need to be injected in some delivery vehicle such as glycerol or dimethylsulfoxide. One disadvantage of this method is its general lack of selectivity for specific cell types or tissues of interest.

For examination of aqueous environments in cells, a common general approach is to take advantage of cell membrane transport systems and intracellular enzymes. One good example of this approach is the uptake and accumulation of fluorodeoxyglucose, a metabolic probe described earlier in this review. Fluorodeoxyglucose is actively transported into brain cells and phosphorylated by hexokinase. Being a poor substrate for glycolysis, an increasing concentration of phosphorylated fluorosugar accumulates, allowing for observation by ^{19}F NMR. This technique was also used to accumulate trifluorothymidine into cells where thymidine kinase has been induced by herpes infection (Rand *et al.*, 1987).

Unfortunately, these strategies cannot be used to introduce many of the nonmetabolized probes. Most of these compounds are fluorinated modified acids and are generally too polar to be passively absorbed by a cell. However, as developed by Tsien (1981), if the acid groups are esterified the probes can diffuse through the membrane. Inside the cell the active compound can be regenerated by cellular nonspecific esterases.

There are two general problems with this strategy. First, the activity and substrate specificity of intracellular esterases vary with cell type. For example, several types of cells would accumulate the methyl esters of different metabolic probes but would not liberate the free acid (Tsien, 1981; Deutsch and Taylor, 1987a). The acetoxymethyl ester appears to be a more generally recognized esterase substrate which enters a larger range of cells. The relatively unstable p-chlorophenylester which will spontaneously hydrolyze has also been employed (Taylor and Deutsch, 1988).

Second, the products of deesterification are potentially toxic to cells. Upon hydrolysis, methoxy esters yield methanol and acetoxymethyl esters produce formaldehyde. Since incorporation of near millimolar quantities of compound into the tissue is necessary for NMR observation, similar amounts of the leaving group are introduced. Direct or subtle changes in metabolic function may result.

3.2. Fluorinated pH Indicators

Fluorinated amino acid analogues, in particular D,L-2-amino-3,3-difluoro-2-methyl propanoic acid or difluoromethylalanine (see Fig. 5 for structure, F_2MeAla abbreviation) (Deutsch *et al.*, 1982; Taylor and Deutsch, 1983), have been developed as probes of intracellular pH. The ^{19}F NMR spectrum of F_2MeAla is an AB pattern resulting from nonequivalence of the two fluorines due to the asymmetric center (see Fig. 4). The difference between the two center lines is pH sensitive, varying by about 2 ppm over the range pH 8 to pH 6. Using F_2MeAla, Deutsch *et al.* (1982) showed that peripheral blood lymphocytes were able to maintain intracellular pH at a steady-state value of 7.2 while varying extracellular pH from 6.6 to 7.6. In a novel experiment, Taylor and Deutsch (1988) loaded the *p*-chlorophenyl ester of F_2MeAla into *Paracoccus denitrificans* and cultured RINm5F cells. These cells cannot hydrolyze methyl esters of F_2MeAla.

Deutsch and co-workers have used F_2MeAla to measure intracellular pH in lymphocytes, hepatocytes, cultured insulinoma cells, frog skin, and rabbit colon (reviewed by Deutsch and Taylor, 1987b). Other laboratories

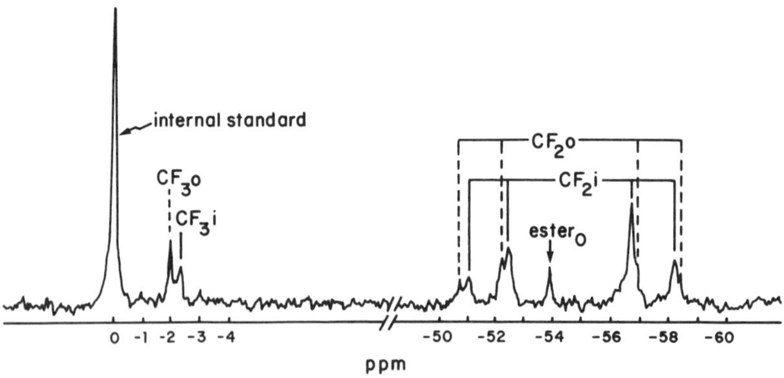

Figure 4. ^{19}F NMR spectrum of fluoroamino acid pH indicators in human lymphocytes. One millimolar difluoromethylalanine methyl ester and 0.17 mM trifluoromethylalanine methyl ester were added to human lymphocytes suspended in Hanks–Hepes buffer, pH 6.6, at 23°C. Resonances marked "o" arise from extracellular species while resonances marked "i" arise from intracellular species. The intracellular pH in this experiment is calculated to be 7.2. The internal standard is aqueous trifluoroacetate. (Reproduced with permission from Taylor and Deutsch, 1983.)

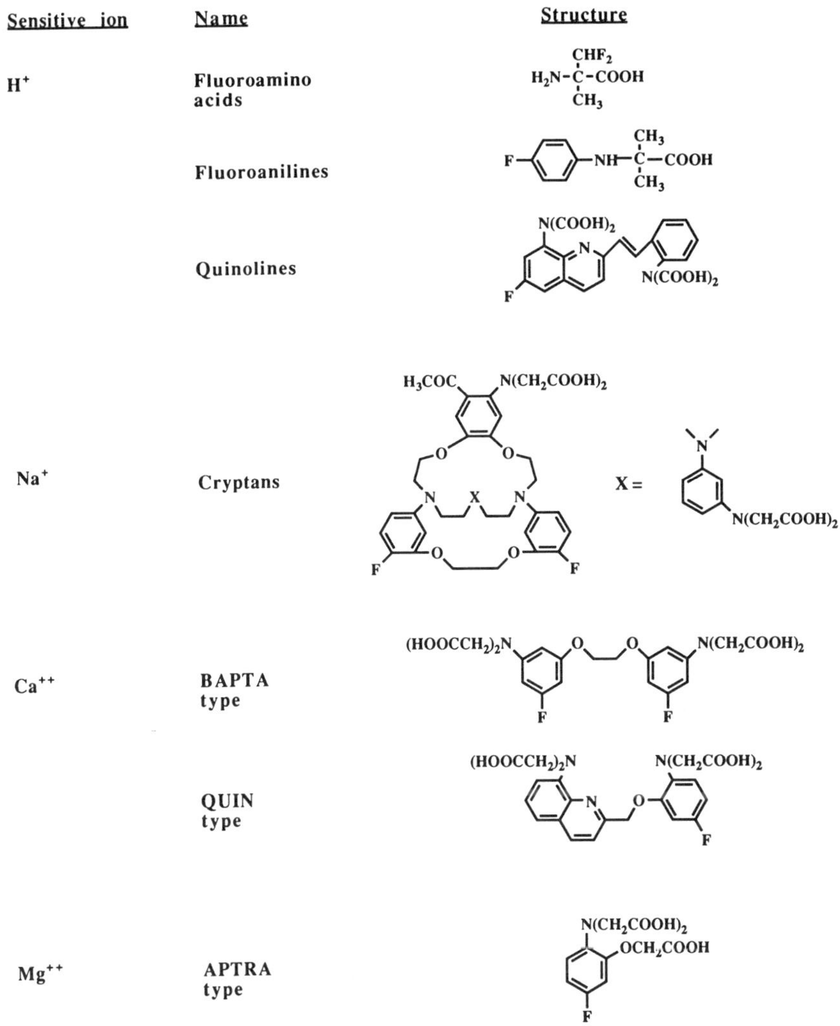

Figure 5. Representative structures of fluorinated ion-selective chelators.

have utilized F_2MeAla for pH determination in perfused liver (Thoma and Ugurbil, 1988) and in hog carotid artery (Grieder *et al.*, 1988). F_2MeAla is preferred over other fluoroamino acid indicators for several reasons. First, the pK_a of F_2MeAla is 7.3; therefore, the change in chemical shift difference between the AB multiplets in F_2MeAla is greatest near normal physiological pH. Second, since pH is determined by the chemical shift difference between two resonances arising from the same indicator, no chemical shift reference is required. Third, F_2MeAla does not appear to interfere with normal cellular metabolism in most tissues.

The major limitation of the fluoroamino acid pH indicators is the insensitivity of their change in chemical shift as a function of pH. The most sensitive indicator has a change in chemical shift of under 2 ppm per pH unit (Deutsch and Taylor, 1987a). The chemical shift of substituted fluoroanilines can be ten times more sensitive to pH changes (Deutsch and Taylor, 1987a). Although still under development, fluoroanilines have the potential to replace F_2MeAla as the NMR pH indicator of choice (Deutsch and Taylor, 1989).

A second alternative to the modified amino acids are fluorinated analogues of fluorescent pH indicators. A fluorinated analogue of the fluorescent indicator Quene-1 (named F-Quene-1; see Fig. 5 for structure) has been synthesized, and its pH is 6.7 (Metcalfe *et al.*, 1985). In a preliminary report, Beech and Iles (1987) used this indicator to measure intracellular pH in rat liver. No details have been published regarding the effect of temperature and cations on the pK of F-Quene-1.

The acid dissociation constants for the fluorinated pH indicators discussed above are provided in Table 4.

3.3. Fluorinated Analogues of Metal Ion Chelators

A major advance in the measurement of the free calcium ion concentration in cells was synthesis of fluorescent calcium ion indicators (Tsien, 1981). Soon afterwards, Smith *et al.* (1983) synthesized various fluorinated analogues of BAPTA [1,2-bis(*o*-aminophenoxy) ethane-*N*,*N*,*N'*,*N'*-tetraacetic acid; see Fig. 5 for structure]. As expected from the structure of the FBAPTA calcium complex (Gerig *et al.*, 1987), the chemical shift of the fluorine nuclei on the chelator is sensitive to cations. Exchange of bound cations from BAPTA derivatives substituted at the 3- and 5-position of the aminophenoxy rings (designated 3F-BAPTA and 5F-BAPTA, respectively) is slow relative to the Larmor frequency (between 94 and 339 MHz, depending on field strength). Slow exchange results in the appearance of two ^{19}F NMR resonances in the presence of a cation, corresponding to cation-bound chelator and uncomplexed chelator. The chemical shift difference between bound and free chelator is dependent on the chemical nature of the chelator and the bound cation. For the derivatives 4F-BAPTA and 6F-BAPTA, cation exchange from the chelator is fast relative to the Larmor frequency. Therefore, in the presence of cations a single ^{19}F NMR resonance is visible whose chemical shift represents the weighted average of the chemical shifts of cation-bound and uncomplexed chelator (Smith *et al.*, 1983; Murphy and London, 1988).

TABLE 4
Dissociation Constants for Fluorinated Ion Chelators

	pK_a	Reference
pH		
Fluorinated amino acids		
α-Monofluoromethylalanine	8.5	Deutsch and Taylor (1987a)
α-Difluoromethylalalinine	7.3	Deutsch and Taylor (1987a)
α-Trifluoromethylalanine	5.9	Deutsch and Taylor (1987a)
α-Monofluoromethylalanine methyl ester	6.4	Deutsch and Taylor (1987a)
α-Difluoromethylalanine methyl ester	5.1	Deutsch and Taylor (1987a)
α-Trifluoromethylalanine methyl ester	4	Deutsch and Taylor (1987a)
α-Trifluoroethylalanine	8.0	Deutsch and Taylor (1987a)
N-Methyl α-trifluoromethylalanine	6.0	Deutsch and Taylor (1987a)
α-Monofluorovaline	8.3	Deutsch and Taylor (1987a)
γ-Hexafluorovaline	6.3	Deutsch and Taylor (1987a)
α-Monofluoromethylhistidine	7.7	Deutsch and Taylor (1987a)
α-Monofluoroalanyl-α-monofluoroalanine	6.54	Deutsch and Taylor (1987a)
Fluorophosphate	4.7	Deutsch and Taylor (1987a)
Fluoroaniline derivatives		
3-Fluoroaniline	3.5	Deutsch and Taylor (1987a)
4-Fluoroaniline	4.6	Deutsch and Taylor (1987a)
2,2,4-Trifluoroanalinine	2	Deutsch and Taylor (1987a)
N-(2-Cyano-isopropyl)-4-fluoroaniline	4.75	Deutsch and Taylor (1987a)
N-(2-Amido-isopropyl)-4-fluoroaniline	3.3	Deutsch and Taylor (1987a)
N-(2-Methylcarboxylate-isopropyl)-4-fluoroanaline	3.5	Deutsch and Taylor (1987a)
N-(2-Carboxy-isopropyl)-4-fluoroaniline	5.8	Deutsch and Taylor (1987a)
Quinoline derivative		
F-quene	6.7	Metcalfe *et al.* (1985)
Na^+		
FCryp-1	1.3	Smith *et al.* (1986)
Ca^{2+}		
3F-BAPTA	6.19	Smith *et al.* (1983)
4F-BAPTA	5.61	Smith *et al.* (1983)
5F-BAPTA	6.15	Smith *et al.* (1983)
	6.47	Marban *et al.* (1987)
	6.20	Levy *et al.* (1987)
6F-BAPTA	4.6	Smith *et al.* (1983)
MF-BAPTA	6.58	Levy *et al.* (1987)
QuinF	6.91	Levy *et al.* (1987)
QuinMF	7.20	Levy *et al.* (1987)
Mg^{2+}		
MF-APTRA	3.21	Levy *et al.* (1988)
5F-APTRA	3.04	Levy *et al.* (1988)
4F-APTRA	2.51	Levy *et al.* (1988)

For determination of intracellular free cation concentrations, a chelator with slow exchange properties is preferred. Free cation concentrations can be determined using the relationship

$$[M^{n+}]_{free} = K_D \frac{[\ln d]_{bound}}{[\ln d]_{free}}$$

where K_D is the dissociation constant of the chelator for the cation of interest. The relative concentrations of free and cation-complexed indicator can be determined by the relative intensities of the corresponding resonances in the ^{19}F NMR spectrum.

Cation chelating agents such as BAPTA contain multiple carboxylic acid groups which contribute to the cation binding site (Gerig *et al.*, 1987). To introduce BAPTA analogues into cells, these carboxylic acid groups must be esterified in a similar manner to that discussed for the pH indicators. Esterification allows passive diffusions of the chelators through the plasma membrane into the cytoplasm, where celluar esterases have the opportunity to liberate the free acid.

Determination of intracellular cation concentrations using chelator probes should be performed with caution. The accuracy of calculated cation concentrations is directly related to the accuracy of the measured cation dissociation constant for the chelator. For example, the reported K_D values for 5F-BAPTA from three different groups differ by a factor of 2 (see Table 4). Determination of chelator K_D is routinely performed by cation titration into chelator solutions which model the intracellular milieu. Since the indicators by design have high affinity for cations, small amounts of accurately measured cations have to be added to the chelator solution. Often, the added cation solutions are buffered by a second chelator, thereby introducing another K_D and potential source of error into the measurement. Also, the viscosity and protein content in the cell are not mimicked in the test solution; therefore, the experimentally determined K_D may not exactly represent the intracellular K_D.

Due to current interest in calcium ions as intracellular messengers, probe development and experimentation has concentrated on calcium ion measurement in cells. To measure calcium ions using a fluorinated probe and ^{19}F NMR, the chelator must be loaded into cells at millimolar concentrations to measure an NMR signal in a reasonable length of time. Cellular free calcium ion concentrations in most resting cells range between 0.1 and 10 μM. Therefore, a large excess of chelator relative to calcium ions must be introduced into the cell, allowing the chelator itself to act as a cellular calcium buffer. This buffering effect is most pronounced in heart muscle;

a prolongation of relaxation time and depression of pressure development are observed after chelator loading (Steenbergen *et al.*, 1987; Marban *et al.*, 1987).

In addition, cell hydrolysis of acetoxymethyl esters has not been carefully controlled. BAPTA analogues contain four carboxylic acid groups per probe molecule. All must be esterified for the chelator to enter the cell. The indicator could possibly be partially deesterified to form a chelator with a different K_D than the free acid form of the indicator. Several pools of indicators with different K_D's would affect the calculation of the intracellular free ion concentration.

Finally, the indicator is assumed to be concentrated in the cytoplasm of target cells. Distribution of the indicator between subcellular organelles, although unlikely, has not been ruled out.

1. *Calcium indicators.* Despite limitations, fluorinated chelators and ^{19}F NMR are uniquely suited for measurement of intracellular free calcium concentrations in perfused tissues and cell preparations. The indicator 5F-BAPTA has been most often used for these measurements. However, fluorinated derivatives of other fluorescent calcium indicators have been prepared (Levy *et al.*, 1987; see Table 4 and Fig. 5). These new indicators vary in their affinity for calcium ions, allowing for the selection of an indicator with a K_D similar to the intracellular free calcium concentration in the system of interest.

The first ^{19}F NMR measurements of free calcium were recorded by Smith *et al.* (1983). Using 5F-BAPTA, the $[Ca^{2+}]$ free in mouse BALB/c thymocytes was calculated to be 250 nM. Interestingly, in a further study by Metcalfe *et al.* (1985) a signal for Zn^{2+} was observed in Ehrlich ascites tumor cells. The free calcium concentration calculated in these cells was higher using 5F-BAPTA than that observed with the fluorescent calcium indicator quin 2. However, at the level of free zinc indicated by ^{19}F NMR there could be significant complexation by the fluorescent indicator, resulting in a higher calculated and hence a lower actual free calcium concentration.

Murphy *et al.* (1986) have used 5F-BAPTA to measure free calcium in red blood cells. There was some leakage of the indicator from their preparation but this was corrected for by the addition of an extracellular paramagnetic shift reagent. Using Quin-MF, an indicator with a lower K_D than 5F-BAPTA, resulted in a calculated $[Ca^{2+}]_{free}$ in normal blood cells of 21 nM (Murphy *et al.*, 1989). The $[Ca^{2+}]_{free}$ in sickle red blood cells was determined to be identical to normal red blood cells. No change in intracellular free calcium was found upon initiating the sickling process (Murphy *et al.* 1989).

In rat heart, a time-averaged free calcium concentration of 630 nM was measured at 37°C using 5F-BAPTA (Steenbergen *et al.*, 1987). The calcium concentration represents the time-weighted average of systole, where the heart is contracting and has elevated cytoplasmic calcium, and diastole, the relaxed state where cytoplasmic free calcium is low. Potassium- or magnesium-arrested hearts had a $[Ca^{2+}]_{free}$ of approximately 250 nM. In globally ischemic (clamped) hearts, the free calcium concentration rose dramatically. In ferret hearts, Marban *et al.* (1987) used paced perfused samples loaded with 5F-BAPTA. The loaded hearts at 30°C yielded a baseline average free calcium of 104 nM. The concentration increased to almost 1 μM after 20 min of global ischemia and then returned to normal during reperfusion. Discrepancies in results between Steenbergen *et al.* (1987) and Marban *et al.* (1987) may arise due to differences in species, protocol (including temperature), or the different K_D for 5F-BAPTA used by the two groups (see Table 4). The time-averaged $[Ca^{2+}]_{free}$ in ferret heart at 30°C was recently found measured to range between 400 and 550 nM (Kirschenlohr *et al.*, 1988).

2. *Magnesium.* The ^{31}P NMR chemical shifts of phosphorus energy metabolites have been used to calculate the intracellular magnesium ion concentration (Gupta *et al.*, 1984). Determinations of this type, however, are subject to error (Garfinkel and Garfinkel, 1984; Burt, 1987).

Several fluorinated magnesium indicators have been developed for measurement of intracellular free magnesium concentrations using ^{19}F NMR. Levy *et al.* (1988) have synthesized three fluorinated magnesium indicators with dissociation constants in the millimolar range, suitable for measurement of cytoplasmic free magnesium. Using these indicators, the free $[Mg^{2+}]$ in human erythrocytes was measured to be 0.25 mM (Levy *et al.*, 1988). In rat heart, the average free $[Mg^{2+}]$ was measured as 0.85 mM independent of the indicator used (Murphy *et al.*, 1989). Global ischemia resulted in a threefold increase in the intracellular magnesium concentration. Cultured chick heart cells have a basal free $[Mg^{2+}]$ of 0.56 mM (Rotevatn *et al.*, 1989).

Metcalfe and colleagues have used fluorocitrate to measure free magnesium concentrations in perfused ferret hearts (Kirschenlohr *et al.*, 1988). The basal free $[Mg^{2+}]$ in ferret heart was calculated as 1.2 mM using fluorocitrate. The difference between this value and the measurement of Murphy *et al.* (1989) in rat heart may reflect differences in species or experimental conditions. Determination of magnesium using fluorocitrate does require a chemical shift standard, which is a potential source of error not found using the indictors of Levy *et al.* (1988).

3. *Other ions.* In their initial report, Smith *et al.* (1983) demonstrated that 5F-BAPTA is sensitive to several different divalent cations. In a preliminary report, Dowd *et al.* (1988) has demonstrated that 5F-BAPTA can be used to measure lead concentrations in osteoblasts, providing information regarding the mechanism of lead toxicity in this system.

Smith *et al.* (1986) have described the synthesis of fluorinated crown ethers specific for sodium ions (see Table 4 and Fig. 5).

4. *Free radicals.* Interest in free radicals and their role in cell injury has recently developed. Unfortunately, as a result of their reactivity and the reducing environment in cells, *in vivo* detection of radicals has been difficult. To address this problem, Selinsky *et al.* (1989) have developed fluorinated "spin-traps" to study radical formation *in vivo.* This method shows some promise; however, the lack of sensitivity of NMR relative to radical concentrations generated will limit its application.

5. *Oxygen.* In NMR studies of perfused organs or cells, a simple method to monitor oxygenation is desirable. The spin–lattice relaxation of perfluorohydrocarbons is strongly dependent on oxygen binding (Hamza *et al.*, 1981; Clark *et al.*, 1985). Taylor and Deutsch (1988) have used this property of perfluorotributylamine to estimate the oxygen saturation in the perfusion buffer of RINm5F cells. This method will be particularly useful when monitoring the effects of hypoxia on another quantity which can be measured by ^{19}F NMR (e.g., [Ca^{2+}]).

3.4. Other Agents Accumulated by Cells Containing Fluorine Markers

The first ^{19}F NMR study on a fluorinated compound in living cells was carried out by Costa *et al.* (1979), who examined the spectra of 2-fluoro-ATP and 4,6-difluoroserotonin in isolated human platelets. Difluoroserotonin in platelets gave no observable NMR signal, indicative of tight binding of the compound to intracellular components. Difluoroserotonin could be observed by NMR in cell extracts. 2-Fluoroadenosine was shown to be taken up by platelets and converted to 2-fluoro-ATP. Diffley *et al.* (1983) were able to observe 6-fluorodopamine loaded into guinea pig nerve microsacs *in vitro.* The antimalarial drug mefloquine has been studied in erythrocytes using ^{19}F NMR (San George *et al.*, 1984).

Neuroleptic drugs would appear to be an attractive choice to be investigated by ^{19}F NMR. A variety of these drugs contain fluorine and many are administered in doses high enough to be optimistic about NMR-observability. Bartels *et al.* (1986) were able to observe the drug fluphenazine in the excised brain of treated rats; however, its low *in vivo* concentration and broad linewidth precluded its *in situ* measurement. A recent abstract indicated that Prolixin, another neuroleptic drug, could be observed and imaged in treated hamsters (Ratner *et al.*, 1987b).

London and Gabel (1989) point out a potential powerful new tool to investigate membrane potential. Following the lead of Labotka and Omachi (1987) who showed that charged phosphorus compounds can be accumulated by erythrocytes, they demonstrated that monovalent negatively charged trifluoroacetate (TFA) also is taken up by erythrocytes. Kirk *et al.* (1988)

had also shown that in erythrocytes intracellular and extracellular ions give resolvable ^{31}P resonances. Since there is an electrical potential difference between the red blood cell and the medium, the TFA must distribute itself according to the Nernst equation. Conveniently, TFA inside and outside the erythrocytes are in NMR slow exchange so that two peaks can be seen. The TFA inside is shifted and broadened due to the intracellular environment compared to the TFA outside. A clever trick then involved using the neutral compound trifluoroacetamide which is shifted as TFA but is distributed uniformly in the aqueous compartment. The area of the peaks under TFA outside and inside compared to the amide inside and outside allows a direct calculation of the potential. They were then able to show that as predicted from the literature, their calculated potential varied as a function of extracellular pH.

4. IMAGING

One exciting area in NMR research and application is magnetic resonance imaging (MRI). The NMR experiment, combined with magnetic field gradients, gives images which reveal the spatial distribution of a particular nucleus, e.g., the distribution of water protons in tissue. As can be noted in many books and reviews, nuclei other than protons are also suitable for imaging (e.g., Mansfield and Morris, 1982; James and Margulis, 1984). Any nuclei can be imaged, if they exist in sufficient quantity and have appropriate spectral characteristics.

Fluorine nuclei in fluorinated compounds provide a tempting target for MRI. Many fluorinated pharmaceutical agents have approval for use from the Food and Drug Administration, and some of these are known to achieve *in vivo* concentrations of up to 1 M (Mattrey *et al.*, 1982). Endogenous fluorine is present only at micromolar concentrations, so background interference is inconsequential. Many fluorinated pharmaceuticals naturally concentrate within specific tissues. Inhalation anesthetics exist at highest concentrations in the lungs during administration, while perfluorinated blood substitutes are essentially confined to the circulatory system. Fluorodeoxyglucose concentrates in the brain as the result of "metabolic localization," due to the brain's preference for glucose as an energy source. In addition, antibodies or other molecules will specifically bind to one type of cell and are often amenable to fluorination without affecting binding affinity.

Obtaining images of fluorine nuclei is complicated by both the broad chemical shift dispersion and the relaxation properties of fluorine. The chemical shift dispersion of organofluorine compounds extends over

200 ppm. If a compound to be imaged has multiple ^{19}F NMR resonances dispersed over a wide chemical shift range, several nonideal options are available. All of the resonance lines can be excited over the spatial region of interest, which is technically difficult especially at high static field strength. A single resonance line can be excited, which would waste the resonance intensity present in the discarded resonance lines. Finally, deconvolution techniques can be used to eliminate chemical shift artifacts from fluorine images obtained using standard imaging procedures (Harrison *et al.*, 1985; Busse *et al.*, 1986).

The relaxation properties of the fluorine nucleus must also be taken into consideration in designing imaging experiments. Fluorine nuclei in perfluorohydrocarbons have a spin–lattice relaxation time (T_1) of between 1 and 3 sec, necessitating either an extended relaxation delay or a short tip angle excitation pulse to achieve a maximal signal-to-noise ratio (Clarke *et al.*, 1984a). The spin–spin relaxation time in perfluorocarbons is relatively short, measured between 10 and 50 msec (Clarke *et al.*, 1984b), while the same quantity in halothane can be as short as 2–3 msec (Evers *et al.*, 1987). The use of traditional imaging sequences, with echo refocus times of 30 msec or so, would result in serious loss of signal. For example, an early attempt to image halothane in brain tissue was unsuccessful, likely the result of a 15-msec echo time (McFarland *et al.*, 1985).

Therefore, the ideal imaging sequence for fluorine nuclei should compensate for chemical shift artifacts, extended relaxation delays, and fast spin–spin relaxation. One procedure toward this end is a driven equilibrium spin–echo sequence described by Freeman *et al.* (1988). The echo time is reduced to 3.5 msec using this procedure, which would make it suitable for quickly decaying nuclei. To minimize the echo time, moderately selective excitation pulses were used. Similar sequences have been implemented by other laboratories (Conboy *et al.*, 1989).

4.1. Oxygen Delivery System

Perfluorinated hydrocarbons (PFCs) have clinical potential as artificial blood substitutes. They possess a high affinity for oxygen, are relatively inexpensive, easily sterilized, and can be stored for extended times. The PFCs being tested as blood substitutes are listed in Table 5.

The large number of fluorine atoms present in PFCs suggested their use as imaging contrast agents if sufficient concentrations of these compounds in tissues could be achieved. PFCs can easily be introduced into the circulatory system up to 20% concentration by volume (Joseph *et al.*, 1985). After time, the PFCs collect in the organs of the reticuloendothelial system. To demonstrate potential tissue concentrations of PFCs, Longmaid

TABLE 5
List of Perfluorocarbons

Meglumine salt of trifluoromethyl sulfonic acid
Perfluorotributylamine (Flurosol 43, PFTA)
Perfluorodecalin
Fluosol DA (14% perfluorodecalin and 6% perfluorotripropylamine)
Perfluorotriproplyamine
Perfluorotributylfuran
Perfluoroacetyl bromide
Perfluoroanthrene

et al. (1985) administered to rats 3 ml of a 16% solution of perfluorotributylamine (PFTA) per 100 g body weight. Two separate injections were given 4 days apart. Five days after the second treatment, the rats were sacrificed, tissues harvested and analyzed for PFTA. The concentrations of PFTA were determined to be about 50 mg/g in liver, 120 mg/g in spleen, and about 25 mg/g in an implanted Walker-256 sarcoma tumor. As 67 mg/g tissue is roughly equivalent to 0.1 M concentration, fluorine imaging should be possible using these dosage conditions.

Initial studies of PFC imaging showed the practicality of the technique and reemphasized that biodistribution of intravenously administered PFCs favored spleen, then liver, and finally abscesses and tumors over other tissues (Joseph *et al.*, 1985; Longmaid *et al.*, 1985; McFarland *et al.*, 1985; Ratner *et al.*, 1987a). In addition, contrast between infarcted and normal myocardium was shown in excised tissue (Nunnally *et al.*, 1985b). As an imaginative alternative to intravenous injection, Thomas *et al.* (1986) loaded animals with PFC by immersion breathing. Aqueous dispersions of PFCs contain enough dissolved oxygen to allow replacement of air with oxygenated PFCc for breathing. After loading with PFCs, the lungs are allowed to drain as fully as possible, and the residual PFC concentration imaged. The spleen again shows good contrast and the lungs are now clearly observed.

Another useful property of PFCs is the sensitivity of the fluorine spin-lattice relaxation upon their degree of oxygen saturation (Hamza *et al.*, 1981; Clark *et al.*, 1985). As described earlier in this review, this relationship between oxygen saturation and T_1 can be used to measure tissue oxygenation *in vivo*. The same property can potentially be used in conjunction with T_1-weighted imaging techniques to generate anatomical images describing regional tissue anoxia. Fishman *et al.* (1987) demonstrated that the quality of a fluorine image of intravascular PFTA improved after changing the respiratory gas from room air to 95% oxygen. Image enhancement resulted from a two- to threefold decrease in the T_1 of PFTA in different tissues, thereby increasing signal intensity.

Eidelberg *et al.* (1988a, b) have performed ^{19}F NMR of PFCs to determine blood oxygenation in cat brain. Some information regarding the oxygen content of different regions of the brain can be delineated.

4.2. Halothane

The mechanism of action of general anesthetics is poorly understood. To address the mechanism of action of the inhalation anesthetics, localization of their target areas in the brain would be useful. ^{19}F MRI would appear to be ideally suited for determining the tissue localization of fluorinated inhalation anesthetics in the brain. Halothane (2-bromo-2-chloro-1,1,1-trifluoroethane), since it has a single ^{19}F NMR resonance and its clinical use is well characterized, was chosen for anesthetic imaging.

A pilot study using halothane injected into green peppers was published to demonstrate that halothane could be observed by fluorine MRI (Wyrwicz *et al.*, 1986). Halothane was demonstrated to accumulate in the seeds of the pepper, which due to their oil content represented a more hydrophobic environment than the flesh of the pepper.

However, attempts to image halothane in the brain of anesthetized subjects initially did not work (McFarland *et al.*, 1985). Subsequently, relaxation measurements of halothane in rat brain found that the fluorine T_2 values for halothane could be as low as 2–3 msec (Evers *et al.*, 1987). Thus, image sequences with longer echo times were not capable of recording the quickly decaying signal from halothane in tissues (e.g., McFarland *et al.* used a spin–echo sequence with a 15-msec echo time). Using imaging experiments with short echo times, halothane has been imaged in rat and rabbit brains using ^{19}F MRI (Matwiyoff *et al.*, 1987; Conboy *et al.*, 1989, Wyrwicz and Conboy, 1989). Unfortunately, image resolution is not sufficient to provide detailed information of halothane localization in anesthetized animals.

4.3. Imaging of Other Fluorinated Compounds

For the same reasons as described for halothane, knowledge of the anatomical localization of neuroleptic drugs would improve our understanding of how these drugs work. Many of the neuroleptic drugs contain fluorine, and hence are potentially suitable for fluorine imaging. Ratner *et al.* (1987b) demonstrated that the neuroleptic agent Prolixin could be imaged in hamsters after 10 days of intraperitoneal administration. Unfortunately, Prolixin could be observed by imaging only in the upper abdomen and not in the brain.

A separate study labeled antibodies to anticarcinoembryonic antigen (CEA) with PFTA and introduced these into a mouse bearing a CEA-producing tumor (Shimizu *et al.*, 1987). In the mouse, the tumor could not be visualized using ^{19}F MRI; however, the excised tumor was imaged. Labeling of the tumor occurred primarily at the exterior of the tumor, with little labeling of the central core.

4.4. Flow

Fluorinated compounds have recently been demonstrated as useful markers of blood flow. Eleff *et al.* (1988) have used Freon (CHF_3) to measure blood flow in cat brain before and after blockage of the carotid artery. Authier (1988) has used the ^{19}F NMR signal from PFTA to monitor blood flow in a rat leg following tourniquet application. In both cases, the fluorinated marker should remain confined to the circulatory system throughout the measurement. This localization, coupled with the sensitivity of the fluorine nucleus and the low natural background, suggests additional flow measurements in the future.

5. CONCLUSIONS

The potential applications of *in vivo* ^{19}F NMR have been demonstrated. The high sensitivity of the fluorine nucleus, coupled with the lack of biological fluorine background, invites the synthesis of fluorinated probes and contrast agents for study by NMR and MRI. Already, *in vivo* ^{19}F NMR is being used to answer basic questions of pharmacology and physiology, such as the mechanism of action of anesthetics and intracellular ion concentration changes in response to cell injury or hormonal stimulation. In the next decade, many of the contrast agents cited in this review will be developed and studied. Improved methods for ^{19}F NMR and MRI will be developed, which will give better-quality fluorine images and clearly demonstrate the utility of fluorinated contrast agents. The tip of the iceberg has been seen; the expanse beyond the tip has yet to be uncovered.

ACKNOWLEDGMENTS. The authors express their gratitude to Dr. Alice Wyrwicz for many helpful discussions, and for supplying preprints of manuscripts before publication. The authors also thank Drs. Robert London, Elizabeth Murphy, and Louis Levy for providing copies of manuscripts before publication. The proofreading efforts of Laura Selinsky and the editorial assistance of Carolyn Milford are gratefully acknowledged.

REFERENCES

Authier, B., 1988, *Magn. Reson. Med.* **8**:80.

Bartels, M., Albert, K., Kruppa, G., Mann, K., Schroth, G., Tabarelli, S., and Zebel, M., 1986, *Psychiatry Res.* **18**:197.

Baxter, R. L., Mackenzie, N. E., and Scott, A. I., 1983, in *Biological Magnetic Resonance*, Vol. 5, Plenum Press, New York, p. 1.

Beech, J. S., and Iles, R. A., 1987, *Biochem. Soc. Trans.* **15**:871.

Berkowitz, B. A., and Ackerman, J. J. H., 1987, *Biophys. J.* **51**:681.

Bernadou, J., Armand, J. P., Lopez, A., Malet-Martino, M. C., and Martino, R., 1985, *Clin. Chem.* (*N.Y.*) **31**:846.

Bessel, E. M., and Thomas, P., 1973, *Biochem. J.* **131**:83.

Bessel, E. M., Foster, A. B., and Westwood, J. H., 1972, *Biochem. J.* **128**:199.

Blomberg, F., and Rüterjans, H., 1983, in *Biological Magnetic Resonance*, Vol. 5, Plenum Press, New York, p. 21.

Brown, F. F., Halsey, M. J., Wardley-Smith, B., and Martin, P. A., 1984, *Proc. Anesth. Res. Soc.* **1984**:806P.

Burt, C. T., 1987, *Biophys. J.* **51**:502a.

Burt, C. T., Eisemann, A., Schofield, J. C., and Wyrwicz, A. M., 1982, *J. Magn. Reson.* **46**:176.

Burt, C. T., Moore, R. R., and Roberts, M. F., 1983, *J. Magn. Reson.* **53**:163.

Burt, C. T., Moore, R. R., Roberts, M. F., and Brady, T. J., 1984, *Biochim. Biophys. Acta* **805**:375.

Busse, L. J., Thomas, S. R., Pratt, R. G., Clark, L. C., Jr., Ackerman, J. L., Samaratunga, R. C., and Hoffmann, R. E., 1986, *Med. Phys.* **13**:518.

Chabner, B. A., 1981, in *Cancer and Chemotherapy* (Crooke, S. T., ed.), Academic Press, New York.

Chang, L.-H., Chew, W. M., Weinstein, P. R., and James, T. L., 1987, *J. Magn. Reson.* **72**:168.

Civan, M. M., and Shporer, M., 1978, in *Biological Magnetic Resonance*, Vol. 1, Plenum Press, New York, p. 1.

Clark, L. C., Ackerman, J. L., Thomas, S. R., Millard, R. W., Hoffman, R. E., Pratt, R. G., Ragle-Cole, H., Kinsey, R. A., and Janakiramen, R., 1985, *Exp. Med. Biol.* **180**:835.

Clarke, L. C., Ackerman, J. L., Thomas, S. R., and Millard, R. W., 1984a, *Adv. Exp. Med. Biol.* **180**:835.

Clarke, L. C., Ackerman, J. L., Thomas, S. R., and Millard, R. W., 1984b, *Magn. Reson. Med.* **1**:135.

Cohen, E. N., Trudell, J. R., Edmunds, H. N., and Watson, E., 1975, *Anesthesiology* **43**:392.

Conboy, C. B., Wyrwicz, A. M., and Nuss, M., 1989, *Biophys. J.* **55**:393a.

Costa, J. L., Dobson, C. M., Kirk, K. L., Poulson, F. M., Valeri, C. R., and Vecchione, J. J., 1979, *FEBS Lett.* **99**:141.

Deutsch, C. J., and Taylor, J. S., 1987a, *Ann. N.Y. Acad. Sci.* **508**:33.

Deutsch, C. J., and Taylor, J. S., 1987b, in *NMR Spectroscopy of Cells and Organisms* (Gupta, R., ed.), CRC Press, Boca Raton, Fla., p. 55.

Deutsch, C. J., and Taylor, J. S., 1989, *Biophys. J.* **56**:799.

Deutsch, C., Taylor, J. S., and Wilson, D. F., 1982, *Proc. Natl. Acad. Sci. USA* **79**:7944.

Diffley, D. M., Costa, J. L., Sokoloski, E. A., Chiueh, C. C., Kirk, K. L., and Creveling, C. R., 1983, *Biochem. Biophys. Res. Commun.* **110**:740.

Diginis, G. A., Hawi, A. A., Hip, H., and Layton, W. J., 1986, *Life Sci.* **38**:2307.

DiVito, M., Podo, F., Torosantucci, A., Carpinelli, G., Whelan, W. L., Kerridge, D., and Cassone, A., 1986, *Antimicrob. Agents Chemother.* **29**:303.

Dowd, T. L., Schanne, F. A. X., Gupta, R. K., and Rosen, J. F., 1988, *Biophys. J.* **53**:217a.

Eidelberg, D., Johnson, G., Barnes, D., Tofts, P. S., Delpy, D., Plummer, D., and McDonald, W. I., 1988a, *Magn. Reson. Med.* **6**:344.

Eidelberg, D., Johnson, G., Tofts, P. S., Dobbin, J., Crockard, H. A., and Plummer, D., 1988b, *J. Cereb. Blood Flow Metab.* **8**:276.

Eleff, S. M., Schnall, M. D., Ligewtti, L., Osbakken, M., Subramanian, H., Chance, B., and Leigh, J. S., Jr., 1988, *Magn. Reson. Med.* **7**:412.

Elgavish, G. A., 1987, in *Biological Magnetic Resonance*, Vol. 7, Plenum Press, New York, p. 1.

Evers, A. S., Berkowitz, B. A., and d'Avignon, D. A., 1987, *Nature* **328**:157.

Evers, A. S., Haycock, J. C., and d'Avignon, D. A., 1988, *Biochem. Biophys. Res. Commun.* **151**:1039.

Fishman, J. E., Joseph, P. M., Floyd, T. F., Mukherji, B., and Sloviter, H. A., 1987, *Magn. Reson. Imag.* **5**:279.

Fox, J. L., 1984, *Science* **224**:143.

Freeman, D. M., Muller, H. H., Hurd, R. E., and Young, S. W., 1988, *Magn. Reson. Imag.* **6**:61.

Garfinkel, L., and Garfinkel, D., 1984, *Biochemistry* **23**:3547.

Gerig, J. T., 1978: in *Biological Magnetic Resonance*, Vol. 1, Plenum Press, New York, p. 139.

Gerig, J. T., Singh, P., Levy, L. A., and London, R. E., 1987, *J. Inorg. Biochem.* **31**:113.

Grieder, T. A., Evans, C. A., Greenberg, S. S., and Diecke, F. P. J., 1988, *Life Sci.* **43**:999.

Gupta, R. K., Gupta, P., and Moore, R. D., 1984, *Annu. Rev. Biophys. Bioeng.* **13**:221.

Halama, J. R., Gatley, S. J., DeGrado, T. R., Bernstein, D. R., Ng, C. K., and Holden, J. E., 1984, *Am. J. Physiol.* **247**:H734.

Hamza, M. A., Serratrice, G., Stebe, M.-J., and Delpuech, J.-J., 1981, *J. Magn. Reson.* **42**:227.

Harrison, C. G., Adams, D. F., and Kramer, P. B., 1985, *Invest. Radiol.* **20**:180.

Hawkins, R. A., and Miller, A. L., 1978, *Neuroscience* **3**:251.

Heidelberger, C., 1974, in *Antineoplastic and Immunosuppressive Agents* (Sartorelli, A. C., and Johns, D. G., eds.), Springer-Verlag, Berlin, p. 193.

Heidelberger, E., and Lauterbur, P. C., 1982, *Abstr. 1st Annu. Meet. Soc. Magn. Reson. Med.* p. 70.

Holland, G. N., Bottomley, P. A., and Hinshaw, W. S., 1977, *Phys. Med. Biol.* **32**:439.

Hollis, D. P., 1980, in *Biological Magnetic Resonance*, Vol. 2, Plenum Press, New York, p. 1.

Hull, W. E., Kunz, W., Port, R. E., and Seiler, N., 1988, *NMR Biomed.* **1**:11.

James, T. L., and Margulis, A. R., 1984, *Biomedical Magnetic Resonance*, Radiology Research and Education Foundation, San Francisco.

Joseph, A., Davenport, C., Kwock, L., Burt, C. T., and London, R. E., 1987, *Magn. Reson. Med.* **4**:137.

Joseph, P. M., Fishman, J. E., Mukherji, B., and Sloviter, H. A., 1985, *J. Comput. Assist. Tomogr.* **9**:1012.

Kanazawa, Y., Momozono, Y., Ishikawa, M., Yamada, T., Yamane, H., Haradahira, T., Maeda, M., and Kojima, M., 1986, *Life Sci.* **39**:737.

Kawashima, A., Seto, H., Kato, M., Yasuda, A., Uchida, K., and Otake, N., 1958, *J. Antibiot.* **38**:1625.

Kawashima, A., Seto, H., Kato, M., Yasuda, A., Uchida, K., and Otake, N., 1986, *J. Antibiot.* **39**:1495.

Kirk, K., Kuchel, P. W., and Labotka, R. J., 1988, *Biophys. J.* **54**:241.

Kirschenlohr, H. L., Metcalfe, J. C., Morris, P. G., Rodrigo, G. C., and Smith, G. A., 1988, *Proc. Natl. Acad. Sci. USA* **85**:9017.

Koehler, L. S., Fossel, E. T., and Koehler, K. A., 1977, *Biochemistry* **16**:3700.

Kojima, M., Kuribayashi, S., Kanazawa, Y., Haradahira, T., Maehara, Y., and Endo, H., 1988, *Chem. Pharm. Bull.* **36**:1194.

Kwee, I. L., Nakada, T., and Rao, G. A., 1986, *Biochem. Arch.* **2**:1.

Kwee, I. L., Nakada, T., and Card, P. J., 1987, *J. Neurochem.* **49**:428.

Labotka, R. J., and Omachi, A., 1987, *J. Biol. Chem.* **262**:305.

Levy, L. A., Murphy, E., and London, R. E., 1987, *Am. J. Physiol.* **252**:C441.

Levy, L. A., Murphy, E., Raju, B., and London, R. E., 1988, *Biochemistry* **27**:4041.

Litt, L., Gonzalez-Mendez, R., James, T. L., Sessler, D. I., Mills, P., Chew, W., Moseley, M., Pereira, B., Severinghaus, J. W., and Hamilton, W. K., 1987, *Anesthesiology* **67**:161.

London, R. E., 1988, *Prog. NMR Spectrosc.* **20**:337.

London, R. E., and Gabel, S. A., 1989, *Biochemistry* **28**:2378.

Longmaid, H. E., III, Adams, D. F., Neirinckx, R. D., Harrison, C. G., Brunner, P., Seltzer, S. E., Davis, M. A., Neuringer, L., and Geyer, R. P., 1985, *Invest. Radiol.* **20**:141.

Malet-Martino, M. C., Martino, R., Lopez, A., Beteille, J. P., Bon, M., Bernadou, J., and Armand, J. P., 1983, *Biomed. Pharmacother.* **37**:357.

Malet-Martino, M. C., Martino, R., Lopez, A., Beteille, J. P., Bon, M., Bernadou, J., and Armand, J. P., 1985, *Biochem. Pharmacol.* **34**:429.

Malet-Martino, M. C., Bernadou, J., Martino, R., and Armand, J. P., 1988, *Drug Metab. Dispos.* **16**:78.

Mansfield, P., and Morris, P. G., 1982, *NMR Imaging in Biomedicine*, Academic Press, New York.

Marban, E., Kitakaze, M., Kusuoka, H., Porterfield, J. K., Yue, D. T., and Chacko, V. P., 1987, *Proc. Natl. Acad. Sci. USA* **84**:6005.

Martino, R., Bernadou, J., Malet-Martino, M. C., Roche, H., and Armand, J. P., 1987, *Biomed. Pharmacother.* **41**:104.

Mattrey, R. F., Scheible, F. W., Gosink, B. B., Leopold, G. R., Long, D. M., and Higgins, C. B., 1982, *Radiology* **145**:759.

Matwiyoff, N. A., Griffey, R. H., Griffey, B. V., and Storey, S. S., 1987, *6th Annu. Meet. Soc. Magn. Reson. Med. Abstr.* p. 67.

McFarland, E., Koutcher, J. A., Rosen, B. R., Teicher, B., and Brady, T. J., 1985, *J. Comput. Assist. Tomogr.* **9**:8.

Metcalfe, J. C., Hesketh, T. R., and Smith, G. A., 1985, *Cell Calcium* **6**:183.

Mills, P., Sessler, D. I., Mosely, M., Chew, W., Pereira, B., James, T. L., and Litt, L., 1987, *Anesthesiol.* **67**:169.

Moore, R. R., and Roberts, M. R., 1985, *Biochim. Biophys. Acta* **844**:346.

Murphy, E., and London, R. E., 1988, in *Reviews in Biochemical Toxicology*, Vol. 9, Elsevier, Amsterdam, p. 131.

Murphy, E., Levy, L., Berkowitz, L., Orringer, E., Gabel, S. A., and London, R. E., 1986, *Am. J. Physiol.* **251**:C496.

Murphy, E., Steenbergen, C., Levy, L., Raju, B., and London, R. E., 1989, *J. Biol. Chem.* **264**:5622.

Nakada, T., and Kwee, I. L., 1986, *Biochem. Arch.* **2**:53.

Nakada, T., and Kwee, I. L., 1987, *Magn. Reson. Med.* **4**:366.

Nakada, T., Kwee, I. L., Rao, G. A., and Conboy, C. B., 1985, *Biochem. Arch.* **1**:163.

Nakada, T., Kwee, I. L., and Conboy, C. B., 1986, *J. Neurochem.* **46**:198.

Nakada, T., Kwee, I. L., Card, P. J., Matwiyoff, N. A., Griffey, B. V., and Griffey, R. H., 1988, *Magn. Reson. Med.* **6**:307.

Nunnally, R. L., Babcock, E. E., and Antich, P., 1985a, *4th Annu. Meet. Soc. Magn. Reson. Med.* p. 810.

Nunnally, R. L., Babcock, E. E., Horner, S. D., and Peshock, R. M., 1985b, *Magn. Reson. Imag.* **3**:399.

Phelps, M. E., Huang, S. C., Hoffman, E. J., Selin, C., Sokoloff, L., and Kuhl, D. E., 1979, *Ann. Neurol.* **6**:371.

Plummer, J. L., Cousins, M. J., and de la M. Hall, P., 1982, *Q. Rev. Drug Metab. Drug Interact.* **4**:49.

Prior, M. J. W., McSheeby, P. M. J., Maxwell, R. J., and Griffiths, J. R., 1987, *6th Annu. Meet. Soc. Magn. Reson. Med.* p. 502.

Raleigh, J. A., Franko, A. J., Treiber, E. O., Lunt, J. A., and Allen, P. S., 1986, *Int. J. Radiat. Oncol. Biol. Phys.* **12**:1243.

Rand, K. H., Raad, I., El Koussi, A., Houck, H. J., Brey, W., Rocca, J., Loftsson, T., and Bodor, N., 1987, *J. Viorol. Methods* **18**:257.

Ratner, A. V., Hurd, R., Muller, H. H., Bradley-Simpson, B., Pitts, W., Shibata, D., Sotak, C., and Young, S. W., 1987a, *Magn. Reson. Med.* **5**:548.

Ratner, A., McCann, U., Simpson, B. B., Muller, H., Sotak, C., Hurd, R., Barchas, J., and Young, S., 1987b, *6th Annu. Meet. Soc. Magn. Reson. Med.* p. 592.

Reivich, M., Kuhl, D., Wolf, A., Greenberg, J., Phelps, G. M., Ido, T., Casella, V., Fowler, J., Hoffman, E., Alavi, A., Som, P., and Sokoloff, L., 1979, *Circ. Res.* **44**:127.

Rotevatn, S., Murphy, E., Levy, L. A., Raju, B., Lieberman, M., and London, R. E., 1989, *Am. J. Physiol.* **257**:C141.

San George, R. C., Nagel, R. L., and Fabry, M. E., 1984, *Biochim. Biophys. Acta* **803**:174.

Sarhan, S., Knodgen, B., Gerhart, F., and Seiler, N., 1987, *Int. J. Biochem.* **19**:843.

Schnall, M. D., Subramanian, V. H., Leigh, J. S., Jr., Gyulai, L., McLaughlin, A., and Chance, B., 1985, *J. Magn. Reson.* **63**:401.

Selinsky, B. S., Thompson, M., Jeffreys, L. M., and London, R. E., 1986, *Biophys. J.* **49**:329a.

Selinsky, B. S., Thompson, M., and London, R. E., 1987, *Biochem. Pharmacol.* **36**:413.

Selinsky, B. S., Perlman, M. E., and London, R. E., 1988a, *Mol. Pharmacol.* **33**:559.

Selinsky, B. S., Perlman, M. E., and London, R. E., 1988b, *Mol. Pharmacol.* **33**:567.

Selinsky, B. S., Levy, L. A., Motten, A. G., and London, R. E., 1989, *J. Magn. Reson.* **81**:57.

Sharma, M., Potti, G. G., Simmons, O. D., and Korytnyk, W., 1987, *Carbohydr. Res.* **163**:41.

Shimizu, M., Kobayashi, T., Morimoto, H., Matsuura, N., Shimano, T., Nomura, N., Itoh, S., Yamazaki, M., Iriguchi, N., Yamamoto, T., Yamai, S., Furuta, T., Maki, T., and Mori, T., 1987, *Magn. Reson. Med.* **5**:290.

Smith, G. A., Hesketh, T. R., Metcalfe, J. C., Feeney, J., and Morris, P. G., 1983, *Proc. Natl. Acad. Sci. USA* **80**:7178.

Smith, G. A., Morris, P. G., Hesketh, T. R., and Metcalfe, J. C., 1986, *Biochim. Biophys. Acta* **889**:72.

Steenbergen, C., Murphy, E., Levy, L., and London, R. E., 1987, *Circ. Res.* **60**:700.

Stevens, A. N., Morris, P. G., Iles, R. A., Sheldon, P. W., and Griffiths, J. R., 1984, *Br. J. Cancer* **50**:113.

Stock, J. G. L., and Strunin, L., 1985, *Anesthesiol.* **63**:424.

Taylor, J. S., and Deutsch, C., 1983, *Biophys. J.* **43**:261.

Taylor, J. S., and Deutsch, C., 1988, *Biophys. J.* **53**:227.

Thoma, W. J., and Ugurbil, K., 1988, *NMR Biomed.* **1**:95.

Thomas, S. R., Clark, L. C., Jr., Ackerman, J. L., Pratt, R. G., Hoffman, R. E., Busse, L. J., Kinsey, R. A., and Samaratunga, R. C., 1986, *J. Comput. Assist. Tomogr.* **10**:1.

Trudell, J. R., and Hubbell, W. L., 1976, *Anesthesiology* **44**:202.

Tsien, R., 1981, *Nature* **590**:527.

Vialaneix, J. P., Chouini, N., Malet-Martino, M. C., Martino, R., Michel, G., and Lepargneur, J. P., 1986, *Antimicrob. Agents Chemother.* **30**:756.

Vialaneix, J. P., Malet-Martino, M. C., Hoffmann, J. S., Pris, J., and Martino, R., 1987, *Drug. Metab. Dispos.* **15**:718.

Wolf, W., Albright, M. J., Silver, M. S., Wever, H., Reichardt, U., and Sauer, R., 1987, *Magn. Reson. Imag.* **5**:165.

Wyrwicz, A. M., and Conboy, C. B., 1989, *Magn. Reson. Med.* **9**:219.

Wyrwicz, A. M., Schofield, J. C., and Burt, C. T., 1982, in *Noninvasive Probes of Tissue Metabolism* (Cohen, J. S., ed.), Wiley, New York, p. 149.

Wyrwicz, A. M., Pszenny, M. H., Schofield, J. C., Tillman, P. C., Gordon, R. E., and Martin, P. A., 1983, *Science* **222**:428.

Wyrwicz, A. M., Murphy, R., Prakash, R. M., Moriarty, R. M., and Dougherty, T., 1985, *4th Annu. Meet. Soc. Magn. Reson. Med. Abstr.* p. 759.

Wyrwicz, A. M., Ryback, K. R., Chew, W., and Hurd, R., 1986, *J. Magn. Reson.* **69**:572.

Wyrwicz, A. M., Conboy, C. B., Nichols, B. G., Ryback, K. R., and Eisele, P., 1987a, *Biochim. Biophys. Acta* **929**:271.

Wyrwicz, A. M., Conboy, C. B., Ryback, K. R., Nichols, B. G., and Eisele, P., 1987b, *Biochim. Biophys. Acta* **927**:86.

6

In Vivo ^{2}H NMR Studies of Cellular Metabolism

Robert E. London

1. INTRODUCTION

The development of *in vivo* spectroscopic methods for the study of cell suspensions, perfused organs, intact animals and humans over the past two decades has relied primarily on the observation of nuclei other than protons. In particular, ^{31}P and ^{13}C NMR studies of intact systems have provided information about the energetics and biochemistry of intact cells, and applications involving these nuclei have been reviewed extensively (Hollis, 1980; Iles *et al.*, 1982; Baxter *et al.*, 1983; London, 1988). A number of other nuclei have also proven useful, including sodium and fluorine; *in vivo* applications utilizing both are reviewed in this volume (Miller and Elgavish, this volume; Selinsky and Burt, this volume). Although deuterium tracers have proven useful for mapping out the biosynthetic pathways of secondary metabolites using *in vitro* NMR methods (Abell, 1986, and references therein), initial consideration of some of the limitations of this isotope—quadrupolar broadening, significant kinetic isotope effects, and small chemical shift dispersion—suggests that it will have limited use as an *in vivo* tracer. Despite these limitations, several factors make deuterium a favorable isotope for *in vivo* metabolic NMR studies in some cases. The low natural abundance (1.56×10^{-2}%) minimizes interference from resonances of endogenous compounds. The relatively short spin-lattice relaxation

---

Robert E. London • Laboratory of Molecular Biophysics, National Institute of Environmental Health Sciences, Research Triangle Park, North Carolina 27709.

Biological Magnetic Resonance, Volume 11: In Vivo Spectroscopy, edited by Lawrence J. Berliner and Jacques Reuben. Plenum Press, New York, 1992.

times due to the quadrupolar relaxation mechanism, lead to important real-time sensitivity gains and minimize the intensity distortions which can arise from overpulsing. In contrast to ^{13}C studies, it is generally not necessary to use decoupling methods which are difficult to implement *in vivo.* The relative ease of synthesis of many deuterated compounds and related low cost of commercially available deuterated materials, particularly in comparison with ^{13}C-labeled compounds, is also an attractive advantage for metabolic studies. In general, *in vivo* studies with ^2H may be useful for situations in which a relatively limited number of spectral lines are present, and in which the molecular size and/or rotational correlation time of labeled metabolites are sufficiently small so that quadrupolar broadening is not excessive. In the experience of this reviewer, the major advantage of using deuterium as an *in vivo* tracer is the extreme technical ease with which studies can be carried out. This factor has led to the suggested use of deuterated metabolites in pilot or feasibility studies even in situations in which it may ultimately prove necessary to use a different labeling strategy (Eng *et al.*, 1990). Deuterium labels provide information about the chemistry of the labeled "proton" which in some cases cannot be obtained using ^{13}C labels, as illustrated in the studies of formaldehyde metabolism (e.g., Mason and Sanders, 1989). Additionally, the presence of deuterium labels in metabolites of interest has been detected based on the perturbations of proton or carbon resonances of the molecule. Recent ^2H NMR studies have utilized D_2O as a freely diffusible tracer to measure blood flow and tissue perfusion (Ackerman *et al.*, 1987a, b), and the potential of deuterium imaging has also been explored (Ewy *et al.*, 1986, 1988; Muller and Seelig, 1987; Link and Seelig, 1990). The present review provides a general consideration of the use of ^2H labels in metabolic studies (Section 2) as well as summarizing some recent metabolic (Section 3) and imaging and perfusion studies (Section 4). The general area of ^2H NMR studies of lipids, which includes some *in vivo* NMR work, has been reviewed elsewhere (Jacobs and Oldfield, 1980; Seelig and Macdonald, 1987) and is not covered here. Finally, we note that for this review we have adopted the following nomenclature: the symbol "D" is used in chemical formulas, e.g., HDO, and "^{2}H" is used to denote labeling when the common name is used, e.g., [*methyl*-$^2H_3$]lactate.

2. GENERAL CONSIDERATIONS

2.1. Sensitivity and Relaxation

The deuteron has spin 1 and a magnetogyric ratio 6.5 times smaller than that of the proton. The inherent sensitivity for detection of an ^2H nucleus at constant magnetic field strength is approximately 0.01 relative to ^{1}H, or 60.6% the sensitivity of ^{13}C (dropping to 20% if a maximal ^{13}C-{^{1}H} NOE is developed) or 14.5% relative to ^{31}P. Although such a calculation would suggest that the concentration threshold for detection of deuterated metabolites would be somewhat higher than for phosphorylated or ^{13}C-labeled metabolites, the comparative deuterium sensitivity is improved by the relatively short spin-lattice relaxation times characterizing the quadrupolar deuterium nucleus, and by the fact that frequently several equivalent or nearly equivalent deuterons will contribute to the observed signal. For example, deuterated methyl groups provide a factor of 3 improvement in sensitivity and there may be 9 equivalent deuterons in *N*-trimethyl metabolites. It is worth noting that although the deuteron spin of 1 leads to an additional quadrupolar relaxation mechanism relative to the proton, the considerably smaller magnetogyric ratio reduces the dipolar interaction with other protons or unpaired electrons by $(\gamma_D/\gamma_H)^2 = 0.024$. Hence, deuterons are subject to a smaller dipolar relaxation contribution than protons, which in some situations can result in narrower resonances for deuterons than for the corresponding protons (Wheeler *et al.*, 1982). This difference has been utilized in a study of the effects of hepatic iron loading on water relaxation (Irving *et al.*, 1987). In this study, HDO deuterium resonances are readily observed under iron loading conditions which broaden the water proton resonance to a degree which makes measurements difficult.

Questions concerning the nature of intracellular water have received considerable attention and been the focus of a number of NMR studies. The deuterium spin-lattice relaxation time of HDO in isolated muscle and liver tissue of mice and in egg white has been found to exhibit significant frequency and temperature dependence (Fung *et al.*, 1975). Perhaps the simplest model which provides a reasonable fit of both proton and deuterium relaxation data is based on the assumption of two rapid exchanging pools of water (Fung *et al.* 1975). A small fraction of water which is associated with cellular macromolecules, the "hydration shell," is characterized by a distribution of correlation times and exhibits a frequency-dependent T_1; it cannot be frozen to the solid state down to −70°C. The bulk of the water molecules exhibit a single correlation time which is sufficiently short so that the T_1 for this pool is frequency independent; this pool freezes at about −8°C. Above −8°C, the observed signal is then a weighted average for the two pools. Analysis of proton and deuteron relaxation data (Fung *et al.*, 1975) led to values of 12% for the fraction of water molecules in the hydration layer in liver and gastrocnemius muscle, and to 5% in egg white. The existence of a hydration shell of water around membranes is further substantiated by studies of deuterium relaxation behavior in the presence of extracted *E. coli* lipids (Borle and Seelig, 1983). At low water/lipid ratios, residual quadrupole splitting of the deuterium resonance was used to estimate the size of the primary hydration layer. A more sophisticated analysis of the behavior in water in a protein solution has recently been developed based on the relaxation behavior of D_2O in solutions containing bovine serum albumin (Kimmich *et al.*, 1990). Fung and co-workers also carried out comparative studies of excised muscle and the distillate, and determined that the intensity of the water signal observed from the muscle corresponds to the entire water pool (Fung *et al.*, 1975). Hence, it was concluded that all of the HDO signal is NMR visible. This result is important in terms of the use of the HDO resonance as an internal concentration standard (see below).

More recently, ^2H spin–lattice relaxation times for the HDO resonance of tissues taken from rats given 20% D_2O in their drinking water were compared with ^1H relaxation times of the same tissues at the same frequency (13.7 MHz), and the relation

$$\frac{1}{T_{1H}} = 0.991\left(\frac{1}{T_{1D}}\right) - 3.48 \quad (1)$$

was empirically derived (Block and Parekh, 1987). Spin–lattice relaxtion data for H_2O and HDO in rat head (Ewy *et al.*, 1988) and mouse liver (Fung *et al.*, 1975) are also consistent with Eq. (1). The existence of a linear relationship with a slope close to 1.0 indicates that the relaxation differences between different tissues are probably not caused by mechanisms which would affect proton and deuterium relaxation behavior very differently. These include intermolecular relaxation due to translational diffusion, interactions with paramagnetic ions, and cross-relaxation effects. Block and Parekh note, however, that the model used by Fung *et al.* (1975) could account for some of the observations.

Relaxation measurements reported for HDO and for deuterated metabolites are summarized in Table 1. Although for small molecules whose motion is expected to fall into the extreme narrowing limit ($\omega\tau \ll 1$) an equivalence of T_1 and T_2 is expected, this has generally not been observed; the one significant exception to this pattern is [*methyl*-$^2H_9$]choline. These discrepancies presumably reflect the more complex dynamics of water and metabolites in cell systems, perhaps arising from exchange between free and macromolecule-associated states as illustrated by the model for cellular water discussed above. The inequivalence between T_1 and T_2 also extends to suspensions of cells in which most of the HDO signal corresponds to extracellular water. The basis for the near equivalence of T_1 and T_2 observed for choline, which allowed this metabolite to be selectively imaged (Eng *et al.*, 1990), is unclear. Clearly more work is required to determine the basis for the relaxation behavior of deuterated metabolites.

TABLE 1
Deuterium Relaxation Parameters

Molecule	T_1 (msec)	T_2 (msec)	Temperature[a] (°C)	Tissue	Frequency (MHz)	Reference
HDO	129	18.2	—	Liver	13.7	Block and Parekh (1987)
HDO	169	22.3		Kidney	13.7	Block and Parekh (1987)
HDO	160	17.4		Muscle	13.7	Block and Parekh (1987)
HDO	158	17.4		Mammary gland	13.7	Block and Parekh (1987)
HDO	188	—		Mammary adenocarcinoma	13.7	Block and Parekh (1987)
HDO	318	72	[37]	Kidney	30.7	Eng *et al.* (1990)
HDO	228	—	[37]	Abdomen	30.7	Brereton *et al.* (1986)
HDO	310	—	[37]	RIF-1 tumor	30.8	Evelhoch *et al.* (1989)
HDO	390	50	37	RBC suspension	76.7	Hotchkiss *et al.* (1990)
HDO	446	—	22	Cell suspension	55.3	Aguayo *et al.* (1988a)
HDO	250	—	[37]	Liver	55.3	Ackerman *et al.* (1987a)
CHD	34		[37]	Abdomen	30.7	Brereton *et al.* (1986)
[*methyl*-$^2H_2$]ethanol	198	—	22	Cell suspension	55.3	Aguayo *et al.* (1988a)
[*methyl*-$^2H_2$]acetate	176	—	22	Cell suspension	55.3	Aguayo *et al* (1988a)
[6,6-$^2H_2$]glucose	290	—	22	Cell suspension	55.3	Aguayo *et al.* (1988a)
[6,6-$^2H_2$]glucose	80	50	37	RBC suspension	76.7	Hotchkiss *et al.* (1990)
[3,3-$^2H_2$]lactate	162	—	22	Cell suspension	55.3	Aguayo *et al.* (1988a)
[3,3-$^2H_3$]lactate	290	40	37	RBC suspension	76.7	Hotchkiss *et al.* (1990)
[*methoxy*-$^2H_3$] misonidazole	210	—	[37]	RIF-1 tumor	30.8	Evelhoch *et al.* (1989)
[*methyl*-$^2H_9$]choline	377	316	[37]	Rabbit kidney	30.7	Eng *et al.* (1990)

[a] Values in brackets correspond to *in vivo* measurements.

2.2. Chemical Shifts and Coupling Constants

The chemical shifts of deuterated molecules are essentially equivalent to those of the corresponding protons expressed in ppm (Mantsch *et al.*, 1977). Although the deuterium resonances in multiply deuterated compounds may be subject to β and γ isotope shifts, such effects are generally very small, particularly in comparison with the deuterium linewidths characterizing the *in vivo* spectra. In contrast, large, one-bond isotope shifts have been used to characterize the distribution of deuterons in a molecule based on observation of the ^{13}C resonances (e.g., Fig. 1a) (Walker and Wageman, 1986; Barrow *et al.*, 1986; Mason *et al.*, 1987a, b; Mason and Sanders, 1989; Mantsch *et al.*, 1977).

Scalar coupling interactions involving deuterons are scaled by the ratio of the magnetogyric ratios relative to the corresponding proton coupling interactions: $J_{DD} = (\gamma_D/\gamma_H)J_{DH}$; $J_{DH} = (\gamma_D/\gamma_H)J_{HH}$; where $(\gamma_D/\gamma_H) = 0.1535$, so that in general, geminal and vicinal proton–deuteron coupling

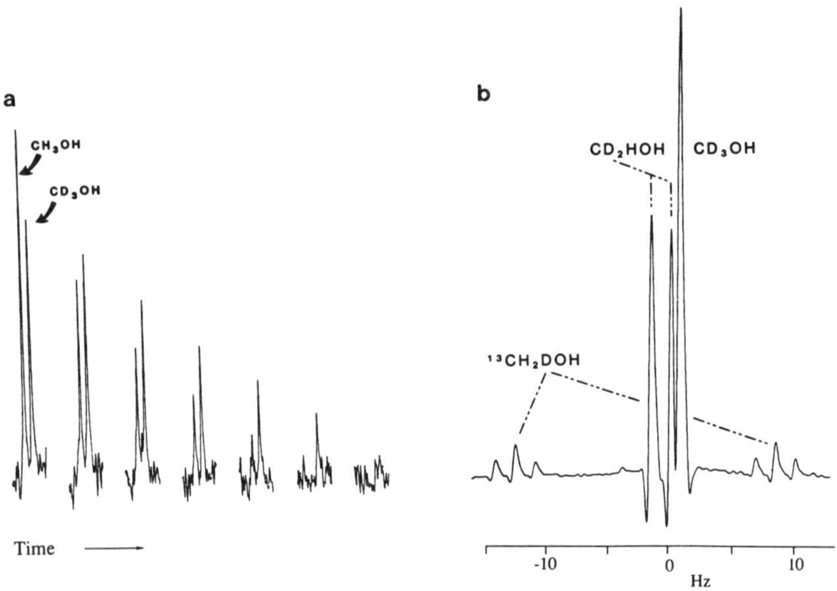

Figure 1. (a) Successive $^{13}C\{^1H,^2H\}$ spectra (100.6 MHz) of a 50:50 mixture of $^{13}CH_3OH$ and $^{13}CD_3OH$ in a suspension of *Methylosinus trichosporium* cells illustrating the isotopic selectivity of methanol consumption. The ^{13}C resonance of the deuterated methanol is shifted upfield by 0.8 ppm due to the deuterium isotope effect. The spectra were acquired at 3-min time intervals and subjected to a 10-Hz line broadening for sensitivity enhancement. (b) ^2H NMR spectrum (76.75 MHz) of methanol isotopomers produced by cross-dismutation of $^{13}CH_2$- and $^{12}CD_2$-formaldehyde by a suspension of *Pseudomonas putida.* The spectrum represents 1500 scans (1.5 hr) subjected to resolution enhancement. (Reprinted from Mason *et al.*, 1987a.)

interactions will not be resolved (Mantsch *et al.*, 1977) (see, however, Fig. 7). One consequence of these relations is that deuterium labeling studies do not directly yield the information about the correlated nature of the labeling of interacting positions as can be obtained through the observation of $^{13}C-^{13}C$ coupling interactions (London, 1988; Malloy *et al.*, 1988). One solution to this deficiency which has been developed by Campbell and co-workers is based on the idea that deuteration will eliminate the homonuclear $^1H-^1H$ coupling interactions (Brindle *et al.*, 1986). Hence, an interacting proton spin system may be "labeled" by deuteration. In contrast to the case of J_{DH} and J_{DD} coupling interactions, $^1J_{DC}$ couplings can sometimes be resolved and observed either by ^{13}C or ^2H NMR (Fig. 1). Observation of $^1J_{CD}$ by ^{13}C NMR requires that the spin–lattice relaxation time of the deuterium be sufficiently long ($T_1 \gg 1/J_{CD} = \gamma_H/(\gamma_D \cdot J_{CH})$), so that the multiplet is not collapsed. Rinaldi and Baldwin (1982) have proposed a $^{13}C\{^2H\}$-INEPT pulse sequence for the selective detection of deuterated carbons, and Gunther and co-workers have utilized spin–echo measurements to distinguish different degrees of deuteration of methine, methylene, and methyl carbons (Schmitt *et al.*, 1983; Wesener and Gunther, 1983; Wesener *et al.*, 1984). Mason *et al.* (1987b) have utilized $^{13}C\{^1H, ^2H\}$ double resonance decoupling in order to compare the intensities of protonated and deuterated methanol in suspensions of *Methylosinus trichosporium* OB3b cells (Fig. 1).

The background ^{2}H spectrum generally observed *in vivo* includes a resonance arising from HDO and, in some cases, from methylene CHD groups of lipids as well. It is probable that only the triglycerides have sufficiently mobile CHD groups to contribute significantly to this resonance. The HDO resonance is particularly useful, as it can provide an internal chemical shift standard, an intensity standard, and, subject to several assumptions, a concentration reference. Since the chemical shift of water exhibits significant temperature dependence of ~0.01 ppm/°C, this factor must be considered when using the HDO peak as a shift reference. We have generaly set $\delta(\text{HDO}) = 4.7$ ppm for *in vivo* studies at 37°C.

2.3. HDO as an Internal Concentration Reference

The use of the deuterium resonance of HDO as a concentration standard has been discussed by several groups (London *et al.*, 1987; Evelhoch *et al.*, 1989). Based on the natural abundance of deuterium, the HDO concentration of pure water is calculated to be 17.3 mM. The use of this figure for determining the level of another deuterated metabolite in the cytosol is based on the assumptions that: (1) the HDO is completely NMR visible, (2) the water NMR signal arises only from the cytosol of the cells containing the labeled metabolite, or the fractional contribution of cytosolic water can be determined or assumed, and (3) the HDO signal is constant. Although the first issue has been somewhat controversial, ^{2}H NMR studies of Fung and co-workers discussed above support the validity of the first assumption. However, it is clear that the water resonance observed *in vivo* arises from extracellular (e.g., interstitial, blood) as well as intracellular pools. In addition, an error can arise as a result of the fact that particular deuterated metabolites may be localized in specific cells or tissues while the water is not. For example, a surface coil experiment may detect deuterated drugs in the liver but HDO from liver as well as muscle or other cells, depending on the coil geometry, localization strategy, and other parameters. Finally, as a result of the low natural abundance of ^{2}H, the intensity of the HDO signal can be significantly increased as a result of the metabolism of deuterated molecules. This effect has been noted in several of the studies summarized here (e.g., London *et al.*, 1987; Aguayo *et al.*, 1988b; Hotchkiss *et al.*, 1990). Alternatively, for molecules which are not water-soluble, tissue (rather than cytosolic) concentrations may be estimated by determining the fractional water content or obtaining values from the literature (e.g., Lentner, 1981).

The use of HDO as a concentration standard for the estimation of drug concentration has been explicitly discussed by Evelhoch *et al.* (1989), who used the relation:

$$[\text{Drug}](\text{in mM}) = 17.3 \cdot I(\text{drug})/I(\text{HDO})/(n \times p) \qquad (2)$$

where $I(\text{drug})/I(\text{HDO})$ is the intensity ratio of the corresponding ^{2}H resonances, n is the number of equivalent ^{2}H nuclei, and p is the fractional enrichment. Data obtained for deuterated [*methoxy*-$^2H_3$]misonidazole (1-(2-hydroxy-3-methoxypropyl)-2-nitroimidazole), a radiation sensitizing agent, were compared with determinations made using tritium-labeled misonidazole with a correction factor of 0.82 g water/g tumor. As discussed above, this determination yields a concentration relative to total tissue water rather than a cytosolic concentration. Misonidazole concentrations determined by ^{2}H NMR were similar, although generally lower than those determined using the radiolabeled drug, and this difference was ascribed to metabolic loss of the deuterium label. As noted above, detection of additional water from other tissues or compartments with lower drug concentrations or an increase in the HDO resonance due to the metabolism of deuterated compounds, will also result in an underestimation of the concentration of the deuterated metabolite.

2.4. Kinetic Isotope Effects

Kinetic ^{2}H isotope effects can be significant (Rose, 1970) and have been observed in several of the studies reviewed here. The existence of such effects presents a problem in studies designed to evaluate metabolic regulation since ideally the NMR label should be nonperturbing. For example, subsequent to an I.V. injection, perdeuterated glucose has been found to be cleared more slowly from the blood and to accumulate more slowly in heart and brain of mice, than nondeuterated glucose (Gatley *et al.*, 1986). The existence of significant isotope effects may not only alter the rate of observed metabolic transformations, but can lead to "metabolic switching" (Foster, 1984) in situations in which the existence of a large isotope effect on one branch of a pathway will favor the flux through a competing pathway with a smaller or nonexistent isotopic selectivity. One approach to the problem presented by significant isotope effects is to compare intracellular enzyme activities determined *in vivo* with activity measured in model solutions using the same deuterated substrates, as has been done in studies of the activity of lactate dehydrogenase in erythrocytes (Brindle *et al.*, 1986). (For the case of lactate dehydrogenase and a 10 mM pyruvate concentration, Simpson *et al.* (1982) find a β-secondary isotope effect >2 upon deuteration of the methyl group.) In complex, multienzyme systems, the observation of isotope effects will be dependent on whether the isotope-sensitive steps are rate limiting. Although a primary isotope effect of 2.4 was observed for glucose-6-phosphate dehydrogenase isolated from *Zymomonas mobilis*, a comparison of the metabolism of [1-^{2}H]glucose by ^{2}H NMR and unlabeled glucose by ^{31}P NMR indicated no significant differences (Barrow *et al.*, 1986). It was therefore concluded that oxidation of glucose-6-phosphate to 6-phosphogluconate was not rate limiting, and that the overall kinetics were determined by glucokinase. Similarly, no differences in glucose consumption or HDO production rates due to isotope effects were observed in the metabolism of [1-^{2}H] glucose by *P. mirabilis*, indicating that the rate-limiting metabolic steps do not involve deuterium transfer (Aguayo *et al.*, 1988a).

Although the existence of deuterium isotope effects limits the use of deuterium as a tracer in studies of metabolic regulation, the observation of kinetic isotope effects can provide considerable insight into chemical and biochemical mechanisms (Rose, 1970; Melander and Saunders, 1980). For example, a comparison of the toxicity of a compound with that of its deuterated analogue has been used to establish the significance of oxidative metabolism in producing the toxic response (Pohl and Gillette, 1985). Several groups have recently derived information about chemical mechanisms and biosynthetic pathways of molecules based on analysis of the distribution of naturally abundant deuterium nuclei (Grant *et al.*, 1982; Martin *et al.*, 1983, 1986; Pascal *et al.*, 1986; Leopold *et al.*, 1988). For example, Martin *et al.* (1986) have analyzed the deuterium levels in a variety of carbohydrates and in ethanol derived from these carbohydrates by fermentation. Since there is no labeling, long accumulation times are required and the studies were carried out on purified materials rather than *in vivo.* Deuterium abundances could not be directly detected for each position of glucose, but for positions 1, 2, 3 + 4, and 5 + 6 + 6′. The authors were able to conclude that the enrichment of the nonexchangeable positions in glucose is correlated with the enrichment of the methyl group of the ethanol product, while the enrichment of the ethanol methylene group is correlated with the initial enrichment of the water.

Mason *et al.* (1987b) have studied the kinetic isotope effect for methanol oxidation in *Methylosinus trichosporium* OB3b. Carbon-13 NMR measurements of cell suspensions containing a mixture of $^{13}CH_3OH$ and $^{13}CD_3OH$ faster rate (Fig. 1a). Based on the different rates of disappearance of the two species, a value of $k_H/k_D = 1.8$ was determined. It was noted that this ratio implies that C–H bond cleavage is not the rate-determining step in overall turnover of methanol, since otherwise a value of 6–8 would have been observed. The kinetic isotope effect inherent in the dismutation of formaldehyde by *Pseudomonas putida* has also been studied by comparing the metabolism of $^{13}CH_2$- and $^{12}CD_2$-formaldehyde (Mason *et al.*, 1987a), and a ratio $k_H/k_D = 3.7 \pm 0.5$ was obtained using computer simulation to determine the ratios of the methanol isotopomers which were formed (Fig. 1b).

2.5. Solvent Exchange Processes

A final aspect of the use of deuteration in metabolic studies which is central to the design of experiments is the inherent lability of particular deuterated sites. For example, the methyl deuterons of pyruvate will exchange with solvent protons due to keto–enol tautomerism (Fig. 2). In studies of [3-$^3H_1$]pyruvate, Rose (1960) found ~3% detritiation at 37°C over a 30 min period. Thus, a significant fraction of the deuterium nuclei at the C-3 position of pyruvate can be lost via enolization during experiments which extend for several hours. Further, the enolization of pyruvate has been shown to be catalyzed by the formation of Schiff's base links with the amino groups of proteins (Simpson *et al.*, 1981). Human serum albumin has been reported to catalyze exchange of pyruvate methyl protons in D_2O (Stepuro *et al.*, 1986). Pyruvate kinase (Rose, 1960), oxaloacetate decarboxylase (Kosicki, 1968), and 6-phospho-2-dehydro-3-deoxygluconate aldolase (Barrow *et al.*, 1986) have been shown to catalyze pyruvate enolization. In the presence of lactate dehydrogenase, the lactate methyl protons will also exchange (Fig. 2). Rognstad and Wals (1976) have concluded that glutamate–pyruvate transaminase exchange plays a major role in the loss of tritium from L-[3-^2H]lactate in hamster liver cells. Conversely, deuterium initially present in the water can label the various metabolites as a result of these processes. In studies of the formation of ethanol from D_2O, Barrow *et al.* (1986) found that metabolic incorporation routes were insufficient to

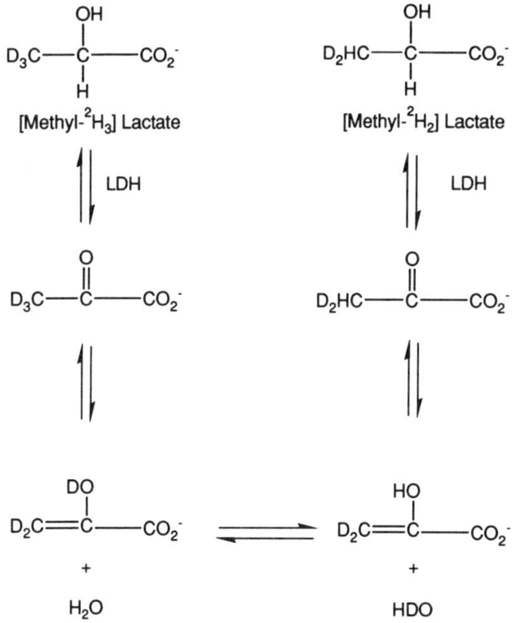

Figure 2. Exchange of lactate/pyruvate methyl deuterons with the solvent. In addition to the involvement of an H_2O molecule in the keto → enol rearrangement as shown, the involvement of a second water molecule in the enol → keto transition could lead to the loss of a second methyl deuteron. LDH, lactate dehydrogenase.

account for the observed labeling of ethanol and that extensive metabolite enolization was therefore occurring. Thus, to some extent the labeling of cellular metabolites can occur by processes which are dependent on the chemical nature of the particular proton and do not reflect metabolic transformations of the molecule. This situation, which also characterizes the use of ^{3}H radiotracers, contrasts with the case of ^{13}C. Despite the apparent analogy between ^{2}H and ^{3}H labeling, the use of deuterated compounds labeled to high specific activity in most ^{2}H NMR studies, compared with the low levels of ^{3}H enrichment used in radiotracer (although not ^{3}H NMR) studies, can lead to significant differences in isotopic enrichment (Abell, 1986, and references therein).

3. METABOLIC STUDIES USING DEUTERATED COMPOUNDS

As noted in the previous section, *in vivo* metabolic studies utilizing deuterium labeling have involved both direct detection of the deuterium isotope and "indirect" detection of the deuteron resulting from the absence of the corresponding proton resonance and associated loss of the corresponding ^1H–^1H coupling interactions in the proton NMR spectrum. Since the latter approach has been reviewed in this series (Brindle *et al.*, 1987), this section focuses on applications involving direct detection. Reported studies have dealt with glutathione metabolism, metabolism of deuterated glucose in a variety of systems, observation of the metabolic incorporation of deuterium from deuterated water, lipid mobilization, the metabolism of choline and methionine, and formaldehyde metabolism.

3.1. Glutathione Metabolism

Perhaps the earliest *in vivo* metabolic studies involving the direct detection of deuterated metabolites dealt with the incorporation of [$^2H_5$]glycine into glutathione. Isab and Rabenstein (1979) and York *et al.* (1982) utilized ^1H and ^2H NMR spectroscopy of intact erythrocytes to follow the exchange between added [$^2H_5$]glycine and the initially unlabeled glycine residue of glutathione in erythrocytes. The use of [$^2H_5$]glycine containing both CD_2 and ND_3^+ groups was motivated primarily by the proton NMR studies which were also carried out, since the amino deuterons will exchange with the water and thus are not detectable by ^2H NMR. As a result of the small shift difference between the free glycine resonance (3.54 ppm) and the glutathione glycine resonance (3.76 ppm), quantitation of the ^2H NMR data was not possible. York *et al.* achieved somewhat improved results by carrying out inversion recovery experiments which exploit the longer spin–lattice relaxation time of the free glycine resonances; however, as in the previous study, quantitation was based on ^1H NMR analyses. Isab and Rabenstein also noted that incorporation of deuterated glutamate into glutathione proceeded much more slowly than the glycine exchange process. These studies also demonstrate the principal limitation of the ^2H approach arising from the limited resolution attainable, which in turn reflects the limited shift range and relatively broad ^2H linewidths. Nevertheless, it should also be noted that in contrast to these ^1H NMR studies, the ^2H NMR experiments did not require the use of deuterated buffers which can compromise the physiological validity of the studies.

3.2. Glucose Metabolism

Since glucose is the ultimate primary organic precursor to the metabolites of microorganisms, plants, and animals, it has not surprisingly been utilized in a number of studies. This is particularly true of *in vivo* NMR studies in which the time and sensitivity limitations restrict observations to major metabolic processes. Nevertheless, it should be noted that there are several problems inherent in the use of deuterated glucose as an *in vivo* NMR tracer. First, the central metabolic role of glucose makes it a precursor to many cellular metabolites. Given the limited chemical shift range of deuterium, this can result in significant resolution problems. Second, deuterium nuclei at several positions in glycolytic intermediates are rapidly lost due to metabolic and exchange reactions. For example, a deuteron at the C-2 position of glucose will be lost upon formation of fructose (with a small probability of transfer to C-1). Deuterons at positions 3 and 4 are lost in glycolytic metabolism to pyruvate. Deuterons at positions 1 and 6 which label pyruvate at C-3 can be lost due to exchange with solvent protons which occurs due to keto-enol tautomerism. These limitations do not prevent the use of deuterated glucose in metabolic studies, but underline the selectivity of the method and the need to consider both the biochemistry and the chemistry of the deuterated metabolites. Indeed, some of the effects noted above can be utilized in the design and interpretation of metabolic studies.

The fermentation of [1-^{2}H]glucose by *Zymomonas mobilis* to ethanol was studied by Barrow *et al.* (1986). Although in most ^{2}H NMR studies the HDO resonance dominates the spectrum, this was not the case in these studies due to the use of an initial concentration of 0.25 M [1-^{2}H]glucose. Time-dependent ^{2}H NMR spectra showed that the β anomer was utilized at 1.6 times the rate of the α anomer, and the principal labeled product was [1-^{2}H]ethanol. In these studies, the anomeric specificity of glucose transport or metabolism is to some extent offset by the mutarotation of glucose which in some cases is accelerated enzymatically by mutarotases (Bailey *et al.*, 1968). Thus, these studies are detecting the inability of the mutarotation process to keep pace with the uptake and/or metabolism of a particular glucose anomer. In the *Z. mobilis* studies, there is relatively little labeling of the water, as judged by the ^{2}H NMR spectra. As a result of the utilization of the Entner–Duodoroff (6-phosphogluconate) pathway by *Z. mobilis*, the 1-^{2}H label is immediately transferred to NAD^+ yielding (*R*)-[4-^{2}H]-NADH, and the label subsequently reintroduced into ethanol by alcohol dehydrogenase (Fig. 3). Carbon-13 NMR studies of the labeled ethanol were utilized to determine the distribution and specific activities of the various deuterated sites.

Aguayo *et al.* (1988a) have recently used deuterated glucose in metabolic studies of five bacterial strains: *Escherichia coli*, *Clostridium perfringens*, *Klebsiella pneumoniae*, *Proteus mirabilis*, and *Staphylococcus aureus*. Deuterated metabolic end products identified included acetate, ethanol, lactate, succinate, and, in studies of *C. perfringens*, butyrate and acetone. Significant differences in the disappearance of the α and β anomers of [1-^{2}H]glucose in studies with *P. mirabilis* were observed, with the β anomer showing an

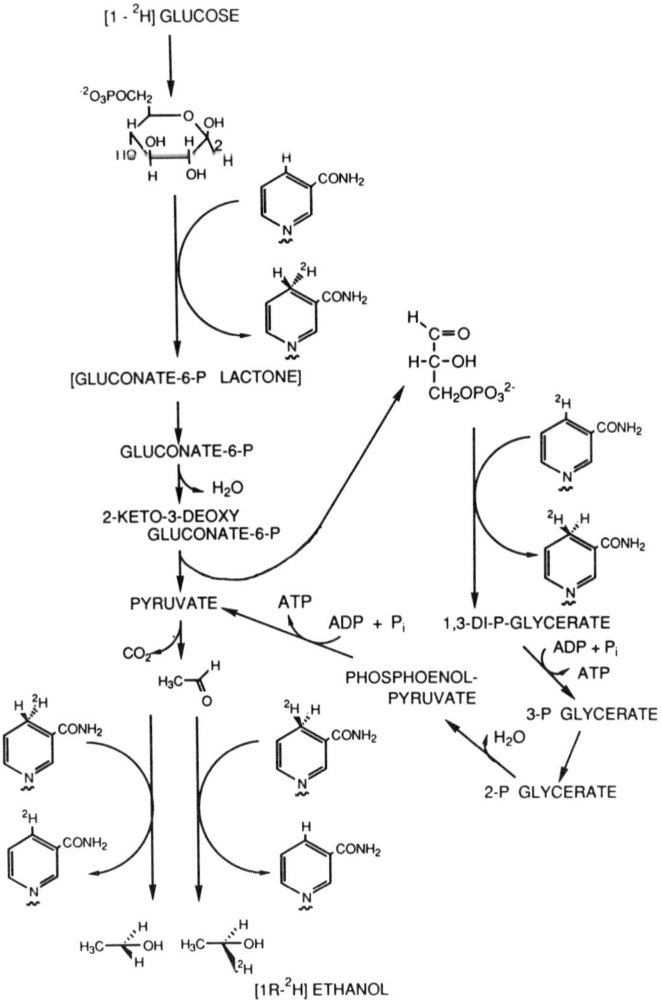

Figure 3. Flow of deuterium derived from [1-^{2}H]glucose through the Entner–Duodoroff pathway utilized by *Zymomonas mobilis* showing catabolism of [1-^{2}H]glucose with the dehydrogenase steps and enzyme stereospecificities indicated. (Adapted from Barrow *et al.*, 1986.)

initial increase in concentration while the α anomer concentration decreased rapidly. Thus, after 20 hr the β-D-glucose signal is ~115% of its initial value, while the α-glucose intensity is ~67% of its initial value. Subsequently, the β-glucose concentration is observed to fall more rapidly than the α-glucose concentration, so that after ~72 hr the α-D-glucose concentration exceeds the β-D-glucose concentration. These data are interpreted to indicate a more rapid uptake and consumption of the β-D-glucose, which is, however, initially overwhelmed by mutarotation of the α anomer to the β. If it is assumed that the initial anomer ratios correspond to the thermodynamic equilibrium, the authors' interpretation appears to imply that the mutarotation process is producing a larger β/α ratio than would exist thermodynamically, which clearly cannot be the case. Potentially complicating factors which could also contribute to these observations include the superposition of intra- and extracellular pools as well as the overlap of resonances corresponding to various metabolites such as glucose-6-phosphate. It was also noted that based on the label distribution in the products, ~20% of the label in [2-^{2}H]glucose is transferred from C-2 to C-1 of fructose-6-phosphate. This migration has been observed previously (Rose, 1970), and is catalyzed by glucose-phosphate isomerase, with the degree of deuteration at C-1 strongly temperature dependent. As in the studies with *Z. mobilis* noted above, no differences in glucose consumption or HDO production rates due to isotope effects were observed, indicating that the rate-limiting metabolic steps do not involve deuterium transfer.

Deuterated glucose has also been used to study metabolism in the lens of the rabbit eye (Aguayo *et al.*, 1987). In this study, both [1-^{2}H]sorbitol and [3-^{2}H]lactate are formed from the [1-^{2}H]glucose precursor. Pulse–chase experiments were carried out in which the lens was initially exposed to the labeled glucose for 24 hr, followed by exposure to unlabeled glucose. Two glucose concentrations, 5.5 and 35.5 mM were used, in order to simulate differences between the normal and diabetic states. [1-^{2}H]Sorbitol formed during the labeling period with the high glucose concentration persisted during the chase period, while labeled sorbitol formed during the labeling period with the low glucose concentration disappeared during the chase period. It was noted that in three previous NMR studies carried out by the same group using ^{13}C labeling, sensitivity was inadequate to observe metabolism of sorbitol by the polyol dehydrogenase pathway under low glucose loading conditions. Aguayo *et al.* (1988b) have subsequently studied the metabolism of [6,6-$^2H_2$]glucose in isolated rabbit corneas. The lactate production/glucose utilization ratio was determined to be 0.6, while glucose oxidation in the Krebs cycle accounted for 40% of the total glucose consumption. Although the analysis is dependent on the degree of label which appears in the lactate and water signals, and relatively long periods (12 hr) were involved, there was no discussion of possible effects due to enolization reactions. Despite the presence of Krebs cycle activity, no [4-^{2}H]glutamate or [3-^{2}H]aspartate was observed.

Hotchkiss *et al.* (1990) have studied the metabolism of [6,6-$^2H_2$]glucose in rat red blood cells obtained from Sprague–Dawley rats subjected to a cecal ligation and perforation protocol to induce sepsis. [*methyl*-$^2H_3$]-Lactate was the only metabolite observed, other than an increase in the intensity of the HDO resonance. Although no significant difference between the glucose metabolism of control rats and rats subjected to the sepsis protocol was observed, it was noted that the production of labeled lactate determined from ^{2}H NMR measurements during the first 4 hr of glucose metabolism did not account for all of the consumption of metabolized glucose, even when the increase of the HDO resonance was taken into account. Although not explicitly discussed, this enrichment most probably is a consequence of the pyruvate enolization noted above. It was proposed that the additional discrepancy resulted from the distribution of the remaining ^{2}H label among a range of other metabolites which did not show up due to low concentration or excessive linewidth. Discrepancies with lactate production rates measured by enzymatic assay were also found to be lower at longer times, and this was ascribed to the initial production of unlabeled lactate from initially unlabeled glycolytic intermediates. A similar effect has been noted based on ^{13}C NMR studies (Oxley *et al.*, 1984).

The question of the extent to which the conversion of glucose to hepatic glycogen follows a direct or indirect pathway, i.e., involves the breakdown of glucose to three carbon intermediates followed by gluconeogenesis, has been the subject of a number of recent studies (Katz and McGarry, 1984). Goodman *et al.* (1989) have studied the deuterium labeling of the hepatic glycogen pool derived from a [6,6-$^{2}H_2$]glucose precursor, and correlated these data with the labeling of venous and arterial glucose pools. In this study, the labeling of the glycogen was determined by ^{2}H NMR subsequent to its hydrolysis to glucose, and hence the question of the NMR visibility of deuterated glycogen was avoided. Rats were fasted prior to the study to reduce contributions from endogenous glycogen to a minimum. Under the conditions of the study, the flux through indirect pathways was determined to be significant.

3.3. Studies Using Deuterated Water

Deuterated water has been used in a number of *in vivo* ^{2}H NMR metabolic studies. In the studies of *Z. mobilis* described above (Barrow *et al.*, 1986), cells were suspended in media which contained D_2O plus glucose in which the hydroxyl protons had been exchanged for deuterons, but the nonexchangeable protons were left in place, and the labeled ethanol produced was analyzed by ^{13}C NMR. Ethanol isotopomers observed (neglecting the exchangeable hydroxyl position) included unlabeled ethanol, ethanol monodeuterated at C-1, and ethanol monodeuterated at C-1 and mono-, di-, or trideuterated at C-2. Labeling at the methylene position arises as a result of the action of pyruvate carboxylase which introduces a deuteron from the solvent to convert pyruvate to acetaldehyde. The acetaldehyde is then stereospecifically reduced to ethanol by alcohol dehydrogenase. Since all of the labeled ethanol isotopomers observed contained only a single methylene deuterium, it was concluded that there is no deuterium exchange between solvent and NAD(H). Alternatively, deuterium incorporated into the methyl group of ethanol was concluded to arise from both enzymatic and nonenzymatic steps, with the former including gluconate-6-phosphate dehydratase, 6-phospho-2-dehydro-3-deoxygluconate aldolase, and pyruvate kinase. It was concluded that the degree of deuteration of the methyl group was greater than could be explained on the basis of these enzymatic steps, and that therefore introduction of label via enolization processes must be significant. Enolization may occur with 6-phospho-2-dehydro-3-deoxygluconate, pyruvate, or acetaldehyde.

Incorporation of deuterium nuclei from labeled water into ethanol by *C. fasciculata* (de los Santos *et al.*, 1985) and into alanine and succinate by *T. cruzi* (Frydman *et al.*, 1990) has also been reported. In the latter study, differential incorporation of deuterium between excreted alanine and alanine derived from perchloric acid cell extracts led to the conclusion that two pools of alanine were present. Only the excreted alanine pool was found to incorporate deuterium (at the C-3 position).

In vivo ^{2}H NMR studies of the abdomens of mice fed a diet containing 10% ^2H_2O in the drinking water have been reported (Brereton *et al.*, 1986). NMR observations of the upper abdominal region were made using a horizontal solenoidal coil of 1.5 cm width tuned to the ^{2}H frequency (30.7 MHz). The deuterium level in the mice was reported to have equilibrated after 3–4 days of exposure to the deuterium-enriched drinking water, and was detected in both the water and lipid methylene signals, presumably corresponding primarily to HDO and to CHD groups. After 28 days of exposure, the loss of deuterium from the body was monitored by ^{2}H NMR. As expected, the rate of loss from water (half life 3–4 days) was significantly faster than the loss from CHD groups (half life 8–9 days). These equilibrium times are considerably shorter than the 4–5 week equilibration time reported by Peng *et al.* (1972) in studies of the deuterium levels of the urine of Sprague-Dawley rats.

Brereton *et al.* (1989) have subsequently carried out a more extensive analysis of lipid metabolism in normal female, Quackenbush mice, and in mice made obese by treatment with monosodium glutamate (MSG). Both groups were given drinking water to which 10% D_2O had been added for a period of 14 days, and the disappearance of the deuterated lipid (triglyceride) resonances after the return to isotopically normal water was subsequently monitored by ^{2}H NMR. For both groups of mice, the disappearance of the lipid CHD/CD_2 signal was found to be biexponential, with $t_{1/2}$ values of 3.1 and 30.2 days for the normal controls, 5.3 and 113.6 days for the obese mice. Both groups were then treated with streptozotocin to produce a diabetic state, and the longer half life corresponding to mobilization of the slow component (hypothesized to correspond to white adipose tissue) dropped by an order of magnitude for both the streptozotocin-treated controls and the streptozotocin-treated obese (MSG treated) mice, indicating dramatically increased rates of lipid utilization. The authors conclude that (1) obesity is associated with decreased utilization of existent body fat stores; (2) there is no significant difference in the change in fat utilization between obese and nonobese mice upon induction of (streptozotocin) diabetes, suggesting that there is no impairment of fat metabolism in obese mice. However, since in the obese mice the total quantity of fat and the corresponding intensity of the lipid CHD/CD_2 resonance are substantially larger than in the normal controls, the number of fat molecules utilized per unit time may not be that different from the value in the controls. Since the HDO resonance is initially increased and may also change as deuterated lipid is utilized, this resonance could not be used as an internal standard in these studies, and an external standard containing 5% D_2O/H_2O was used.

3.4. Metabolism of D- and L-Methionine

Although methyl groups represent an attractive target for deuterium labeling due to the linewidth reduction which can be calculated to result from the internal motion of the methyl group, relatively little work has been done using methyl labeled precursors. The hepatic metabolism of excess [*methyl*-$^2H_3$]-L- or -D-methionine has been studied in anesthetized rats using surface coils (London *et al.*, 1987; London and Gabel, 1988). Excess dietary methionine has been reported to be the most toxic of the naturally occurring amino acids, and the catabolic pathway has been the object of some controversy. Catabolism initially involving either transamination to give the corresponding α-keto acid, or transmethylation via the formation of *S*-adenosylmethionine have been proposed to be significant (Fig. 4). Deuterium NMR spectra of the liver, obtained by placing a surface coil over the liver of rats in which the portion of muscle directly over the organ had previously been removed and the skin subsequently sutured (London *et al.*, 1985), exhibit resonances corresponding to background HDO and CHD peaks. Subsequent to I.V. administration of [*methyl*-$^2H_3$]-L-methionine a resonance is observed at the expected shift of 2.1 ppm (using the HDO signal as an internal reference), and a second metabolite, subsequently identified as [*methyl*-$^2H_3$]sarcosine (2.7 ppm), is rapidly formed (Fig. 5). On a slower time scale, an additional deuterium resonance at 3.15 ppm arising from *N*-methyl metabolites is observed, so that after a period of 24 hr, only this resonance remains. These studies established a rapid flux through the transmethylation pathway for the catabolism of excess methionine. A gradual increase in the intensity of the HDO resonance was observed, presumably reflecting the oxidation of the deuterated sarcosine by mitochondrial sarcosine dehydrogenase. As discussed in the study,

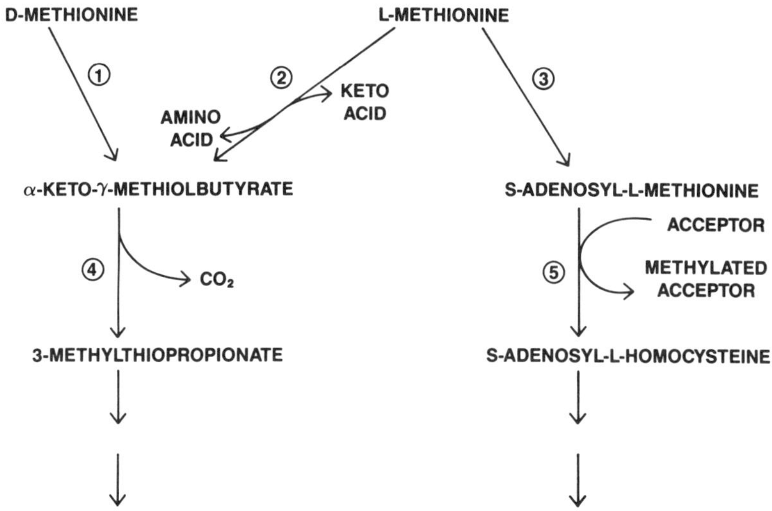

Figure 4. Proposed metabolic pathways for the catabolism of D- and L-methionine. Enzymatic steps indicated correspond to: (1) D-amino acid oxidase, (2) transamination (glutaminase II), (3) methionine adenosyl transferase, (4) oxidation of α-keto-γ-methylthiobutyrate, and (5) glycine methyl transferase. (From London and Gabel, 1988, reprinted with permission from *Biochemistry* **27**:7864–7869, copyright © 1988 American Chemical Society.)

deuterium isotope effects are unlikely to significantly alter the rate of the transmethylation reactions, but probably are significant for sarcosine dehydrogenase. If the availability of glycine becomes rate limiting, this will significantly retard the overall kinetics. It was further demonstrated that pretreatment of the rats with ethionine, the *S*-ethyl analogue of methionine, dramatically reduced the rate of transmethylation of [*methyl*-$^2H_3$]-L-methionine.

Subsequent studies with [*methyl*-$^2H_3$]-D-methionine were carried out (London and Gabel, 1988). Since D-methionine is not a substrate for methionine adenosyl transferase (Lombardi and Talalay, 1970), it must initially be deaminated to give the α-keto acid via D-amino acid oxidase (Fig. 4). Nevertheless, analysis of the hepatic metabolism by ^{2}H NMR yielded spectra which were essentially identical with those observed in the labeled L-methionine study. This result was interpreted to indicate that although D-methionine is initially deaminated to the α-keto acid, it is not subsequently catabolized to a significant extent, but is transaminated to yield L-methionine. This process occurs with sufficient flux that it is apparently not rate limiting, since the kinetics of disappearance of the labeled methionine peak and formation of the labeled sarcosine and HDO are essentially identical to the observations in the L-methionine study. As a further test of this conclusion, rats were pretreated with sodium benzoate,

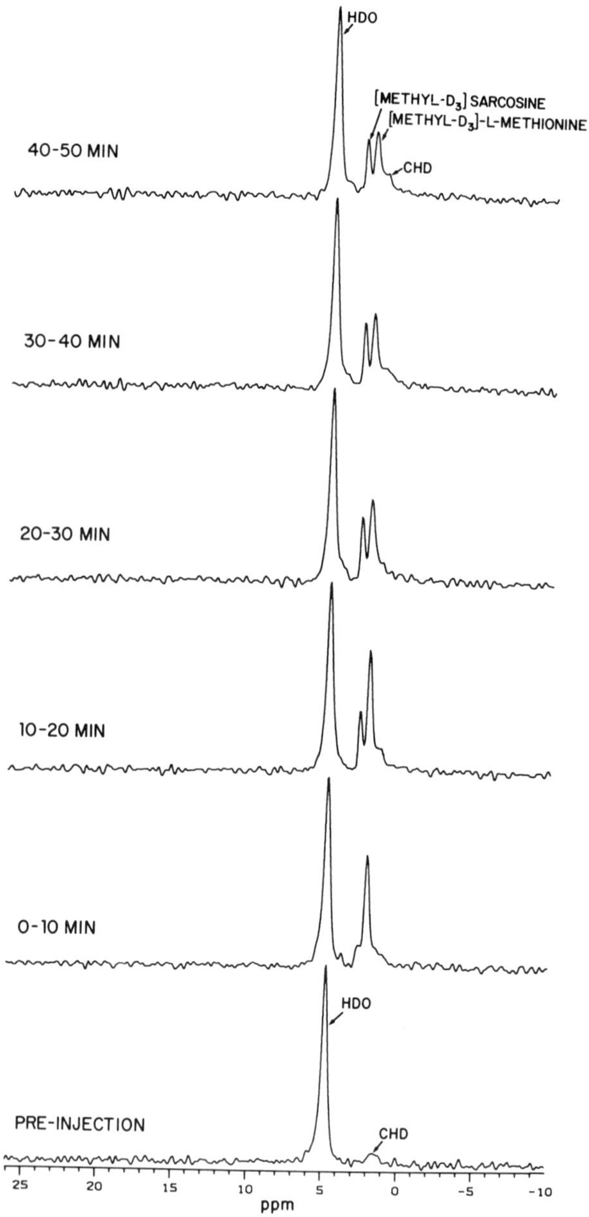

Figure 5. A series of ^2H NMR spectra obtained by placing a surface coil over the liver of a rat surgically modified as described by London *et al.* (1985, 1987). The lower trace corresponds to the background resonances assigned as indicated to HDO and CHD groups of lipids. The [*methyl*-$^2H_3$]-L-methionine was introduced via an intravenous cannula at time zero. All spectra have been resolution enhanced using double-exponential multiplication and a parameter of 10 Hz. (From London *et al.*, 1987, reprinted with permission from *Biochemistry* **26**:7166–7172, copyright © 1987 American Chemical Society.)

an inhibitor of D-amino acid oxidase, and subsequently given [*methyl*-$^2H_3$]-D-methionine. The disappearance of the methionine methyl resonance, and the appearance of labeled sarcosine and HDO were all greatly retarded relative to the untreated animals (Fig. 6). It was also found that rats pretreated with sodium benzoate and subsequently dosed with [*methyl*-$^2H_3$]-L-methionine showed a rapid formation of labeled sarcosine, but slower clearance of labeled methione and no observable progressive enrichment of the water with deuterium. The differences in the rates of formation of sarcosine in these studies are consistent with the action of sodium benzoate as an inhibitor of D-amino acid oxidase, and with the metabolic flux of D-methionine noted above. However, the differences in L-methionine clearance and water labeling which are found between the control and sodium benzoate-treated rats indicate that the sodium benzoate exerts additional metabolic effects on the hepatic metabolism of the rat. It is interesting to note that in combination, these studies show that while the hepatic catabolism of L-methionine in the rat does not involve a significant flux in the transamination pathway, this pathway is extremely active in the reamination of the α-keto acid formed as a result of the oxidation of D-methionine. Thus, the flux through this transamination pathway under physiological conditions is essentially one way. Such metabolism is presumably a consequence of the involvement of the glutaminase II system which transfers the α-amino group of glutamine (Cooper and Meister, 1972). As noted in Section 2, concentrations can be estimated from data such as those shown in Fig. 6 by comparison of the intensities of the metabolite and initial HDO resonances, using a correction factor of 1/3 for the methionine and sarcosine, and 1/9 for the *N*-trimethyl-labeled metabolites based on the assumption that these are formed from three methyl groups derived from the labeled methionine.

3.5. Choline Metabolism

The distribution of deuterated choline in mice fed a diet containing [*methyl*-$^2H_9$]- or [1,2-$^2H_4$]-choline was studied by Hanin and Schuberth (1974), but no NMR data were obtained. The ratio of deuterated to unlabeled choline in plasma was found to increase from 0.22 on day 1 after initiation of a diet containing deuterated choline, to 0.57 on day 14. The mole ratio of deuterated/unlabeled acetyl choline in brain was slightly lower than that of plasma. Oldfield *et al.* (1976) reported ^{2}H NMR studies of chemically transformed mouse fibroblast LM cells, and of mitochondria from Sprague-Dawley rats fed a diet containing [*methyl*-$^2H_9$]choline. Two sets of signals, one corresponding to free labeled choline plus water, and a second to the choline headgroups of phospholipids, were observed. The free choline

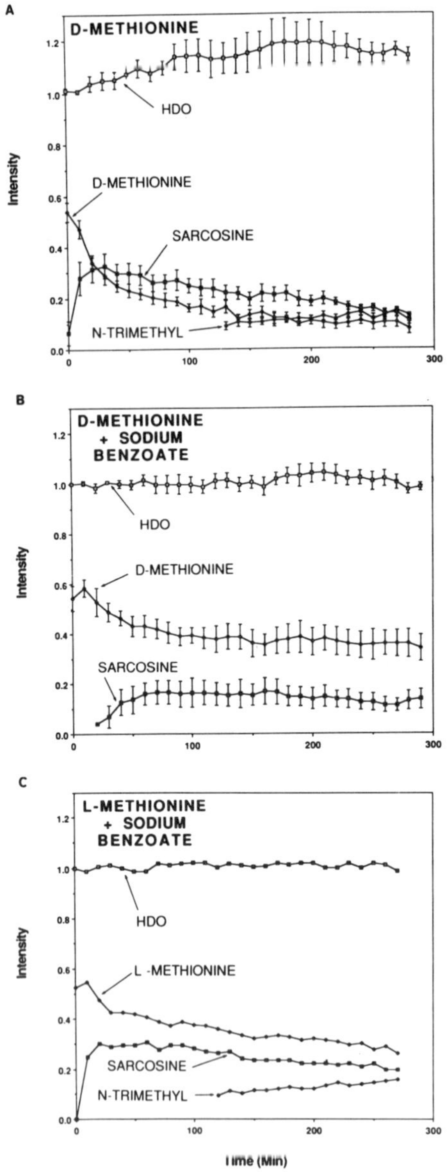

Figure 6. Relative intensities, normalized with respect to the intensity of the initial HDO resonance in each study, of the ^{2}H resonances of observed metabolites plotted as a function of time after the injection of deuterated methionine via intravenous cannula. (A) Rats dosed with [*methyl*-$^2H_3$]-D-methionine; (B) rats dosed with [*methyl*-$^2H_3$]-D-methionine 1 hr subsequent to receiving an intraperitoneal dose of 0.9 mg/kg sodium benzoate; (C) rats dosed with [*methyl*-$^2H_3$]-L-methionine 1 hr subsequent to receiving an intraperitoneal injection of sodium benzoate. Observed metabolites are total (D + L) methionine (◆), sarcosine (■), *N*-trimethyl-deuterated compounds (○), and HDO (□). (Adapted from London and Gabel, 1988.)

observed in his study may actually correspond to other small metabolites, particularly phosphorylcholine, as observed in ^{13}C NMR studies of suspended CHO cells grown in media containing [*methyl*-^{13}C]choline (London *et al.*, 1976) and in ^{13}C NMR studies of *T. cruzi* grown in a medium containing [1-^{13}C]glucose (Frydman *et al.*, 1990). Eng *et al.* (1990) have studied the metabolism of [*methyl*-$^2H_9$]choline in the rabbit kidney. Deuterium NMR studies indicated a linear relationship between choline uptake and the rate of choline infusion, with a detection sensitivity limit of 100 μM over a 2-min data collection period. The much longer T_2 of the labeled deuterated methyls compared with HDO allowed selective imaging of these metabolites by choosing an echo time intermediate between the water and methyl T_2 values. Deuterium imaging studies of the kidney *in vivo* after infusion of deuterated choline showed high trimethylamine concentrations in both cortex and inner medulla, but not in outer medulla. The ^{2}H methyl resonances corresponding to betaine and choline could not be resolved *in vivo*, but were analyzed in extracts. Deuterium images obtained following infusion of furosemide (a loop diuretic) show a washout of inner medullary (deuterated) trimethylamines. Analysis of extracts revealed that cortex betaine:choline ratios increased from 0.69 ± 0.05 to 3 ± 1 after furosemide treatment, indicating that much of the choline observed was present in the urine. Thus, the ^{2}H image after furosemide administration provided a map of kidney betaine distribution.

3.6. Metabolism of Formaldehyde

Although *in vivo* deuterium NMR should be a useful approach to study xenobiotic metabolism, relatively little work has been described. Sanders and co-workers have carried out an elegant series of studies on the metabolism of formaldehyde (a generally potent antibacterial agent) by microorganisms. Proton, ^{13}C, and ^{2}H NMR studies carried out on suspensions of *E. coli* exposed to ^{13}C-labeled formaldehyde or to [$^2H_2$]formaldehyde have demonstrated the formation of methanol, formate, CO_2, and the three-carbon alcohols glycerol, 1,2-propanediol, and 1,3-propanediol (Hunter *et al.*, 1984, 1985). Deuterium NMR studies demonstrated that formate and methanol are produced in nearly equal quantities, and that the labeling pattern obtained corresponded to DCO_2^- and CD_2HOH, where in the latter the coupling of the single methyl proton to the two deuterons could be observed (Hunter *et al.*, 1984). Subsequent ^{2}H NMR studies in *Pseudomonas putida* F61 and *Staphylococcus aureus* have focused on a search for a formaldehyde dismutase or "Cannizzarase" activity in the metabolism of formaldehyde (Mason and Sanders, 1989). The dismutase activity transferring a hydrogen or deuterium atom from one formaldehyde molecule to another would generate [^{2}H]formate and [$^2H_3$]methanol from two molecules of [$^2H_2$]formaldehyde according to

$$2CD_2(OH)_2 \rightarrow DCO_2H + CD_3OH + H_2O \quad (3)$$

where, based on the observed spectral shift of 4.8 ppm, formaldehyde is written as the hydrated form. In contrast, the methanol produced by the independent action of oxidizing and reducing enzymes, each using NADH or a related cofactor, would contain only two deuterons:

$$CD_2(OH)_2 + NAD^+ \rightarrow DCO_2H + NADD \quad (4)$$

$$CD_2(OH)_2 + NADH \rightarrow CD_2HOH + NAD^+ \quad (5)$$

provided that the pool of NADH is large enough not be be significantly perturbed by the NADD produced in the first reaction (4) above. Data obtained with *E. coli*, *P. putida*, and *S. aureus* are shown in Fig. 7, and demonstrate the dominance of pathway (4, 5) for *E. coli*, and pathway (3) for the latter two organisms. Analogous studies of the metabolism of formaldehyde by the potato (*Solanum tuberosum*) support contributions of both pathways, in a ratio of 3 : 1 (Mason *et al.*, 1986). Studies in which both [$^2H_2$]formaldehyde and a second aldehyde were present, demonstrated deuteride transfer to the second aldehyde via a generalization of pathway (3):

$$CD_2(OH)_2 + RCHO \rightarrow DCO_2H + RCDHOH \quad (6)$$

In these studies, [1-^{2}H]propanol, but not propionate, was formed from propionaldehyde, demonstrating that formaldehyde is heavily favored as the hydride donor and that propionaldehyde is the preferred acceptor. Pyruvaldehyde was reduced exclusively at the aldo and not the keto group to yield [1-^{2}H]hydroxyacetone. Cross dismutation was also observed for benzaldehyde, *n*-butyraldehyde, and acrolein, but not for acetone, crotonaldehyde, trichloroacetaldehyde, DL-glyceraldehyde, or isobutyraldehyde (Mason and Sanders, 1989).

A mixed isotope experiment using ^{13}C and [$^2H_2$]formaldehyde was carried out in order to detect cross dismutation (Fig. 8). Three of the predicted isotopomers were observed by ^{2}H NMR, and the fourth by ^{13}C NMR. Carbon-13 NMR analyses of studies using equimolar ^{13}C and [$^2H_2$]formaldehyde indicated that [$^{13}CH_3OH$] > [$^{13}CH_2DOH$] and ^{2}H NMR indicated that [CD_2HOH] > [CD_3OH] = [$^{13}CH_2DOH$], indicating a significant kinetic isotope effect. Computer simulations gave a value of 3.7 ± 0.5 for k(H transfer)/k(D transfer).

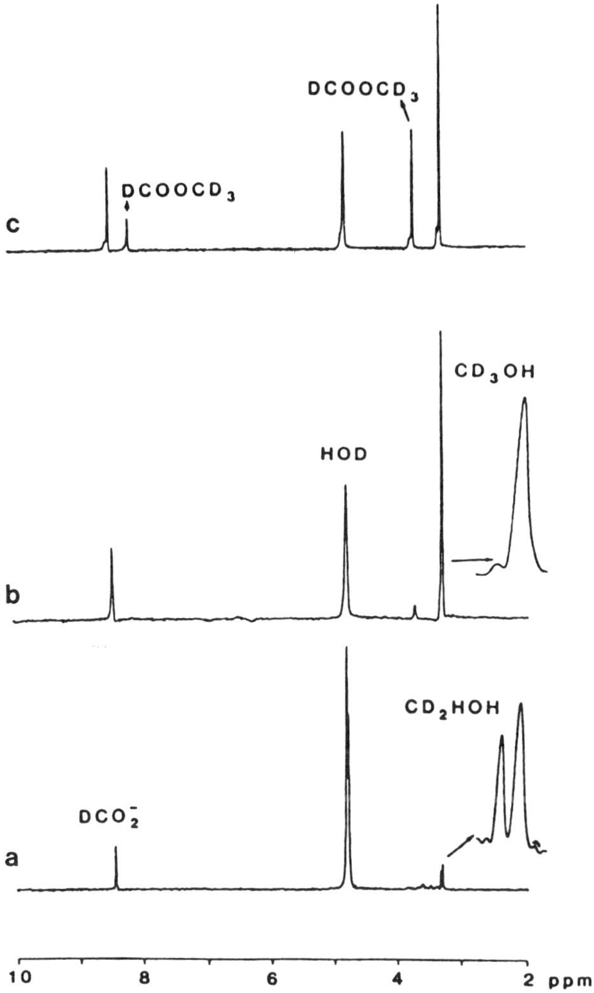

Figure 7. Proton-decoupled ^{2}H NMR spectra (61.4 MHz) of suspensions of (a) *E. coli*, (b) *P. putida*, and (c) *S. aureus* acquired during anaerobic metabolism of 10 mM D_2-formaldehyde. One hundred transients were acquired over 5 min. Gaussian resolution enhancement was applied. (From Mason and Sanders, 1989, reprinted with permission from *Biochemistry* **28**:2160–2168, copyright © 1989 American Chemical Society.)

4. MAGNETIC RESONANCE IMAGING AND TISSUE PERFUSION

The potential utility of deuterium for magnetic resonance imaging studies and for studies of blood flow and tissue perfusion have been evaluated recently (Muller and Seelig, 1987; Ewy *et al.*, 1986, 1988). Ewy

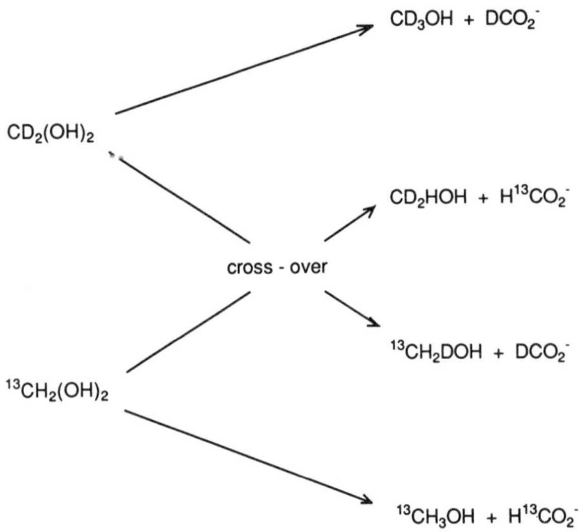

Figure 8. Cross-dismutation reaction scheme for measurement of k_H/k_D for hydride transfer in the metabolism of a mixture of D_2- and H_2-formaldehyde. (Adapted from Mason and Sanders, 1989.)

et al. (1986) presented an ^2H image of the hyperkalemia-arrested rabbit heart perfused with a cardioplegic solution in D_2O. Muller and Seelig (1987) obtained a deuterium image of the head of an anesthetized rat which had been fed drinking water containing 5% (v/v) D_2O for a period of 5 weeks, and Ewy *et al.* (1988) obtained ^2H images of a cat brain *in vivo* after an analogous enrichment of the ^2H level to 4–5%. As pointed out by these authors, the sensitivity loss resulting from the lower intrinsic sensitivity of ^2H (1% relative to a proton at constant field) and lower abundance is partially offset by the shorter T_1 values of HDO. In order to fully exploit this shorter T_1 value, Muller and Seelig used a steady-state free precession imaging technique developed by Frahm *et al.* (1986). Subsequent studies of plants aimed at optimizing deuterium imaging strategies have shown that concentrations as low as 20 mM can be imaged at a spatial resolution of 5 mm (Link and Seelig, 1990). Ewy *et al.* (1988) suggest that deuterium T_2 variations for different tissues may be significant for HDO and that T_2 weighted images may provide contrast which differs from that provided by ^1H imaging. The selective imaging of deuterated trimethylamines, which was accomplished based on their relatively long T_2 value, was noted above (Eng *et al.*, 1990).

The use of D_2O as a tracer to study flow and tissue perfusion has also been discussed (Ackerman *et al.*, 1987a, b; Muller and Seelig, 1987). In order to deal with the problem of recirculation of the deuterium label, a

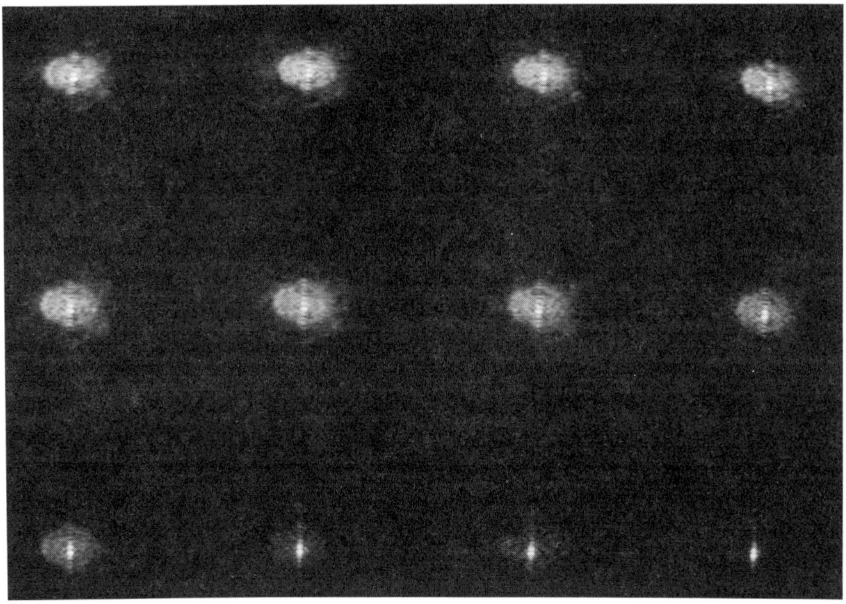

Figure 9. *In vivo* deuterium images of a rat's head subsequent to an intraperitoneal infusion of 20 ml D_2O. The sequence is from upper left to lower right, with image acquisition beginning every 10 min and corresponding to a total acquisition time of 6.6 min per image. The bright spot in the center is an artifact. (Reprinted with permission from Muller and Seelig, *J. Magn. Reson.* **72**:456–466, 1987.)

two-compartment series flow model (Ackerman *et al.*, 1987b), and ultimately multicompartment models (Kim and Ackerman, 1988) have been developed. Deuterium images obtained subsequent to D_2O administration (presumably, containing physiological levels of salts to avoid osmotic shock) can potentially provide spatially localized information on blood flow and tissue perfusion (Fig. 9). Since tissue oxygenation is required for normal aerobic metabolism, the availability of flow and tissue perfusion data can provide information which is useful for the interpretation of other NMR observations of metabolic activity, particularly in tumors or injured areas which are poorly perfused.

5. FUTURE DEVELOPMENT

In vivo deuterium NMR measurements have been made on a broad range of systems and provided structural, metabolic, distribution, and flow-related information. The ease with which *in vivo* ^{2}H NMR studies may be carried out is the most important advantage of this approach, and in the experience of this reviewer has resulted in data which are quantitatively more reproducible than studies using either ^{31}P or ^{13}C. The use of ^{2}H NMR in metabolic studies is somewhat less general than the analogous use of ^{13}C, as a consequence of the inherent lability of some positions, the relatively broad linewidths which can be obtained, the limited chemical shift range, and the existence of significant deuterium isotope effects. As noted by Mason and Sanders (1987b), the latter effects not only can alter the apparent rates derived from studies of deuterated molecules, but can also alter the relative flux through competing pathways by favoring conversions with no or smaller isotope effects. Data outlined above suggest that the use of methyl labels can be a particularly useful strategy, since the internal diffusion of the methyl group will reduce the line broadening problem. As is generally the case, the limitations noted above on the use of deuterium as a metabolic tracer also provide opportunities for obtaining other kinds of information. The analysis of ^{2}H isotope effects can provide useful information related to enzyme mechanism and the identification of rate-limiting enzymatic steps. The loss of tritium from tritiated metabolites such as [2-^{3}H]glucose (Katz *et al.*, 1979) and [3-^{3}H]lactate (Rognstad and Wals, 1976) at specific metabolic conversions has been utilized to quantitate flux through these pathways, and analogous applications of deuterated molecules exist. Deuterated glucose has been utilized to study glucose turnover in humans, utilizing GC-MS analysis (Martineau *et al.*, 1985).

New NMR approaches may also extend the range and information content of *in vivo* metabolic studies. Rinaldi and Baldwin (1982) have provided a $^{13}C\{^{2}H\}$-INEPT pulse sequence for the selective detection of deuterated carbons, and an experimentally less demanding spin–echo sequence has been developed by Wesener *et al.* (1984) for distinguishing carbon resonances on the basis of the deuterium labeling. Recent *in vivo* ^{3}H NMR metabolic studies utilizing tritiated metabolites (Newmark *et al.*, 1990) offer an analogous information content with considerably improved resolution compared with ^{2}H, although with the added difficulties resulting from high specific activity and the associated radioactivity.

ACKNOWLEDGMENTS. Useful discussions with Drs. Bruce Berkowitz, Jim Brainard, Elizabeth Murphy, and Don Davis on the subject of this review are greatly appreciated.

REFERENCES

Abell, C., 1986, *Modern Methods Plant Anal.* **2**:60.

Ackerman, J. J. H., Ewy, C. S., Becker, N. N., and Shalwitz, R. A., 1987a, *Proc. Natl. Acad. Sci. USA* **84**:4099.

Ackerman, J. J. H., Ewy, C. S., Kim, S.-G., and Shalwitz, R. A., 1987b, *Ann. N.Y. Acad. Sci.* **508**:89.

Aguayo, J. B., McLennan, I. J., Aguiar, E., and Cheng, H.-M., 1987, *Biochem. Biophys. Res. Commun.* **142**:359.

Aguayo, J. B., Gamcsik, M. P., and Dick, J. D., 1988a, *J. Biol. Chem.* **263**:19552.

Aguayo, J. B., McLennan, I. J., Graham, C., Jr., and Cheng, H.-M., 1988b, *Exp. Eye Res.* **47**:337.

Bailey, J. M., Fishman, P. H., and Pentchev, P. G., 1968, *J. Biol. Chem.* **243**:4827.

Barrow, K. D., Rogers, P. L., and Smith, G. M., 1986, *Eur. J. Biochem.* **157**:195.

Baxter, R. L., Mackenzie, N. E., and Scott, A. I., 1983, *Biol. Magn. Reson.* **5**:1.

Block, R. E., and Parekh, B. C., 1987, *Magn. Reson. Med.* **5**:286.

Borle, F., and Seelig, J., 1983, *Biochim. Biophys. Acta* **735**:131.

Brereton, I. M., Irving, M. G., Field, J., and Doddrell, D. M., 1986, *Biochem. Biophys. Res. Commun.* **137**:579.

Brereton, I. M., Doddrell, D. M., Oakenfull, S. M., Moss, D., and Irving, M. G., 1989, *NMR Biomed.* **2**:55.

Brindle, K. M., Campbell, I. D., and Simpson, R. J., 1986, *Eur. J. Biochem.* **158**:299.

Brindle, K. M., Campbell, I. D., and Simpson, R. J., 1987, *Biol. Magn. Reson.* **7**:81.

Cooper, A. J. L., and Meister, A., 1972, *Biochemistry* **11**:661.

de los Santos, C., Buldain, G., Frydman, B., Cannata, J. J. B., and Cazzulo, J. J., 1985, *Eur. J. Biochem.* **149**:421.

Eng, J., Berkowitz, B. A., and Balaban, R. S., 1990, *NMR Biomed.* **3**:173.

Evelhoch, J. L., McCoy, C. L., and Girj, B. P., 1989, *Magn. Reson. Med.* **9**:402.

Ewy, C. S., Babcock, E. E., and Ackerman, J. J. H., 1986, *Magn. Reson. Imag.* **4**:407.

Ewy, C. S., Ackerman, J. J. H., and Balaban, R. S., 1988, *Magn. Reson. Med.* **8**:35.

Foster, A. B., 1984, *Trends Pharmacol. Sci.* **5**:524.

Frahm, J., Haase, A., and Matthaei, D., 1986, *Magn. Reson. Med.* **3**:321.

Frydman, B., de los Santos, C., Cannata, J. J. B., and Cazzulo, J. J., 1990, *Eur. J. Biochem.* **192**:363.

Fung, B. M., Durham, D. L., and Wassil, D. A., 1975, *Biochim. Biophys. Acta* **399**:191.

Gatley, S. J., Wess, M. M., Govoni, P. L., Wagner, A., Katz, J. J., and Friedman, A. M., 1986, *J. Nuclear Med.* **27**:388.

Goodman, M. N., Masuoka, L. K., deRopp, J. S., and Jones, A. D., 1989, *Biochem. Biophys. Res. Commun.* **159**:522.

Grant, D. M., Curtis, J., Croasmun, W. R., Dalling, D. K., Wehrli, F., and Wehrli, S., 1982, *J. Am. Chem. Soc.* **104**:4492.

Hanin, I., and Schuberth, J., 1974, *J. Neurochem.* **23**:819.

Hollis, D. P., 1980, *Biol. Magn. Reson.* **2**:1.

Hotchkiss, R. S., Song, S.-K., Ling, C. S., Ackerman, J. J. H., and Karl, I. E., 1990, *Am. J. Physiol.* **258**:R21.

Hunter, B. K., Nicholls, K. M., and Sanders, J. K. M., 1984, *Biochemistry* **23**:508.

Hunter, B. K., Nicholls, K. M., and Sanders, J. K. M., 1985, *Biochemistry* **24**:4148.

Iles, R. A., Stevens, A. N., and Griffiths, J. R., 1982, *Prog. NMR Spectrosc.* **15**:49–200.

Irving, M. G., Brereton, I. M., Field, J., and Doddrell, D. M., 1987, *Magn. Reson. Med.* **8**:88.

Isab, A. A., and Rabenstein, D. L., 1979, *FEBS Lett.* **106**:325.

Jacobs, R. E., and Oldfield, E., 1980, *Prog. NMR Spectrosc.* **14**:113.

Katz, J. J., and McGarry, J. D., 1984, *J. Clin. Invest.* **74**:1901.

Katz, J., Golden, S., and Wals, P. A., 1979, *Biochem. J.* **180**:389.

Kim, S.-G., and Ackerman, J. J. H., 1988, *Magn. Reson. Med.* **8**:410.

Kimmich, R., Gneiting, T., Kotitschke, K., and Schnur, G., 1990, *Biophys. J.* **58**:1183.

Kosicki, G. W., 1968, *Biochemistry* **7**:4310.

Lentner, C., 1981, *Geigy Scientific Tables*, Vol. 1, Ciba-Geigy Corp., West Caldwell, N.J., p. 217.

Leopold, M. F., Epstein, W. W., and Grant, D. M., 1988, *J. Am. Chem. Soc.* **110**:616.

Link, J., and Seelig, J., 1990, *J. Magn. Reson.* **89**:310.
Lombardini, J. B., and Talalay, P., 1970, *Adv. Enzyme Regul.* **9**:349.
London, R. E., 1988, *Prog. NMR Spectrosc.* **20**:337.
London, R. E., and Gabel, S. A., 1988, *Biochemistry* **27**:7864.
London, R. E., Hildebrand, C. E., Olson, E. S., and Matwiyoff, N. A., 1976, *Biochemistry* **15**:5480.
London, R. E., Galvin, M. J., Thompson, M., Jeffreys, L., and Mester, T., 1985, *J. Biochem. Biophys. Methods* **11**:21.
London, R. E., Gabel, S. A., and Funk, A., 1987, *Biochemistry* **26**:7166.
Malloy, C. R., Sherry, A. D., and Jeffrey, F. M., 1988, *J. Biol. Chem.* **263**:6964.
Mantsch, H. H., Saito, H., and Smith, I. C. P., 1977, *Prog. NMR Spectrosc.* **11**:211.
Martin, G. J., Zhang, B. L., Martin, M. L., and Dupuy, P., 1983, *Biochem. Biophys. Res. Commun.* **111**:890.
Martin, G. J., Zhang, B. L., Naulet, N., and Martin, M. L., 1986, *J. Am. Chem. Soc.* **108**:5116.
Martineau, A., Lecavalier, L., Falardeau, P., and Chiasson, J. L., 1985, *Anal. Biochem.* **151**:495.
Mason, R. P., and Sanders, J. K. M., 1989, *Biochemistry* **28**:2160.
Mason, R. P., Sanders, J. K. M., and Gidley, M. J., 1986, *Phytochemistry* **25**:1567.
Mason, R. P., Sanders, J. K. M., and Cornish, A., 1987a, *Biochem. Soc. Trans.* **15**:148.
Mason, R. P., Sanders, J. K. M., and Cornish, A., 1987b, *FEBS Lett.* **216**:4.
Melander, L., and Saunders, W. H., 1980, *Reaction Rates of Isotopic Molecules*, Wiley, New York.
Muller, S., and Seelig, J., 1987, *J. Magn. Reson.* **72**:456.
Newmark, R. D., Un, S., Williams, P. G., Carson, P. J., Morimoto, H., and Klein, M. P., 1990, *Proc. Natl. Acad. Sci. USA* **87**:583.
Oldfield, E., Meadows, M., and Glaser, M., 1976, *J. Biol. Chem.* **251**:6147.
Oxley, S. T., Porteous, R., Brindle, K. M., Boyd, J., and Campbell, I. D., 1984, *Biochim. Biophys. Acta* **805**:19.
Pascal, R. A., Jr., Baum, M. W., Wagner, C. K., Rodgers, L. R., and Huang, D.-S., 1986, *J. Am. Chem. Soc.* **108**:6477.
Peng, S.-K., Ho, K.-J., and Taylor, C. B., 1972, *Arch. Pathol.* **94**:81.
Pohl, L. R., and Gillette, J. R., 1985, *Drug. Metab. Rev.* **15**:1335.
Rinaldi, P. L., and Baldwin, N. J., 1982, *J. Am. Chem. Soc.* **104**:5791.
Rognstad, R., and Wals, P., 1976, *Biochim. Biophys. Acta* **437**:16.
Rose, I. A., 1960, *J. Biol. Chem.* **235**:1170.
Rose, I. A., 1970, in *The Enzymes*, 3rd ed. (Boyer, P. D., ed.), Vol. II, Academic Press, New York, p. 281.
Schmitt, P., Wesener, J. R., and Gunther, H., 1983, *J. Magn. Reson.* **52**:511.
Seelig, J., and Macdonald, P. M., 1987, *Acc. Chem. Res.* **20**:221.
Simpson, R. J., Brindle, K. M., Brown, F. F., Campbell, I. D., and Foxall, D. L., 1981, *Biochem. J.* **193**:401.
Simpson, R. J., Brindle, K. M., Brown, F. F., Campbell, I. D., and Foxall, D. L., 1982, *Biochem. J.* **202**:581.
Stepuro, I. I., Moroz, A. R., and Piletskaya, T. P., 1986, *Biokhimiya* **51**:729.
Walker, T. E., and Wageman, W. W., 1986, *Magn. Reson. Chem.* **24**:157.
Wesener, J. R., and Gunther, H., 1983, *Org. Magn. Reson.* **21**:433.
Wesener, J. R., Schmitt, P., and Gunther, H., 1984, *J. Am. Chem. Soc.* **106**:10.
Wheeler, W. D., Kaizaki, S., and Legg, J. I., 1982, *Inorg. Chem.* **21**:3250.
York, M. J., Kuchel, P. W., Chapman, B. E., and Jones, A. J., 1982, *Biochem. J.* **207**:65.

7

Some Applications of ESR to *in Vivo* Animal Studies and EPR Imaging

Lawrence J. Berliner and Hirotada Fujii

1. INTRODUCTION

Paramagnetic centers, such as naturally occurring free radicals and paramagnetic metal ions, are of crucial importance in many biological functions. Some examples are flavoenzymes, vitamin B12, aging, oxygen radicals, and melanins. While work in the early days of ESR was limited by the size and electronic constraints of the X-band cavity, as well as the need, frequently, to freeze-clamp the sample or excised tissue, an "ideal" *in vivo* ESR spectrometer is one in which a live animal can be directly accommodated (Swartz *et al.*, 1972). These latter experiments were virtually impossible with commercial X-band (9.5 GHz) instruments as the large electric field losses through water and ions would result in "cooking the animal" besides the extreme difficulty in matching the sample. Conventional X-band EPR spectrometers present several technical limitations to measurements on live animals, principally microwave heating. Hence, the first attempts at EPR imaging were limited to very small nonliving samples.

The requirements for animal studies are thus twofold: a resonant device (analogous to the X-band cavity) with dimensions suitable to fit a whole animal; and a choice of an operating frequency that does not produce dielectric heating. An important methodological breakthrough involved the

---

Lawrence J. Berliner and Hirotada Fujii • Department of Chemistry, The Ohio State University, Columbus, Ohio 43210. *Present address of Dr. Fujii:* Tokyo Metropolitan Medical Research Institute, Tokyo, Japan.

Biological Magnetic Resonance, Volume 11: In Vivo Spectroscopy, edited by Lawrence J. Berliner and Jacques Reuben. Plenum Press, New York, 1992.

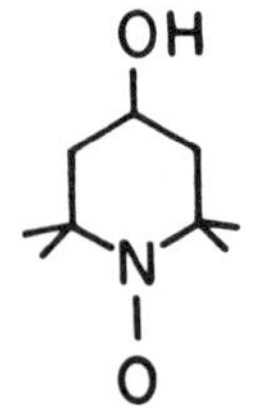

TEMPOL

Figure 1. TEMPOL (4-hydroxy-2,2,6,6-tetramethyl piperidine-1-oxyl).

utilization of low-field, low-frequency, instruments operating in the 1–2 GHz (L-band) region or lower where dielectric losses were reduced significantly (Lukiewicz and Lukiewicz, 1984; Fujii and Berliner, 1986). Second was the range of resonant devices developed, from surface coils to resonant cavities to loop gap resonators (reviewed in Koscielniak and Berliner, 1991).

The field of *in vivo* ESR and EPR imaging blossomed with these instrumental advances beginning in 1984–1985 with the reports of Lukiewicz and Lukiewicz (1984), and Berliner and Fujii (1985). Besides the Laboratory of In-Vivo Electron Spin Resonance Spectroscopy (LIVERS) at Ohio State, significant developments have also come out of the Illinois ESR Center (Harold M. Swartz), The Medical College of Wisconsin ESR Center (James S. Hyde), and The University of Chicago Medical School (Howard Halpern). Most of the work reviewed here comes out of the LIVERS laboratory; however, the reader is encouraged to refer to a recent text which addresses progress through the end of 1990 by most of the laboratories noted above, as well as others (Eaton *et al.*, 1991).

Two-dimensional EPR images of plant materials have been reported using single-turn flat-loop surface coils (Berliner and Fujii, 1985; Fujii and Berliner, 1985, 1986; Nishikawa *et al.*, 1985). The first report of the imaging of diffusional processes by EPR was also shown at L-band for an aqueous solution of the nitroxide TEMPOL (shown in Fig. 1) diffusing into a polyacrylamide gel rod (Berliner and Fujii, 1986). This led to the realization that similar experiments on live animals should be feasible (Berliner, 1991; Berliner *et al.*, 1987). The experiments outlined here represent extensions of these feasibility experiments to problems relevant to metabolic and pathological states in animals.

2. TUMOR IMAGING: SUCCESSES, PROBLEMS, AND PITFALLS

It has been of much interest in experimental oncology to unravel the internal structure of *in situ* growing tumors. The first report on *in vivo* ESR on implanted tumors originated with the report of Lukiewicz and Lukiewicz (1984), then at The Medical College of Wisconsin where the loop gap resonator had been developed, at X-band and subsequently at lower frequencies. In these experiments, the paramagnetic material detected was a naturally occurring product, i.e., the radical melanin species biosynthesized in the implanted melanoma tumor (Lukiewicz and Lukiewicz, 1984). In order to both increase sensitivity and visualize vascularization (i.e., aqueous environment) by EPR, it was necessary to introduce stable nitroxide "imaging agents." In initial feasibility studies, Berliner *et al.* (1987) implanted a dense suspension of Cloudman S-91 melanoma cells into the tail of DBA-2J mice (20 g average weight). After 4 to 6 weeks, the tumor size approached 6–9 mm in diameter. The animal was subsequently "prepared" by five serial 0.2 ml injections of an ~50 mM solution of the nitroxide imaging agent CTPO (3-carboxamido-2,2,5,5-tetramethylpyrroline-1-oxyl) or its saturated pyrrolidine analogue over a 6- to 8-hr period. The serial injections were necessary over several hours in order to build up a sufficient level of nitroxide in the tumor for detection and image reconstruction over a 1- to 2-hr period. The low-field ESR imaging spectrometer operated at 1.55 GHz with a 10-mm-i.d. flat-loop single-turn coil which was placed concentrically over the tumor by threading the loop over the tail. The ESR spectrum (i.e., zero gradient) of the CTPO in the melanoma tumor in the anesthetized rat is shown in Fig. 2. This spectral lineshape was identical whether measured before, during, or after collection of the gradient projection spectra. This spectrum (Fig. 2) reflects a highly mobile nitroxide

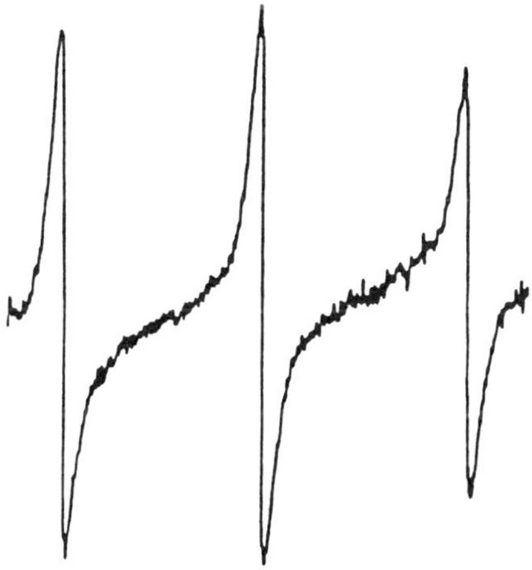

Figure 2. Zero gradient EPR spectrum of CTPO in the melanoma tumor. The tumor was surrounded cross-sectionally by a 10-mm-i.d. flat-loop coil. The microwave frequency was 1.55 GHz; applied microwave power 100 mW; applied static field 0.056 T; modulation frequency 100 kHz; scan time 2 min. (From Berliner *et al.*, 1987.)

molecule dissolved in aqueous solvent; i.e., it is not interacting with the tumor tissue, but represents instead the fluid distribution in the tumor. This also allowed us to assume a single (noninteracting) lineshape throughout the object. Figure 3 depicts gradient projection spectra for four orientations of the tumor in a single field gradient. The arrow denotes the orientation with respect to the **z**-field gradient. While the first feasibility study developed an image produced from only four 45° projections, it was the intention to go to higher resolution with an 18-projection image. Considering that each projection requires two or more time-averaged 2-min scans for the CW ESR measurement, it required almost 1 hr to collect the full 18 projections. Consequently, six 30° projections (18-min) were measured, then staggered by 10° for the second and third sets of six projections each. Each set of six projections was reconstructed by filtered back-projection as shown in Fig. 4A, B, and C, respectively. Figure 4D, which is the composite 18-projection image, does not reflect the individual "time images," 4A, B, C. Note how the lower left intensity feature in 4A, B, and C (presumably a highly vascularized region) decreases more rapidly with time versus the upper right

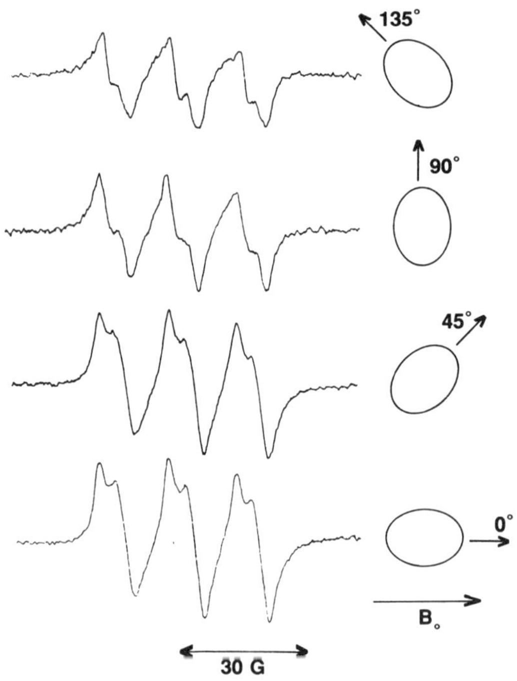

Figure 3. EPR gradient projection spectra of the tumor in Fig. 2. The arrow denotes the tumor orientation with respect to the **z**-field gradient direction. The **z**-field gradient strength was 1.1 G/mm in this example. All other conditions were as in Fig. 2. (From Berliner *et al.*, 1987.)

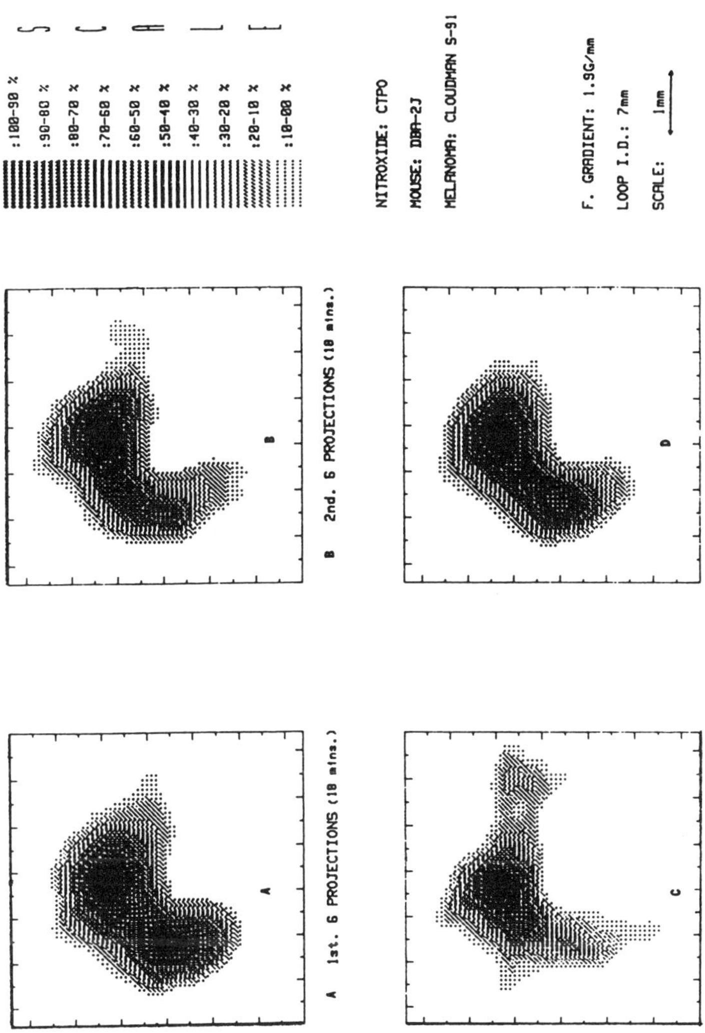

Figure 4. EPR images of a Cloudman S-91 melanoma tumor with time. Six projections were measured during an 18-min period for each image A, B, C, successively. Image D represents a composite of the entire 18 projections collected over the 54-min period. The **z**-field gradient strength was 1.9 G/mm. All other conditions were similar to those in Fig. 3. (From Berliner, 1991.)

feature. Some aspects that might contribute to differences in nitroxide concentrations in various parts of this tumor are:

1. Different mechanisms of nitroxide transport (via blood circulation and diffusion)
2. Different vascularization within the tissue
3. Differential metabolic and nitroxide reduction/oxidation rates within the tissue.

It is suggested that the first two contributions above do not contribute significantly to the differences observed in the time images 4A, B, and C since the animal was essentially "saturated" with the nitroxide imaging agent and especially since diffusion, transport, and recirculation processes are quite rapid compared with the ~18-min periods between each image. On the other hand, the predominant bioreduction of the nitroxide moiety to the diamagnetic hydroxylamine occurs with half-times of the order of 12–15 min *in vivo* as we shall discuss in more detail below (Berliner and Wan, 1989). Thus, even with the current time resolution of the *in vivo* ESR technique, one can distinguish dynamic physicochemical processes within a tumor image.

Some of the problems, however, are resolution and sensitivity. The details of these images are far from the resolution and discrimination of a histological slide. This would require enormously high field gradients, about 10–100 times larger. While these have been achieved at X-band, the physical requirements at L-band present some obstacles for *in vivo* experiments. Some of these sensitivity problems should be overcome soon with development of higher-performance spectrometers and the use of isotopically labeled nitroxides (^{2}H, ^{15}N) which result in much narrower linewidths, hence larger signal-to-noise ratios.

3. *IN VIVO* ESR PHARMACOKINETICS

The use of nitroxide reagents has been demonstrated previously as a physical handle for studying biological materials by ESR. However, as alluded to earlier in Fig. 4, there are nitroxide metabolic processes that occur, namely reduction of the *paramagnetic* nitroxide moiety to its diamagnetic hydroxylamine. The reduction of nitroxides has been well documented in cells (Swartz *et al.*, 1986) and is believed to be mediated by several processes intracellularly, including ascorbic acid, glutathione, and electron transfer enzymes of the mitochondria. In another approach to *in vivo* ESR, we may take advantage of the information available in the nitroxide reduction behavior *in vivo* and observe localized oxidation–reduction metabolism *in vivo*. In order to understand this redox chemistry *in vivo*, Berliner and Wan (1989) followed the metabolism and distribution of the paramagnetic probe, CTPO, in the tail blood circulation of Sprague–Dawley rats (250–300 g) with time.

In a typical pharmacokinetics experiment, a bolus of a "drug" is introduced intramuscularly or intravenously (I.V.) after which it is distributed over the animal into one or more "compartments." The distribution phase requires up to 15–20 min for an intramuscular or intraperitoneal injection in the rat, while an I.V. bolus is evenly distributed within 1–2 min. The data for the decrease in concentration of the drug (i.e., nitroxide) with time are plotted in a semilog fashion. When multiple injections (i.e., consecutively spaced injections) are made the pattern repeats itself, presumably since the compartment approximates an infinite reservoir. A semilog plot for a multiple-injection single-compartment model thus gives a sawtooth ("seesaw") pattern.

The rats were catheterized in the jugular vein and then given serial doses of CTPO nitroxide. The results are shown in Figs. 5–8. Figure 5 depicts the direct ESR spectral data obtained for an anesthetized rat after

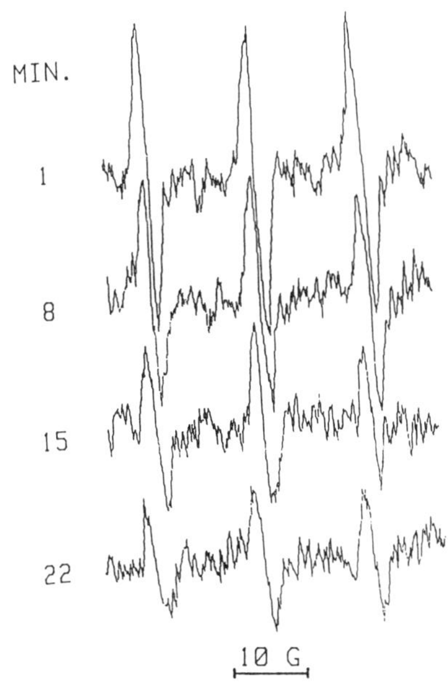

Figure 5. A typical time course of the *in vivo* L-band ESR spectrum for a single (30-sec) I.V. 1.0-ml injection of 75 mM CTPO just prior to spectral collection. The anesthetized rat was placed with the tail in the sample arm of a rectangular two-arm one-entrance cavity resonator. The microwave frequency was 1.4 GHz, applied power 100 mW, applied field 500 G, modulation frequency 100 MHz, scan width 100 G, scan time 1 min. (From Berliner and Wan, 1989.)

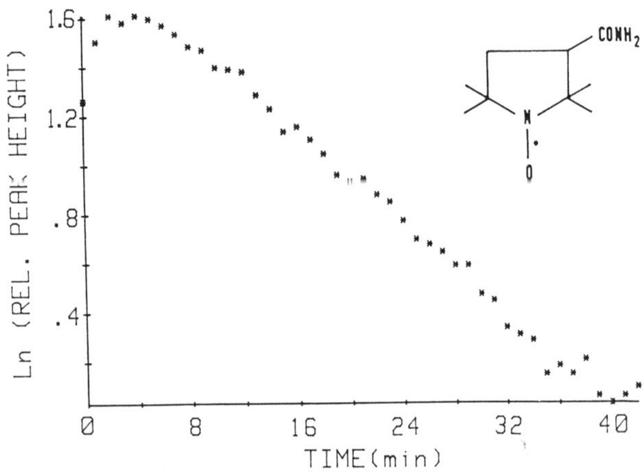

Figure 6. Semilog plot of the ESR signal intensities versus time (averaged over the three lines in each spectrum in Fig. 5). The $t_{1/2}$ was 18 ± 2 min for a typical single-injection study of 15 rats. The data fit a one-compartment model. All other conditions are identical to those in Fig. 5. (From Berliner and Wan, 1989.)

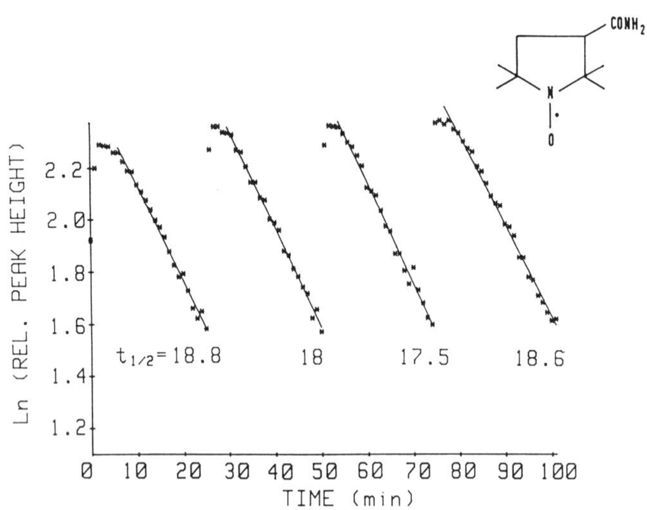

Figure 7. Semilog plot of the ESR signal intensity for multiple (15-sec) doses of CTPO. An initial priming dose of 1.0 ml (30 sec) was applied, followed by maintenance doses of 0.5 ml. The $t_{1/2}$ was 18 ± 2 min for a study of 13 rats. The data are consistent with a simple one-compartment model. All other conditions are identical to those in Fig. 5. (From Berliner and Wan, 1989.)

a single 1.0-ml injection of 75 mM CTPO. The ESR signal intensity is estimated by averaging the peak heights of each three-line spectrum since the lineshape is constant throughout the measurement, reflecting a free aqueous nitroxide molecule with no interaction with any macromolecular material. The semilog plot in Fig. 6 is identical to that found in a typical, simple one-compartment pharmacokinetics study by the standard array of analytical techniques used in pharmaceutical testing. However, this represents one of the first *in vivo* ESR pharmacokinetic plots. The multiple-injection curve shown in Fig. 7 further confirms single-compartment behavior as depicted by the repetitive "seesaw" behavior. In both Figs. 6 and 7, the rapid rise in the signal height just after injection (ca. 2 min) reflects the rapid distribution phase enabled by virtue of the jugular I.V. bolus. Lastly, Berliner and Wan (1989) were also able to maintain a steady ESR signal in the tail blood circulation by setting up a continuous injection of the CTPO solution (Fig. 8). Here a 24-μl/min injection of 75 mM CTPO was sufficient to "balance" the rate of nitroxide decrease. Thus, an imaging experiment requiring a constant level of nitroxide could be maintained by such a continuous injection protocol.

The observed overall elimination rate constant for a simple one-compartment model is composed of linear pseudo-first-order contributions from biotransformation (oxidation–reduction, metabolism), and excretion through the bile, urine, and exhalation. The bile and urine contributions accounted for 10–15 and $\leqslant$10% of the total dose, respectively. Excluding exhalation as a viable pathway, this means that the major contribution to

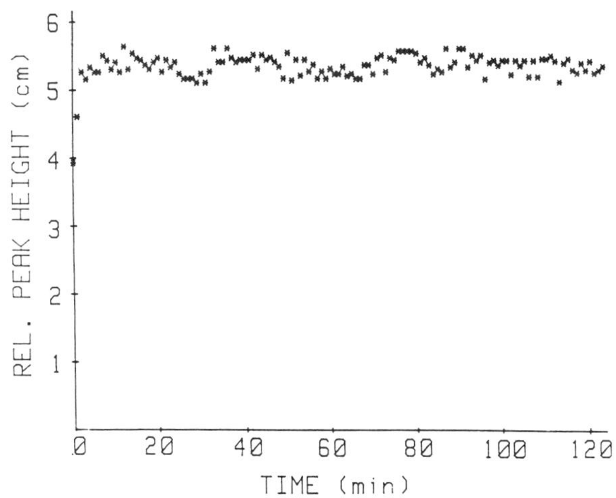

Figure 8. Semilog plot of the ESR signal intensities versus time for a continuous injection of CTPO (24 μl/min) after a single priming dose (0.6 ml). All other conditions are identical to those in Fig. 5. (From Berliner and Wan, 1989.)

this first-order decay was biotransformation with the most likely mechanism as reduction to the hydroxylamine. It should be noted that CTPO is nonpolar (hydrophobic) and contains no charged groups. Thus, this compound may passively diffuse across the cell membrane into the more highly reducing intracellular cytoplasm. We would therefore expect longer lifetimes for charged nitroxides as observed with *in vitro* cell suspensions (Swartz *et al.*, 1986).

4. *IN VIVO* SPIN TRAPPING OF DRUG INTERMEDIATES

The previous examples have addressed the use of nitroxide "imaging agents" or "labels" as the vehicle for observing ESR signals *in vivo.* The most desirable use of *in vivo* ESR would be where naturally occurring, metabolically produced free radicals could be observed. The challenge for *in vivo* detection has been great. While most "bioradicals" have been observed on rapidly frozen samples, this obviously creates controversy if attempted on living animals! An example of direct observation of melanin radicals in a melanoma tumor was discussed earlier by Lukiewicz and Lukiewicz (1984). Overall, this has been a very difficult problem due to the level of radicals *in vivo,* their lifetime, and the sensitivity of low-frequency (i.e., L-band) resonators compared with commercial X-band cavities. That is, biologically important radicals, such as superoxide or hydroxyl radical, are probably well below the $\sim 1\ \mu M$ detection limits of current low-field (L-band) *in vivo* spectrometers. There are several disproportionation and radical scavenging reactions which occur *in vivo* as part of the cells' antioxidant defense mechanisms. The only approach for overcoming these severe sensitivity limitations involves "accumulating" a steady-state generation of these radicals with some type of radical trap which is itself paramagnetic and longer-lived.

The use of nitrone spin traps has opened up new possibilities. That is, short-lived free radicals may be "trapped" and accumulated as stable nitroxide adducts (Mottley and Mason, 1989). An example with drug metabolites was shown by Maples *et al.* (1988) with the compound phenylhydrazine, a common pharmaceutical two decades ago (Maples *et al.*, 1988). Unfortunately, this compound undergoes a series of electron transfer and radical reactions, which are shown for the case of the rat in Fig. 9. Fe(II) hemoglobin is oxidized to the met- form, eventually resulting in production of a phenyl radical, and subsequently a (buried) thiyl radical in the hemoglobin is trapped by the stable spin trap, DMPO (5,5′-dimethylpyrroline nitrone). The proposed reaction scheme is depicted in Fig. 9. The upper spectrum in Fig. 10 depicts the X-band ESR of an *ex vivo* blood sample while the lower spectrum reflects a direct *in vivo* observation of the tail

$$C_6H_5NHNH_2 + HbFe(II)\text{-}O_2 \rightarrow C_6H_5\dot{N}NH_2 + HbFe(III)\text{-}O_2H \tag{1}$$

$$2C_6H_5\dot{N}NH_2 \rightarrow C_6H_5N{=}NH + C_6H_5NHNH_2 \tag{2}$$

$$C_6H_5N{=}NH + O_2 \rightarrow C_6H_5N{=}N\cdot + (H^+ + O_2^{\dot{-}}) \tag{3}$$

$$C_6H_5N{=}N\cdot \rightarrow \cdot C_6H_5 + N_2 \tag{4}$$

$$\cdot C_6H_5 + DMPO \rightarrow DMPO/\cdot C_6H_5 \tag{5}$$

$$HbSH + R\cdot \rightarrow HbS\cdot + RH \tag{6}$$

$$HbS\cdot + DMPO \rightarrow DMPO/HbS\cdot \tag{7}$$

Figure 9. Proposed sequence of reactions resulting from the radical oxidation of rat hemoglobin by phenylhydrazine. The last two steps represent the formation and subsequent trapping of a hemoglobin thiyl radical by the nitrone spin trap, DMPO. (From Maples *et al.*, 1988.)

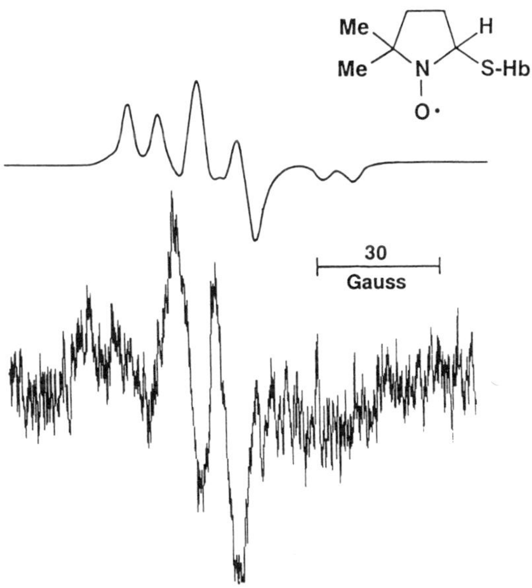

Figure 10. (Upper spectrum) X-band EPR spectrum obtained from the blood of a rat 2 hr after I.P. administration of 188 mg/kg phenylhydrazine and 50 μl/kg CMPO. (Lower spectrum) L-band *in vivo* EPR spectrum of the tail blood circulation of a rat administered as above. (From Berliner, 1991, and Maples *et al.*, 1988.)

blood circulation in the same L-band spectrometer utilized in Fig. 5 (Berliner, 1991; Maples *et al.*, 1988). There are two aspects of the spectra in Fig. 10 that should be noted. First, the spectrum reflects a strongly immobilized spin–trapped–spin-labeled hemoglobin. The broader linewidths of the strongly immobilized spin label (i.e., the nitroxide tumbling rate is essentially that of the protein) mean that significantly higher concentrations of nitroxide are required for the same signal-to-noise ratio as that of a freely tumbling nitroxide (e.g., Figs. 2 or 5). Second, this signal remained constant over long time periods, despite the fact that the spin-labeled protein is located in the intracellular space of the red blood cell. This remarkable stability occurs, as pointed out by Maples *et al.* (1988), since the spin trap adduct was on a unique buried thiol group in rat hemoglobin which is inaccessible to the bulk solvent. The lower spectrum in Fig. 10, which was measured *in vivo,* was actually remarkably strong when compared with spectra such as in Figs. 2 or 5, despite the poor signal-to-noise ratio relative to the X-band spectrum. Most significant was the identical lineshapes in both spectra, confirming the feasibility of direct *in vivo* measurements of spin-trapped drug intermediates.

5. CONCLUSIONS

Several aspects of the results presented above have far-reaching implications for the feasibility and applications of *in vivo* ESR and EPR imaging to animal systems. First, tumor imaging has been demonstrated utilizing only a water-soluble nitroxide. Nitroxides have been proposed as NMR contrast agents (Brasch, 1983; Swartz *et al.*, 1986) and the complementary EPR image serves as a valuable confirmation. Since nitroxide concentrations in different tissue regions are based partly on the local metabolism of the nitroxide label, a second level of contrast (i.e., chemical discrimination) is possible! Lastly, we can observe drug metabolism *in vivo* by trapping transient short-lived radicals created during the biochemical transformation(s) of the pharmaceutical.

ACKNOWLEDGMENTS. This work was supported, in part, by a grant from the USPHS (RR03126).

REFERENCES

Berliner, L. J., 1991, in *EPR Imaging and In-Vivo EPR* (Eaton, G., Eaton, S., and Ohno, K., eds.) CRC Press, Boca Raton, Fla., p. 291.

Berliner, L. J., and Fujii, H., 1985, *Science* **227**:517.

Berliner, L. J., and Fujii, H., 1986, *J. Magn. Reson.* **69**:68.
Berliner, L. J., and Wan, X., 1989, *Magn. Reson. Med.* **9**:430.
Berliner, L. J., Fujii, H., Wan, X., and Lukiewicz, S. J., 1987, *Magn. Reson. Med.* **4**:380.
Brasch, R., 1983, *Radiology* **147**:781.
Eaton, G., Eaton, S., and Ohno, K., (eds.), 1991, *EPR Imaging and In-Vivo EPR*, CRC Press, Boca Raton, Fla.
Fujii, H., and Berliner, L. J., 1985, *Magn. Reson. Med.* **2**:275.
Fujii, H., and Berliner, L. J., 1986, *J. Magn. Reson.* **68**:377.
Koscielniak, J., and Berliner, L. J., 1991, in *EPR Imaging and In-Vivo EPR* (Eaton, G., Eaton, S., and Ohno, K., eds.), CRC Press, Boca Raton, Fla., p. 65.
Lukiewicz, S. J., and Lukiewicz, S. G., 1984, *Magn. Reson. Med.* **1**:297.
Maples, K. R., Jordan, S. J., and Mason, R. P., 1988, *Mol. Pharmacol.* **33**:344.
Mottley, C., and Mason, R. P., 1989, in *Spin Labeling: Theory and Applications*, Biological Magnetic Resonance, Vol. 8 (Berliner, L. J., and Reuben, J., eds.), Plenum Press, New York, p. 489.
Nishikawa, H., Fujii, H., and Berliner, L. J., 1985, *J. Magn. Reson.* **62**:79.
Swartz, H., Bolton, J. R., and Borg, D. C. (eds.), 1972, *Biological Applications of Electron Spin Resonance*, Wiley, New York.
Swartz, H. M., Chen, K., Pals, M., Sentjurc, M. and Morse, P. D., 1986, *Magn. Reson. Med.* **3**:169.

Contents of Previous Volumes

VOLUME 1

VOLUME 3

VOLUME 4

VOLUME 6

VOLUME 7

VOLUME 8

Index